FRANCIS WATKINS AND THE DOLLOND TELESCOPE PATENT CONTROVERSY

Science, Technology and Culture, 1700–1945

Series Editors

David M. Knight
University of Durham

and

Trevor Levere
University of Toronto

Science, Technology and Culture, 1700–1945 focuses on the social, cultural, industrial and economic contexts of science and technology from the 'scientific revolution' up to the Second World War. It explores the agricultural and industrial revolutions of the eighteenth century, the coffee-house culture of the Enlightenment, the spread of museums, botanic gardens and expositions in the nineteenth century, to the Franco-Prussian war of 1870, seen as a victory for German science. It also addresses the dependence of society on science and technology in the twentieth century.

Science, Technology and Culture, 1700–1945 addresses issues of the interaction of science, technology and culture in the period from 1700 to 1945, at the same time as including new research within the field of the history of science.

Also in the series

Acid Rain and the Rise of the Environmental Chemist in Nineteenth-Century Britain
Peter Reed

Matter and Method in the Long Chemical Revolution
Laws of Another Order
Victor D. Boantza

Jesse Ramsden (1735–1800)
London's Leading Scientific Instrument Maker
Anita McConnell

Making Scientific Instruments in the Industrial Revolution
A.D. Morrison-Low

Francis Watkins and the Dollond Telescope Patent Controversy

BRIAN GEE

edited by

ANITA MCCONNELL *and* A.D. MORRISON-LOW

The completion of the book was made possible by funding from the Scientific Instrument Society, which chose to devote a legacy from the late author to ensure an appropriate memorial to one of its long-serving members.

Routledge
Taylor & Francis Group
LONDON AND NEW YORK

First published 2014 by Ashgate Publishing

Published 2016 by Routledge
2 Park Square, Milton Park, Abingdon, Oxfordshire OX14 4RN
711 Third Avenue, New York, NY 10017, USA

First issued in paperback 2016

Routledge is an imprint of the Taylor & Francis Group, an informa business

British Library Cataloguing in Publication Data
A catalogue record for this book is available from the British Library.

The Library of Congress has cataloged the printed edition as follows:
Gee, Brian.
Francis Watkins and the Dollond telescope patent controversy / by Brian Gee ; edited by Anita McConnell and A.D. Morrison-Low.
pages cm. -- (Science, technology, and culture, 1700-1945)
Includes bibliographical references and index.
ISBN 978-1-4094-6643-7 (hardcover)
1. Watkins, Francis, 1723-1782. 2. Dollond, Peter, 1730-1820. 3. Dollond, John, 1706-1761. 4. Telescopes--Great Britain--History. 5. Telescopes--Patents. I. McConnell, Anita. II. Morrison-Low, A. D. III. Title.
QB88.G44 2014
681'.4123092241--dc23

2013047643

ISBN 13: 978-1-138-27954-4 (pbk)
ISBN 13: 978-1-4094-6643-7 (hbk)

Contents

Appendices

List of Figures and Tables

Figures

Tables

List of Colour Plates

Foreword

'Nothing tends so much to the advancement of knowledge as the application of a new instrument' commented Humphry Davy when introducing his discussion of the Voltaic pile in his *Elements of Chemical Philosophy* in 1812. It was Brian Gee's ambition to show how scientific instruments and their makers had played essential roles in the advancement of knowledge. A Yorkshireman through and through, Brian Gee was born in Barnsley on 19 December 1939, the son of Walter Gee (a shop assistant) and his wife Eva (née Turner). Although he came from a working-class family, he was amused in later life to discover, when tracing his ancestry, that many of his forebears had been clergymen, including a bishop commemorated in York Minster. Like many wartime working-class children his prospects were transformed by the 1944 Education Act and its '11-plus' examination that gave him access to further education. In principle, the Butler Act allowed parents a choice of secondary education between a traditional grammar school education, a trades and industrial-oriented education in a technical school, and a 'modern' secondary education for those who would enter manual and service industries and trades. Leeds, where Brian and his parents had settled by the time he was ten, was one of the few educational authorities to have taken technical education seriously in the form of its splendid Leeds Central High School. Founded in 1885 as a Higher Grade (secondary) school, in 1944 it became one of two technical schools in Leeds. Here, at the end of the second year at the school, boys had a choice of following different aspects of technical education, one of them being engineering which involved the study of applied mathematics, wood and metalwork and technical drawing as well as physics and chemistry.

This was the pathway chosen by Brian and the skills he acquired at school were to serve him well. In the eyes of the school's headmaster, John Whitehead (a former physics teacher), a technical school offered 'a type of education more comprehensive in nature than that of the average grammar school, with the road to university wide open, but with an atmosphere of commerce and industry to whet the appetites of the potential industrial and commercial executives.'[1] In the event Brian chose the university option by studying mathematics, physics and chemistry at Goldsmiths College, University of London between 1958 and 1961, emerging from there with his teaching certificate in 1962. From 1962

1 Jenkins (1985), p. 209.

until 1968 he taught mathematics and physics at Hackney Downs School where one of his older colleagues, who was teaching English, was the school's later headmaster, John Kemp (1929–2009). While teaching at Hackney Downs, though under what stimulus is unclear, he embarked upon a part-time MSc course in the history and philosophy of science at University College London and gained his MSc in 1968 for a thesis supervised by Angus Armitage entitled *The Harvard Studies on Stellar Astronomy, 1840–1890.*[2] He was to remain interested in astronomy throughout his life and sometimes used the telescope at the University of Plymouth with his friend Percy Seymour.

With the thesis complete he left Hackney Downs School and took up a science teaching post at the International School in Geneva where he became fluent in French. (He would often write to his friends in this language to keep in practice, and also made a translation of Watkins's *L'Exercise du Microsope* (1754) for his own use.) After some years in Switzerland, which gave him a taste and opportunity for world travel and listening to Bach's organ music, he returned to England as a Senior Lecturer in Science Education at the College of St Mark & St John in Chelsea ('Marjon') – the oldest teacher training college in the country and certainly the earliest college to take the teaching of science by practical work seriously under Henry Moseley, Britain's first Inspector of Schools. When Marjon moved from Chelsea to Plymouth in 1973, Brian made his bachelor home on the other side of the Tamar at Landrake in Cornwall. Already intrigued by the physical apparatus used in his degree course at Goldsmiths and by telescopes, he now found that much of his teaching involved showing students how to demonstrate physical experiments. He soon acquired a reputation among his Marjon colleagues of being an outstanding teacher of physics. Inspired further by David Layton's *Science for the People* (1972), and by Jacob Bronowski's television series 'The Ascent of Man' the following year, Brian began to see how the history of science he had learned at UCL could be used in teaching would-be science teachers how to put across scientific concepts in the classroom. He became especially interested in the history of eighteenth- and nineteenth-century electrical and optical apparatus and how the history of their development might be used in teaching physics. He bequeathed his own collection of nineteenth- and twentieth-century instruments to the Museum of History of Science in Oxford.

I first met and corresponded with Brian in 1980 when I was involved in attempts by the British Society for the History of Science (BSHS) and the Association for Science Education (ASE) to persuade science and history teachers to use aspects of the history of science in their teaching. Brian had joined the Institute of Physics in 1963 and he had convened a small working

---

2 The only surviving copy is the one presented to the library of Harvard Observatory in 1968.

group of its Education Committee to edit a series of interactive articles in the Institute's monthly journal, *Physics Education.* His guidelines to authors, written in 1978, were extremely effective, as one paragraph will illustrate.

> Historical Case Studies should provide the reader with the feeling that he/she has been:
>
> (1) intrigued/challenged/ pleasantly informed;
>
> (2) motivated to follow up the study – possibly as a project – either as a piece of literary research in physics, and/or with some practical experimentation;
>
> (3) [the reader should find themselves] provided with a sense that science is essentially humanistic, involving curiosity, and helped in developing a sense of the past.[3]

Not surprisingly he became a founding member of the Institute's History of Physics Group in 1984 and he served on its Steering Committee until 1987. The Group's very first meeting at the Institute's headquarters in Belgravia Square on the theme of 'experiments and instruments: the interaction between experimental skills and instrumental craftsmanship' owed much to Brian's personal interests and persuasive powers. In July 1983 the Group held an ambitious three-day meeting at Oxford on 'The history of physics for the physicist' to which Brian contributed a paper on the role of the instrument maker in the development of physics.[4]

He also became an enthusiastic member of the Scientific Instrument Society on its foundation in 1983, the year that he first registered for a part-time doctorate under my supervision at the University of Leicester. Despite the huge geographical distance between us, he was a model postgraduate student who was always punctilious in sending material for comment and discussion before attending supervisions once a term, and meticulous in his writing and referencing. His outstanding thesis on philosophical instrument makers from 1750 to 1900 was defended in 1988 and led to several important publications. In 1990, together with Robert Anderson and John Burnett at the Royal Scottish Museum, Brian produced a useful Handlist of scientific instrument makers' trade catalogues which firmly addressed the significance of these ephemeral documents as historical evidence.[5] The original conception was Brian's, though much of the donkey work for this National Museums of

3 For a good example of what Brian had in mind, see Gee (1983).

4 The papers were published as Roche (1990). See Gee (1990c), 202–222.

5 Anderson, Burnett and Gee (1990).

Scotland/Science Museum publication, which involved a painstaking search for copies of these ephemera in libraries all over the United Kingdom as well as correspondence with overseas repositories, was done by Burnett. It was during this period that Brian discovered the Watkins & Hill archives in a disused air-raid shelter in Lewisham, and persuaded the firm's successor, Elliott Brothers, to transfer them to its headquarters in Chatham for safer preservation.[6]

After taking early retirement from Marjon in 1993 (thanks to the political policies of the Department of Education and Science, as he told me), the scientific instruments trade became his abiding interest, though his devotion to his elderly invalid parents, whom he moved down from Wakefield in 1993, soon began to restrict his bachelor freedom to travel and to conduct his research. In retirement Brian tutored for the Open University and set up the 'Tamar Literary Services', copy-editing technical journals like the *Journal of Applied Electrochemistry* and *Sports Science*. This 'pocket money' enabled him to pay for his parents' care whenever he was able to find time for trips to distant archives or to meetings of the Scientific Instrument Society.

At first, his parents lived a few doors away in their own bungalow, but as they grew increasingly frail, he moved them into his own home, to the further detriment of his research time. His sole recreation was relaxation by playing Bach on his piano. The isolation in Cornwall and his full-time caring meant that Brian was greatly obliged to many distant helpers like Peta Buchanan and John Millburn. He was always profoundly grateful for the titbits of information they provided him. 'You see,' he wrote, 'being trapped here has some advantages – I can examine and consider in fine detail all the little bits that [they] find.' Nevertheless, the strain told upon his health. 'I am very tired and strained by it all,' he wrote (21 January 2000). The death of his father in 1998 provided only momentary relief since his mother almost immediately became bed-ridden and needed 24-hour attention. In 2005 Brian underwent seven weeks of radiotherapy for prostate cancer while still juggling with care for his mother. Unfortunately, it did not work and the following year he underwent cryosurgery at the Royal Surrey Hospital in Guildford. Brutally, although living some 200 miles away, his consultant arranged for him to have a pre-operation inspection at Guildford on Christmas Eve 2007 at 4.00 pm. The logistics of his journey and making arrangements for his mother's care during the festive season proved a nightmare, but his stoicism and sense of humour remained unbroken by the experience. There were further post-operative blows: he had become diabetic and his mother died three weeks after his operation.

By the time of his own death in 2009, he had drafted an 11-chapter monograph on the activities of the rival optical instrument makers John Dollond and Francis Watkins. All of his spare time when not involved in

[6] Subsequently transferred to the Museum of History of Science, Oxford.

copy-editing and acting as a carer for his parents had been spent in following up the fine detail of the Charing Cross optical business of Francis Watkins and his legal war with the rival firm of Dollonds. It was not until the end of 2002 that he was free for a few days' visit to the National Archives at Kew (the then PRO). On this occasion he examined the court case records of Peter Dollond v Francis Watkins and the latter's petition to George III that forms the heart of the present posthumous monograph. 'I must say,' he wrote (8 December 2003) 'there was a certain *thrill* at untying the knots and seals on the proceedings of Lord Mansfield's court at Kings Bench, Michaelmas 1764 – a document never before examined! My white gloves quickly became filthy with 250 yrs of muck.' Never given to despair, he viewed the restrictions placed upon him in retirement with equanimity and even as a blessing in disguise. He confided (14 December 2006):

> My salvation has been in working 'backwards' from our days with my thesis. I suppose when Anita McConnell invited me to write the *ODNB* article on Francis (I) Watkins I was forced to work backwards into the 18th century. Glory be – what rich pickings here! The bits and bobs (often seen churned out in books when instruments are mentioned) really tell very little until contexts are constructed. But I have been amazed how much I have managed to winkle out and have taken the greatest pleasure in spending thousands of hours interpreting everything. I suppose the four chapters that I devote to legal history – litigation with the Dollonds (rages and acrimony of mammoth proportions); petitions to the King by the Spectaclemakers' Company; tracing the upheaval throughout the nineteenth century, too, have given me the most to think about. The National Archives Library and (luckily) the Record Office Library at Hereford (where I have spent several respite breaks), and Law Society Library in Chancery Lane have all proved vital; even this week Anita McConnell (who has occasionally dug out 'extras' for me at the Cambridge University Library), has now helped to clinch another mystery of local history and the instrument makers via Parliamentary papers. All in all, the 'inside-out' story emerging is a good view on the instrument trade and its optical history from Newton's denial of achromatism.

Brian was the most modest of scholars and hated contention in the literature, especially if it involved himself. He had a keen sense of human foibles. Witness this reaction to the nation's fears of the Millennium Bug on 1 January 2000.

> Apart from the fact that my computer tells me you wrote your e-mail in 1900 (and perhaps you did, courtesy of some post-Babbage electronic miracle!) my old machine rumbles on without any hitches. I had imagined that it would have 'awoken' on 1st January 1900 and, since it is an intelligent machine, [told] itself

> that it was back in time by a century. It might then have uttered some remark like: 'Today is 1 January 1900. But computers of my kind didn't exist in 1900; therefore I don't exist' ... and then promptly disappeared in a puff of smoke. [But] It didn't and so I'm here too!

Brian died in Derriford Hospital, Plymouth on 3 December 2009 a few days short of his seventieth birthday, and was cremated at Glyn Valley Crematorium, Bodmin, three weeks later. Subsequently, on 21 April his Marjon colleagues and friends held a memorial service for him at the Chaplaincy Centre of University College Plymouth. His scholarship will be sorely missed. We must all be grateful for his determination, as well as his historical skills, and that Anita McConnell has managed to bring the manuscript of his book into a publishable form after such adverse circumstances.

W.H. Brock
University of Leicester

A Note on Prices and Dimensions, and on the Spectaclemakers' Company

Prices throughout the eighteenth century and well into the nineteenth century were generally quoted in guineas, for which gold coins circulated. The guinea was equivalent to one pound (£) and one shilling (s). For accounting purposes guineas, half-guineas, and so on, were converted into £, s and d (pence). No one-pound coins were produced until the 'Great Recoinage' of 1816, the gold sovereign being minted from 1817.

All the instrument makers mentioned here, including the Dollonds, Watkins, Watkins & Hill, and Elliotts, measured their telescopes and other instruments in feet and inches. The metric system, increasingly employed on the continent, was largely ignored in Britain. Likewise, temperatures continued to be measured in degrees Fahrenheit, rather than the Réaumur, centigrade, or Celsius scales found on instruments from continental makers.

The religious guilds to which various trades belonged in medieval times were abolished in 1557, being replaced by secular fellowships or companies. Following an application by several London opticians, who were then members of other companies, King Charles I granted a royal charter, setting up 'The fellowshippe of Spectacle makers of London' in 1628. The company's early records were destroyed in the Great Fire of 1666 but have survived thereafter, with occasional gaps, continuing to the present time. Over the years, its title has been variously expressed as Spectacle Makers' Company, Spectaclemakers Company or Spectaclemakers' Company, which is the form now preferred.[1]

---

1 Clifton (1993a).

Author's Acknowledgements

I am grateful to all those archivists, librarians, information officers, archives managers, collections co-ordinators, local historians and museum curators who have patiently assisted me in seeking out material. Answers to my numerous questions were not always possible but each contribution made, however small, has enriched the overall account, often in ways quite unexpected. Among those who have been of assistance are: Peter Abrahams (Portland, Oregon), David Allen (Royal Society of Chemistry, London), Juliet Bankes (Corporation of London Records Office, City of London), Jane Baxter (Local Studies Librarian, Richmond), Helga Beez (formerly of the Optisches Museum der Ernst-Abbe-Stiftung, Jena), Ron Bristow (Scientific Instrument Society), Larysa Bolton (Barclays Bank Archives), Neil Brown (formerly of the Science Museum, London), Vera Brown (Powys Family History Society), the late Mary Brück (Royal Observatory, Edinburgh), Rosalind Caird (Hereford Cathedral Library), James Caplan (Observatoire de Marseille), Marcus Cavalier (Scientific Instrument Society), Peter de Clercq (Antiquarian Horological Society), Gloria Clifton (formerly of the National Maritime Museum, Greenwich), Diana Crawforth-Hitchins (formerly of Project Simon, Oxford and now of the Antique Metals Society), Elizabeth H. Cuthbert and Allison Derrett (The Royal Archives, Windsor), Teresa Gray (Surrey County Archivist, Richmond), Liz Green (Cheshire Record Office), Håkan Hallberg (Uppsala University Library), Neil Handley (British Optical Association Museum at the College of Optometrists, London), the late Peter Hingley (Royal Astronomical Society, London), Jo Hopkins (Royal Society Picture Library), Giles Hudson (formerly of the Museum for the History of Science, Oxford), Rhian James (Powys Archives), Ruth Jenkins (Southwark Local History Library, London), the late James Legon (Docklands Ancestors and Parish Registers.com), Jane Lewis (Surrey History Centre), Alison Morrison-Low (National Museums Scotland), Allan Lloyd (Lady Hawkins' School, Kington, Herefordshire), Darryl Moody (Swindon Reference Library), Tony O'Brien (Newchurch, Powys), D.J. Pohl (formerly of the Huguenot Society Library, London), Sue Hubbard (Hereford Record Office), Catherine Richards (Powys County Record Office, Llandrindod Wells), Jeremy Smith (London Metropolitan Archives), Rosemary Smith (formerly of the Science Museum Library, London), Emmanuelle Russier (Bibliothèque Municipale du Puy-en-Velay, France), Tony Simcock (Museum of the History of Science, Oxford), Lorna Standen (Hereford Record Office), John Underwood (Science Museum Library, London), Lesley Whitelaw (The Law Society and Temple Archives, London), Jane Sewell (formerly of the City of London

Records Office), Stacey Gee (Guildhall Library, London), Catriona West (formerly of the Whipple Museum, Cambridge), Britta Wijkander (Uppsala University Library), Norman E. Wright (Brigham Young University), Mariam Yamin (Royal Bank of Scotland Archives, Islington), and Rolf Willach (Switzerland).

During the course of the research I have been guided by several very special individuals who gave me considerable encouragement. Some, sadly, are no longer with us. These include Dr John Chaldecott (formerly curator at the Science Museum) and Victor Kenneth Chew (formerly curator and research scholar at the Science Museum). I am, beyond all measure, also indebted to the late John R. Millburn whose biographies of Benjamin Martin, James Ferguson and George Adams are an inspiration to all who care about the role of the instrument maker in the development and diffusion of science. His patience and thoroughness in solving my more abstruse queries mark a level of scholarship that is rarely surpassed. Not only have I drawn inspiration from Millburn's studies but also, more recently, from the magnificent studies by Anita McConnell and Alison Morrison-Low respectively, whose works on Jesse Ramsden and provincial instrument makers reinforce the importance of instrument makers and their trade when considering Britain's past as an industrial, manufacturing nation.

Numerous friends have volunteered or submitted to my coercion to read through various drafts. Allan Mills (Leicester University) checked my physics in a crucial chapter on Newton's optics, Diana Crawforth-Hitchins improved upon grammatical niceties, Marcus Cavalier cast his legal eye over issues of eighteenth-century litigation, Neil Handley (British Optical Association Museum at the College of Optometrists) added useful suggestions, and Anita McConnell invited me to write the entry for Francis Watkins for the *Oxford Dictionary of National Biography*, and so began the problem of finding out about an individual whom many have felt was of importance without actually knowing why.

My deepest gratitude must go to Peta Buchanan without whose expert understanding of family genealogy this project might well have been closed down at an early stage. Her familiarity with political, economic history and social history has led to a more rounded account. For a dozen years she has faithfully offered comments and advice and has taken on the task of locating and extracting complex material from the depths of many strange sources. [Editors' note: The history of science community greatly regrets that Peta Buchanan died in 2011.]

Without these very tolerant friends, none of the discoveries, hypotheses and explanations would have surfaced. It is interesting to note that through the process of constructing this unique story, an invisible college has developed which has shown that worthwhile and accurate scholarship can still be pursued to good effect, even in the absence of any obligation of institutional academia.

Brian Gee
Landrake, Cornwall, 2009

Abbreviations

BC	Birkbeck College, University of London
BCC	Buckinghamshire County Council Archives
BL	British Library
BM	British Museum (Department of Prints and Drawings)
CLO	Clarendon Laboratory, Oxford
CLRO	Corporation of London Record Office
CUL	Cambridge University Library
DNB	*Dictionary of National Biography*, all issues up to 1993. These memoirs can be found as 'Archive' under the matching *OxDNB* online entries.
DSA	Department of Science and Art
GL	Guildhall Library, London
HCL	Hereford Cathedral Library
HRO	Herefordshire Record Office
LMI	London Mechanics' Institute
MHSO	Museum of the History of Science, Oxford.
NLW	National Library of Wales
NMAH	National Museum of American History
NMSI	National Museum of Science and Industry
OxDNB	*Oxford Dictionary of National Biography*
PC	Private communication
PCA	Powys Council Archives
PP	Parliamentary Papers
PRS	President of the Royal Society
RAS	Royal Astronomical Society
RBS	Royal Bank of Scotland
RCRL	Richmond Central Reference Library
RGO	Royal Greenwich Observatory
RI	Royal Institution
RS	Royal Society
RSA	Royal Society of Arts
SDUK	Society for the Diffusion of Useful Knowledge
SHC	Surrey History Centre
SMC	Spectaclemakers' Company
SML	Science Museum Library

TNA	The National Archives. All documents cited are from PRO, Public Record Office, Kew
UCL	University College London
UUL	University of Uppsala Library
WA	Westminster Archives
WCRO	Worcestershire County Record Office

Editors' Introduction

Brian Gee believed that the instrument-maker's role should never be ignored or excluded from the record of scientific and technical development. He was aware that such recognition is often ignored, largely on the grounds that there is some disjunction between the instrument maker and the scientist. It was with the intention of developing a model of the relationship between the maker and the user of scientific instruments that he sought examples by which to develop a model of this relationship, noting that in the eighteenth century the elder Francis Hauksbee, 'demonstrator' to the Royal Society, had teamed up with John Theophilus Desaguliers to provide lectures in the coffee houses. In the nineteenth century there was co-operation between mathematician and civil engineer Peter Barlow and the chemist James Marsh at the Royal Military Academy. Much the same role for the instrument maker was defined 100 years later when the microscope makers Watson & Co. instructed their leading employees to associate with professional users of scientific instruments, recognising that such men might be able to advise and suggest modifications or new designs.

With the intention of providing a detailed study, Brian had hoped to seek out instances where members of the Watkins family played some part in the advance of science. The choice of this particular family was not unreasonable since they had held a business at Charing Cross for over a century and there were clear instances of their success in business having depended upon more than incidental trade of the kind where the maker merely acts as technician exercising his skills. And yet, the search for evidence towards this end had largely failed; nor had there been sufficient evidence to produce a traditional account of the 'life and work' variety such as M.J.G. Cattermole and A.F. Wolfe's account of Horace Darwin and his Cambridge workshop, which had developed into the Cambridge Scientific Instrument Company.

Nevertheless, sufficient material had come to hand to provide a remarkable insight to the life of Francis-I Watkins (1723–1791), the first instrument maker of this family, whose 40 years of working life covered an era stretching from a very traditional background to a time when less heed was paid to such values. Successive members of the same family with the same forename are distinguished by the customary addition of I, II and III. For present-day historians of science, the most important aspect of Watkins's career was his collaboration with John Dollond on the patent for achromatic lenses, followed by the controversy

– one might say battle – with his son Peter over John Dollond's right to be considered as the 'true inventor'. For the first time, Brian's thorough investigations have cleared up the confusion over the two court cases, which many historians have mistaken as a single hearing.

The survival of business archives relating to the later stages of Watkins & Hill shows the part played by the firm in the expansion of education, both at school and university level, and in the Great Exhibition of 1851. The subsequent decline led to the takeover by Elliott & Sons, a firm with its own history, to which Brian himself made significant contributions.

Brian died in December 2009, leaving a substantial proportion of the text for this book, but residence in Cornwall, and his personal circumstances (as carer for an invalid mother), had made it difficult for him to come to London to check and complete the bibliography. Under the auspices of the Scientific Instrument Society, the editors have made some reorganisation of the text and dealt with the bibliographic lacunae.

Editors' Acknowledgements

We are extremely grateful to the staff of the Museum of the History of Science, Oxford, for their generosity in providing many of the illustrations at a minimal cost; to National Museums Scotland for a similar gesture; and to a number of private individuals who allowed photographs of items in their collections, or who sponsored the illustrations used here.

The completion of the book was made possible by funding from the Scientific Instrument Society, which chose to devote a legacy from the late author to ensure an appropriate memorial to one of its long-serving members.

Anita McConnell
Suffolk
August 2011

Alison Morrison-Low
Edinburgh
August 2011

Chapter 1
On Becoming an Optical Instrument Maker

This chapter is an account of the recruitment and training of an optical instrument maker. It is just one example of the many accounts that might be told about young people who were dispatched by their parents from the provinces to the capital in the hope of a better and more rewarding life than would have been possible had they stayed at home. In the early decades of the eighteenth century, London was recovering from a disastrous century of political unrest, a plague, and an immense conflagration that devastated the capital. A more settled period brought with it a sense of the future. By the mid-century, prosperity showed through and London was once more attracting foreign visitors. For those migrating from the provinces, their journeys were less to do with seeking fame or fortune than an expedient often adopted by large families in the hope of finding a trade or career for their offspring. In the case of Francis Watkins (bapt. 1723) – on whom this work is focused – and his brother William, of Newchurch (*Llannewydd*) in Elwell (*Elfael*), Radnorshire, their parents were of good stock, not without money and local influence and, most importantly, they followed an ordered way of life. Thus, the attraction of sending their two youngest offspring to the capital was the opportunity afforded for a successful life among the hard-working artisan élite and shop-owning traders of the great and expanding Georgian capital.

The Welsh had been in London in considerable numbers since Tudor times and by this time comprised about six per cent of the capital's population. Unlike the Huguenots, who were an easily distinguishable immigrant group, the Welsh were fully integrated and had been since the English Government – by an Act of Union in 1536 – made itself responsible for Welsh affairs. Those migrating to the capital, once settled, gladly adopted the English tongue in order to reap the benefits of a working life.[1] In the first half of the eighteenth century, these 'antient' Britons sought an identification of their own, forming in 1716 the Honourable and Loyal Society of the Antient Britons and organising a Welsh charity school in Clerkenwell.

1 Jones (2001), p. 33, passim.

Migration to the Capital

The majority of apprentices came from the home counties, lesser numbers being drawn from elsewhere in the southern half of England. Very few came from Wales or the northern counties. Travel between Radnorshire–Herefordshire and London in the eighteenth century was common enough although for those who wished to make the journey cheaply and safely there was always the possibility of travelling in numbers with the drovers bound for the meat market at Smithfield, a journey that would have taken three weeks or more. More probably, William (and later Francis) made the two or three-day journey on one of the carriers or stage coaches that relayed regularly from Abergavenny, Brecknock or Hereford, into London. Routes such as these had been in operation for over 100 years and are described in John Taylor's *The Carriers Cosmographie* (1637). Indeed, travel to and from the capital by carriage was remarkably well organised by the time that the Watkins brothers made their respective journeys, as may be witnessed by the schedules of departures listed in *The Intelligencer*, a contemporary guide for all people on the move.[2]

Uprooted from their families and familiar surroundings, these boys, or young lads, were sent off to the capital to learn their respective trades. At least, their chances for developing latent abilities were far better placed in London, well away from the rural economy of the timeless, sparsely populated Welsh Marches. Once there, they would quickly adapt to London life and, with a little luck and a good deal of hard work, reap a profitable living. In Francis's case, he set in train a scientific instrument business that was to last over a century in the hands of his family successors; by contrast, William's life remains obscure although there is evidence that his progeny remained in contact with their uncle Francis and also made some separate headway in the instrument trade. As it happened, Francis arrived in the capital at a good time as far as the trade itself was concerned. For a lad with a modest intelligence and an ability to comprehend quickly, the instrument trade assured him a moderately secure future; for a youth with determination and flair – additional qualities evidently very much a part of young Francis's nature – such a choice of trade was not only apt but, as will be seen, it turned out to be his making.

Family Background

Francis and William Watkins were the youngest surviving children of Jeremiah-I Watkins. His good fortune in marriage to Mary Whitney in 1713 paved the way for his elevated status as one of the newly emerging breed of

---

2 *The Intelligencer* (edition consulted: 1738). WA [microfilm].

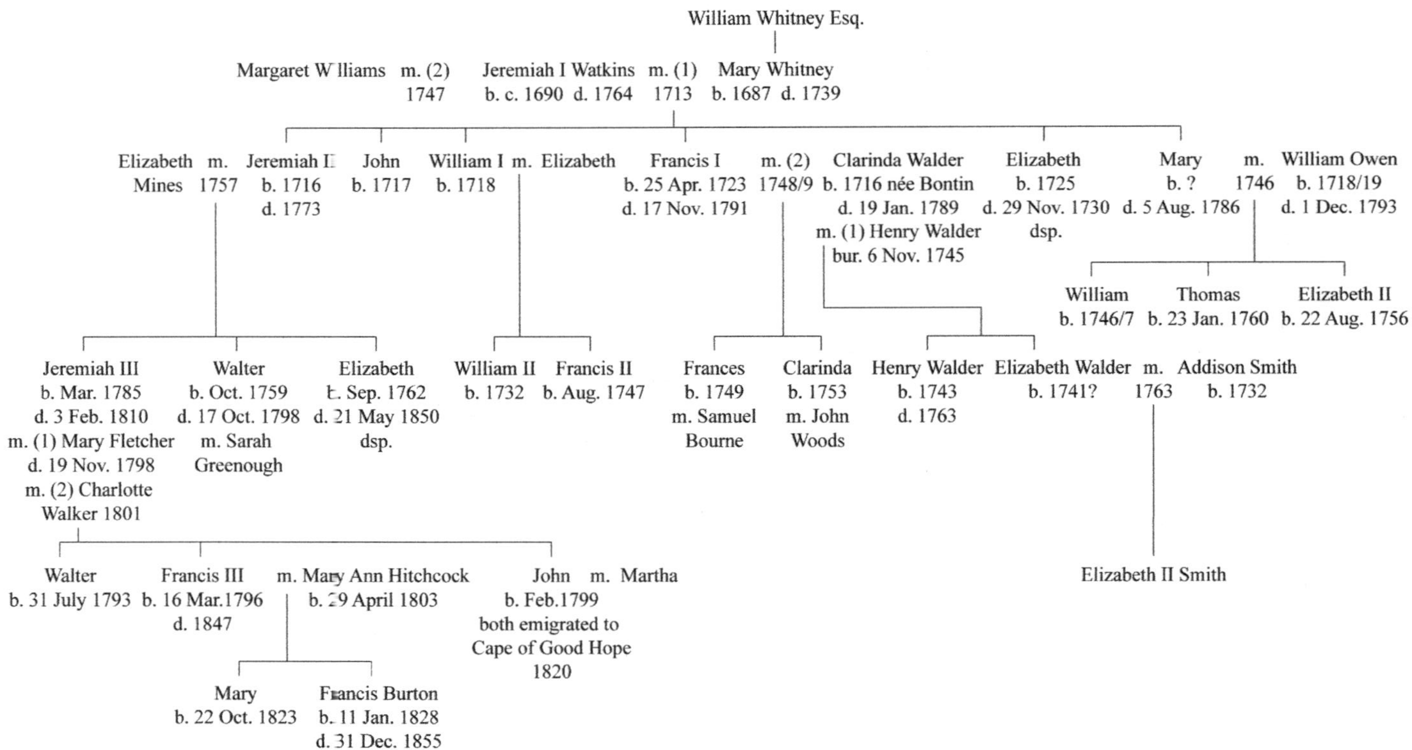

Figure 1.1 The Watkins family tree

gentlemen landowners. That is, he was always addressed in legal documents as 'gentleman', thus signifying that he was of a standing above that of a 'yeoman' although not quite an 'esquire' as was his father-in-law.[3] When Mary died on 28 May 1739, aged 52, a memorial tablet to her life was placed on the north wall of the chancel of the parish church of St Mary. Again, this tablet reflected the importance of the Whitney family by recording the father's name alongside that of his deceased daughter.[4] From this background it may safely be assumed that the children of this union began their lives in modest comfort and very secure surroundings.

As head of the household, Jeremiah-I Watkins was a responsible man and it is clear – even from the little that is known – that he did his best to enable his sons to pursue a satisfactory and profitable life. Education was important as a preparation to any future and it is probable that he employed a good tutor, at least in their early years. For the eldest son, Jeremiah-II, there was also the possibility of a good education in Kington at Lady Hawkins' School, situated just a few miles distant from Newchurch and therefore within easy daily reach on horseback.[5] This school had been founded almost 100 years earlier in 1632 and, in accordance with the Rules laid down at the foundation, boys from certain parishes within the county of Herefordshire, together with those from the parish of Michaelchurch in the county of Radnorshire, received their education entirely free. Newchurch boys, however, were not so privileged and, as fee-paying pupils, they were termed 'foreigners' although made welcome nevertheless. The rates, drawn up at the foundation – probably changing little over the years – were five shillings a quarter for a gentleman's son, two-shillings and six-pence for a yeoman's son and 12 pence for 'the poorest sort.' Jeremiah-I Watkins would have paid at the quarterly rate of five shillings and, for this fee, his eldest son would have received the experience and tutelage of the Reverend John Griffin, headmaster since 1717. Griffin was an Oxford graduate and one qualified – as stipulated by the foundation Rules – as a person 'well able to instruct and train up children in the said school till they be fit for other studies in the universities.'[6]

Apparently, Jeremiah-II – at a mere 12 years old – was very much a candidate ripe to be 'trained up' for the university. It was through this process of coaching that he came to exchange the Radnor hills for the Oxford spires when, in 1730, at only 14 years old, he left for Jesus College to read Divinity. Jeremiah-I's second and third sons, John and William, also attended Lady Hawkins' school and, like their elder brother, appear to have remained there for three years, receiving a thorough grounding in reading, writing and arithmetic together with some

3 Bidgood (1981), p. 27.

4 Davies (1905), p. 264.

5 Lloyd (1998), passim.

6 Lloyd (1998), p. 23. Griffin remained headmaster for 49 years.

Latin and Greek, geography (ancient and modern), along with extra object lessons on topics such as 'the use of the globes'. Francis, the youngest son on whom this work focuses, does not appear on the relevant extant school registers, and therefore he may not have received the same privileged start as his elder brothers although it is likely that he received tuition of some sort, possibly from the local priest, Francis Wadely, or even his eldest brother who, by this time, had returned from Oxford.

Jeremiah-II matriculated from Jesus College in 1734 and returned to Newchurch.[7] On 27 August 1739 he was ordained priest and licensed as curate of St Mary's for a stipend of £8 a year although, prior to his full-time incumbency there, he also served as curate at nearby Clyro for a stipend of £12 per year. Eventually, on 19 December 1748, he returned to Newchurch where he was appointed rector following the death of Francis Wadely. Meanwhile, his brother John's path seems to have been planned as inheritor to his father's estate. Transactions of land by 'lease and release' were the norm and the first of note for John occurred in August 1753 which was stipulated jointly with his younger brother Francis even though, by this time, Francis was fully fledged as an optician and working in London (§ Chapter 8).[8] This deed was for the property and estate of Pontvane farm (*Pont-faen*, Welsh, meaning stone bridge) in Newchurch (Figure 1.2).

About the time that John and Francis received entitlement to the farm, Jeremiah-II appears to have benefited by a separate endowment from his father because, in 1752, he was able to raise half of the £400 sale price for Cae Higgin, another farm at Clyro.[9] Both these farms, occupying the same land, are still in existence today under the same names. However, at some point, Jeremiah-II moved to Hereford and it was there, at the parish church of All Saints, that he married Elizabeth Mines on 11 April 1757. It was also at this church that the sons of this union, Jeremiah-III and Walter (later successors to Francis's instrument business in Westminster) were baptised. Jeremiah-II later moved to Claines in Worcestershire where his death is recorded in 1773.

---

7 Foster (2000), vol. 4, p. 1508. 'Jeremiah, s Jer. of Clirow [Clyro], co. Radnor, pleb Jesus Coll., matric 27 May, 1734, aged 18, B.A. 18 Jan 1737–8, M.A. 1749.'

8 The assignment of mortgage to 'Francis Watkins of Saint Martin in the Fields, Liberty of Westminster, co. M'sex, optician' (along with one David Morgan) was 'dorsed' for a consideration of £298. PCA, R/D/GNX/153–156). The map is 'An Exact Survey of Pontvane being part of the estate of Jn. Watkins Gent situate in the Parish of Newchurch in the County of Radnor taken and delineated in the Year of our Lord 1754. By Meredith Jones [Land Surveyor of Brecon].' National Library of Wales, Department of Prints and Maps, Map 7605, 134/2/10.

9 O'Brien (1991), p. 65. Renting the 45 acres of Cae Higgin brought in £47 annually which, interestingly, became a part of Jeremiah-II's stipend.

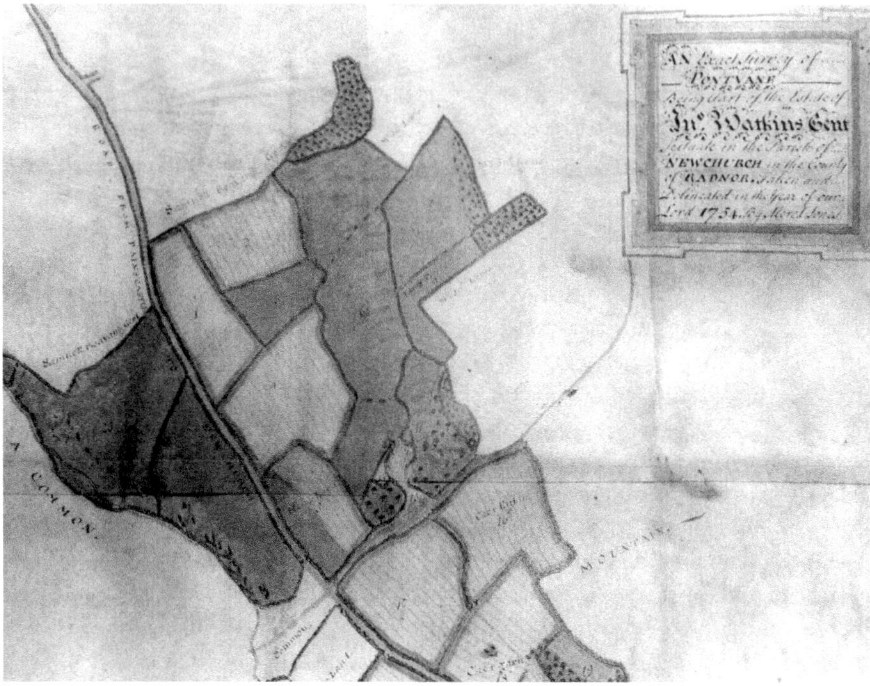

Figure 1.2 Plan of Pontvane estate by Meredith Jones, estate and land surveyor, 1754. By permission of Llyfrgell Genedlaethol Cymru/ The National Library of Wales

Jeremiah-I Watkins remarried in 1747 – that is, after a respectable period had passed following the death of his first wife, Mary, to a widow from Llandeilo Graban, just a few miles south of Newchurch.[10] However, from the absence of any mention of his second wife (Margaret Williams) in his Will of 1764, it is assumed that she predeceased him.[11] Because Jeremiah-I's Will is unusual in that it provides relational links over the two succeeding generations – that is, including his grandchildren who were nephews and nieces of Francis Watkins and who eventually held some loose connection with the instrument business at Charing Cross – it is worth pausing to consider the several bequests therein in order to capture the nature of this strong head of household whose actions set the course of two succeeding generations of his progeny.

[10] That is, according to the Index to pre-1837 marriage bonds. NLW: Department of Manuscripts & Records, SD/Z/46/37. This bond was dated 18 May 1747 although the Bishop's Transcripts record the date as 21 May 1747.

[11] Will of Jeremiah-I Watkins (1764). TNA: PROB 11/899.

The Will began by stipulating that three of his sons, John, Jeremiah-II and Francis, were to receive 'one shilling a-piece' – in common parlance they were 'cut-off with a shilling' – a legal ploy applied to prevent any future rivalry or dispute over the Will, the implication here being that these brothers had been sufficiently catered for during their father's lifetime. Given the aforementioned dealings in land during the early 1750s this may well have been the case. William-I, however, did not receive 'the shilling' and so this implies that he received nothing comparable to his brothers during his father's lifetime. Perhaps to compensate for this, he did, however, receive a specific pecuniary bequest of £20. Also, in William's case, the Will reveals that his children by his wife, Elizabeth, namely, William-II and Francis-II, were each to receive £30 upon their attaining the age of 21 years. To provide for this, John, as executor, was instructed to redeem a bond for £60 at the appropriate time.

Other grandchildren were also to benefit from this Will. The sons of Jeremiah-II (Jeremiah-III and Walter) were each to receive £20 on reaching 21 years. However, Francis-I's children, both girls, Frances and Clarinda, by his union with the widow Clarinda Walder, are absent. It is, of course, possible that Jeremiah-I never met these London grandchildren but it is more likely that he was reluctant to provide for granddaughters whose prospects depended on future husbands rather than the price of an apprenticeship binding. However, there was another granddaughter – Elizabeth-II by his daughter's (Mary's) marriage (on 4 October 1746) to William Owen of Ashperton, Hereford – who was bequeathed £80, a curiously generous sum, for her twenty-first birthday (otherwise towards her marriage) while her brothers, William and Thomas, were each to receive £30 on their twenty-first birthdays. Moreover, their father (William Owen) was also to receive a mortgage bond on a house. Such highly favoured bequests could only be indicative of Jeremiah's desire to ensure a happy and profitable marriage for Mary (his daughter) and her husband who started out their married life in London – just a little after William and Francis's move to London – as printers and publishers.[12] Other bequests included relatives and trusts for the poor people of Aberedw (the place of his birth) and Newchurch (the place of his home).

Jeremiah-I's care in spreading his assets to benefit these succeeding generations characterises a man of deep conscience who thoroughly understood the need to plan a secure start for his offspring. Certainly, the sons were dealt with fairly in their upbringing although – for the sake of the estate and family name – only

[12] As a printer, at the outset of his own business, Owen printed Francis Watkins's Electrical Experiments (1747); then, as the business developed, he became Benjamin Martin's principal publisher: Millburn (1976), p. 62. A later connection between Owen and William and John Watkins (later generations of the Watkins family) is visible through his Will. TNA: PROB 11/1239 (1793 December) f. 611, ff. 395–402. See also Millburn (2004).

Jeremiah-II and John were needed to perpetuate the prime positions within the community at Newchurch; it was, therefore, far more practical to allow William and Francis a chance to find their own feet in the expanding capital. At least, given a little encouragement, they had the chance to make a profitable and respectable life through their respective trades. In short, Jeremiah-I's four sons were split geographically but their contact with each other was never lost. Not only did John liaise with Francis over the lease of the Pontvane estate at Newchurch but Jeremiah-II also maintained contact with Francis and, eventually, his sons (Jeremiah-III and Walter) succeeded to their uncle Francis's instrument business.

Choosing an Apprenticeship

One outcome of the Statute of Labourers and Apprentices (1563), drawn up in the time of Elizabeth-I, was that no person could exercise any of the 'crafts or mysteries' practised in England unless an apprenticeship had been served. This Statute laid down a number of basic rules, including the minimum age at binding. For example, a boy destined for a life at sea with a ship owner could be less than ten years old but no younger than seven years old, while in market and corporate towns apprentices could be as young as ten years old. More typically, apprentices were older than 13 or 14 years but no older than 24 years (the maximum age being lowered to 21 years in 1766–1767). Until a statute of 1710 made stamp duty payable on apprentice indentures, there was no central record of apprentices in England and Wales but after this year registers of money received from the duty on indentures were kept by the Commissioners of Stamps.[13] Thus, around the time that young William and Francis were ready for the world of work, these rules were fully in force and so their presence in the capital can be made visible.

In terms of choice of apprenticeship, a father had to consider how best to match the skills and abilities of his offspring to the opportunities presented by a trade. For parents not knowing which way to turn for advice, an insight to the London trades and types of apprenticeship could be gained through Robert Campbell's compendium *The London Tradesman* (1747). Although this work was not yet published when the Watkins brothers arrived in the capital, it is this text that best provides a contemporary view to how the necessary choices were made.

Briefly, Campbell considered the scope of apprenticeships across five principal domains: namely, religious, judicial, clerical, mechanical and medical. Within these, he distinguished between those trades requiring *strength* and

[13] The Apprenticeship Books are TNA: Class IR 1, but not all apprenticeship bindings are recorded. Compulsory bindings were abolished in 1814.

those requiring *education*. For example, a glass grinder might be 'any [sort of] common Laborour,' a pump-maker required 'more Strength in his Hands than judgement in his Head,' and a clockmaker needed 'no great Strength nor much Education.' To become an instrument maker, particularly in those branches demanding numeracy, Campbell remarks that: 'He ought to have a Mathematically turned Head, and be acquainted with the Theory and Principles upon which his several Instruments are constructed; as well as the practical Use of them.' Indeed, as Campbell pointed out, the choice of trade should be based according to the youth's 'capacity', 'disposition' and 'constitution'.[14] All too often, the author warned the 'fond Mother' and 'doating Father, ... the Child is bound, that is, chained to a Trade, to which Nature never designed him [and hence] furnishes the Pulpit with Coblers [sic] instead of Divines.'[15] Mistakes of this kind were hardly likely to occur with a father who took pains to provide nothing but the best for his family. Whatever else Jeremiah and Mary Watkins may have contemplated for their youngest two sons, they did, it transpires, make effective choices of apprenticeships through masters within the Vintners' and Spectaclemakers' Companies.

Attainment of a company freedom – that is, allowing an individual to practise his trade or craft anywhere within the jurisdiction of the City companies – was a long process that entailed an apprentice being bound to a master who was, by a very strict rule, a freeman of a particular company. The binding, nominally, was for seven years although exceptions to this rule often occurred. Other routes to a freedom were through 'patrimony' (through a parent, usually the father, who was already a company freeman) or 'redemption' (by purchase) or, occasionally, as a 'foreign brother' (member of another company). Neither of these latter options was applicable to William or Francis; therefore, the only course available to them was through 'servitude'. It was by such a course that their respective future masters, Joseph Thorpe (citizen and Vintner) and Nathaniel Adams (citizen and Spectaclemaker), had earned their freedoms and so were qualified to take apprentices.[16] How initial connections were made between these masters and the father may only be imagined but we know that William arrived in London to begin his apprenticeship on 11 April 1733 and Francis took up his apprenticeship on 18 August 1737.[17] At least, young Francis would have received some comfort and reassurance from William's experience of the ways and customs of the capital.

---

14 Campbell (1747), pp. 173, 247, 252, 254 and passim.

15 Campbell (1747), p. 2.

16 Joseph Thorpe (fr. Vintners' Company, 1723). Index to register of freemen, 1721–1733. GL, COL/CHD/FR/03/02/004; Nathaniel Adams (fr. SMC in June 1730). GL: SMC Minute book CLC/L/SH/B/001/2, p. 134.

17 CLRO (Indentures and Freedom Records): ELJL/667/127 (William Watkins) and ELJL/747/63 (Francis Watkins).

Francis's first master – Nathaniel Adams – was a Soho man, being born and bred within his local parish of St Anne. When Francis first met him, Adams was 29 years of age.[18] His own apprenticeship from 1722 had been served in the workshop of the local and very respectable optician and instrument maker Edward Scarlett, whose shop in Macclesfield Street was within sight of St Anne's church. Nathaniel's father, a local glover, probably considered it a great honour that his son should be apprenticed to such an upstanding person who, quite apart from his trade as optician, was a churchwarden, an overseer of the poor and an administrator of vestry matters, including the church accounts. Such were the duties and responsibilities of educated persons of the parish, particularly those competent with figures and calculations.

By the time that Adams was free of Scarlett's binding he had become aware of far more about life than the making of optical and philosophical instruments.[19] Although this was irregular, he married before his apprenticeship had ended on 3 June 1730 and by March 1730 the first child was born.[20] By 1734, he was sufficiently equipped and capable to set up shop on his own account; this he did from premises in Cranbourne Street, near Leicester Fields, but still within St Anne's parish. Also, while living within the confines of Soho, he took over some of the parish responsibilities, including the overseeing of the maintenance of roads. By mid-1737, however, he moved to the adjacent parish of St Martin-in-the-Fields where he traded from a shop adjacent to the Golden Cross tavern on the north side of the statue of Charles-I at Charing Cross.[21] In fact, Francis Watkins's arrival at Charing Cross in August 1737 had immediately followed Adams's move there.

Before a boy could be introduced into the ways of his apprenticeship, there were matters of business to be settled between the master and the father. If Jeremiah-I had accompanied either of his sons to London, then it is likely that he would have met with their respective masters to discuss his expectations and seek out information concerning their prospects and reassurances concerning their well-being. Meetings such as this would have been informal; conversations may have run from explaining the skills required to matters of practicality concerning clothes and food. Most important of all, there would be the need to agree the costs and terms of the apprenticeship. In the case of William, a

---

[18] Nathaniel Adams, bapt. 14 December 1707. WA: Parish register for St Anne, Soho, Westminster. Nathaniel Adams was not related to the well-known instrument maker George Adams (1709–1772).

[19] SMC court minutes 28 November 1722. GL: Minute Book CLC/L/SH/B/001/2, p. 134.

[20] Nathaniel and Hannah had two girls: Hannah (bapt. 5 March 1730) and Elizabeth (bapt. 4 August 1732). WA: Parish register for St Anne, Soho, Westminster.

[21] WA: Poor Rates: Books & Ledgers (St Martin-in-the-Fields). [Microfilm F499 (April 1737)].

binding fee – called 'the consideration' – of 30 guineas was levied whereas for Francis this was somewhat cheaper at 20 guineas.[22] Beyond these gentlemanly agreements, however, there would be the formality, undertaken by the master, to seek permission from the relevant company (the Vintners' in the case of William and the Spectaclemakers' for Francis) to take on the apprentice and thence the task of procuring an indenture form to register the new apprentice.

The Indenture

The form of words employed on the indenture – legally binding upon both master and apprentice – was almost as ancient as the system of apprentices itself. The form supplied to Nathaniel Adams for Francis was nearly identical to the standard format of the previous century. The document itself was doubly printed with blank portions to be filled in appropriately. The due process entailed that both parts be signed and sealed. On agreement and signature by both parties, the clerk would then divide the sheet by cutting (*indenting*) in a wavy or saw-tooth fashion; thus, the separated copies could be matched for authenticity should a need ever arise. Finally, the master was handed one copy and the other copy was retained by the apprentice. The latter copy would then act akin to a passport or transcript by including (on the reverse side) all changes of arrangement that occurred during the apprenticeship.[23]

To conclude the business of starting out into the world of work, some masters or fathers would give the apprentice boy a copy of a pocket-sized tract entitled *A Present for an Apprentice, or a sure Guide to win esteem and an Estate, with Rules for his Conduct to his Master, and in the World.* This compact work, first published in 1737 and originally from the pen of John Barnard, a much-respected Lord Mayor of London at the time of Francis's arrival, was intended as a guide to duties and responsibilities.[24] Five editions appeared in the period from Watkins's binding (1737) until his setting up shop (1747). Thereafter, it became a tradition (into the following century) for a youth to receive a copy of this very practical, if somewhat moralistic, pocket book. Thus, the forbidding tone of the traditional indenture, in which it was abundantly clear that youthful spirits were to be curbed, all temptation of gaming by Tables (backgammon) or Queck (chequer board)

22 Cf. Campbell (1747), 'Table of trades', pp. 331–340, p. 337.

23 Indentures and Freedom Records. CLRO: ELJL/747/63.

24 (Sir) John Barnard, merchant, politician, Quaker, sheriff (1735), lord mayor (1737), president of Christs' Hospital (1740), a powerful speaker and reputedly a thorn in Walpole's side; always popular as a 'City man', advocating the rights of the ordinary liverymen in common hall against the oligarchy of the court of aldermen. Sir John Barnard (c. 1685–1764) www.oxforddnb.com/view/article/1456 accessed 10 June 2011.

This Indenture witnesseth That Francis Watkins Son of Jeremy Watkins at the Parish of Newchurch in the County of Radnor Gentleman *doth put himself Apprentice to* Nathaniell Adams Citizen and Spectacle Maker *of* London to learn his Art and with him (after the manner of an Apprentice) to Serve from the Day of the Date hereof unto the full end & Term of Seven Years from thence next following to be fully Compleat and ended: during which Term the said Apprentice his said Master faithfully shall serve his Secrets keep his lawful Commandments every where gladly do: He shall do no damage to his said Master, nor see to be done of others, but that he is to his power shall let, or forthwith give warning to his said Master of the same He shall not waste the goods of his said Master nor lend them unlawfully to any He shall not commit Fornication nor contract Matrimony within the said Term He shall not play cards Dice Tables or any other unlawful Games whereby his said Master may have any loss With his own goods, or others, during the said Term without License of his said Master he shall nether [sic] bay nor fell, He shall not haunt Taverns or Play-houses, nor absent himself from his said Master's service day nor night unlawfully: But in all things as a faithful Apprentice, he shall behave himself towards his said Master and all his during the said Term: And the said Master his said Apprentice in the same Art which he useth by the best means that he can, shall Teach and Instruct or cause to be Taught and Instructed finding unto the said Apprentice Meat Drink Apparel Lodging and all other Necessaries, according to the Custom of the City of London, during the said Term And for the true performance of all and every the said Covenants and Agreements Each of the said Parties bindeth himself unto the other by these presents, In Witness whereof, the Parties abovementioned to these Indentures interchangeably have put their Hands and Seals, the eighteenth day of August in the eleventh Year of the Reign of our Sovereign Lord George the second of Great Britain Anno Dom 1737.

Sealed & delivered in the Presence of us
(after a Consideracon of twenty guineas paid [?this day])

[signed] Thos Lincoln
[signed] Francis Watkins
[signed] Na^{ll} Adams

Figure 1.3 Apprenticeship indenture of Francis-I Watkins, 1737. Transcript courtesy of the City of London, London Metropolitan Archives

or Cards dissuaded, and rioting against Catholics (which still prevailed in post-Restoration London) cautioned, was moderated by Barnard's more enlightened approach. Nevertheless, it reminded the apprentice of appropriate decorum for all conceivable situations of daily life. For most of the capital's youth, who worked from dawn to dusk, the nine o'clock evening parish church bell signalled a curfew, after which there was little chance to create mischief. Not all of Barnard's tract was restrictive however, and a good deal of common sense was proffered which an intelligent, conscientious youth could apply to good effect, even into his early years as journeyman or master when he would need to know how to manage servants, choose a wife, and deal with matters of money, politics and religion.

The business of setting up the indentured agreement was still much respected in the mid-eighteenth century and, in addition to its major function of binding master and apprentice within a livery company, the indenture was also of interest to the City fathers (the aldermen) to whom it provided a first step for a fully fledged apprentice becoming a freeman of the City. However, there were penalties for those who infringed any part of the agreement. For example, an apprentice who married before the end of his binding without the knowledge of or application to the court of aldermen might not be granted his freedom by servitude and, therefore, he would be forced to obtain it by redemption (that is, by purchase). This would not have mattered to Francis Watkins because he married after receiving his freedom.

A Disturbed Apprenticeship

On 18 August 1737 a court minute of the company recorded: 'Francis Watkins, son of Jeremy, of the Parish of New Church in the County of Radnor, Gentleman, was this day bound Apprentice to Nathaniel Adams for Seven Years from the date ... Consideracōn 20 Guineas.'[25]

This simple record marks the start of young Francis's life in the instrument trade. He was then left in the hands of Nathaniel Adams and his wife who, in *loco parentis*, would have taken their responsibility very seriously although this was not always the case for less reputable masters. If Jeremiah Watkins had accompanied the boy from Radnorshire in the first place then, at least, he could return home comfortable in the knowledge that his son was bound to a master already with the credential of being optician to Frederick, the eldest son of George-II, and therefore the Prince of Wales.[26]

---

25 SMC Court minutes GL: Minute book, CLC/L/SH/B/001/2.

26 Nathaniel Adams was optician to the Prince of Wales according to an inscription on the mirror of a Culpeper microscope, c.1735. Court and von Rohr (1930), p. 81, entry 142; Turner (1989), pp. 34–35.

Any home comforts would now be lost; indeed Francis may not have liked his new surroundings, nor the sleeping arrangements for which his master had provided a bedsettee in the shop (§ Chapter 3). He would, however, have had a companion in John Margas, an older apprentice taken on by Adams two years earlier. It is even possible that he was made welcome by the Margas family whose shop and home was nearby at the corner of Cecil Court in St Martin's Lane. John's father, Jacob Margas, was a second generation Huguenot who traded as a goldsmith, as had his Normandy ancestors in Rouen before coming to England.[27] Here, perhaps for the first time, Francis would have heard spoken French in conversation between members of the family; at least, the language was common enough following the tremendous influx of Protestant families in the district over the preceding decades (§ Chapter 5).[28] Wealthy merchants and bankers (including wigmakers, shoemakers, food and drink vendors, clockmakers, jewellers and smiths of various kinds from gunsmiths to goldsmiths) made their homes in the Leicester Fields and Soho region of Westminster. To cater for their religious and social needs, some 14 French-speaking churches flourished in Westminster alone as well as the Savoyard French church in the Strand, already sanctioned in 1661. Certainly, this lively backdrop was to have its impact on young Francis who later married a Huguenot and also published a work in French (§ Chapter 3).

Watkins was only about half way through the seven-year term when calamity struck with the unexpected death of his master in March 1741 (§ Chapter 3). Under these exceptional circumstances, it became incumbent upon the widow, Hannah Adams – as administratrix of her deceased husband's estate – to arrange for the apprentices to be 'turned over' to some other suitable master according to 'the custom of London'.[29] Naturally, perhaps, she turned to the elder Edward Scarlett – her deceased husband's former master and employee – for guidance. It so happened that in Scarlett's employ at the Soho workshop there was a freeman capable of training young Watkins. This was Henry Walder, son of a Southwark distiller, a man of about 27 years, initially trained under William Radford,

[27] Murdoch (1985), pp. 74–76. There is some disagreement of locations between Murdoch (1985) and Heal (1935). Heal places Jacob Margas in St Martin's Lane between 1706–1720 and King Street, Covent Garden, in 1720. Jacob had two other sons, Samuel and Jacob, who may have joined the goldsmith trade. See also Grimwade (1982).

[28] By 1700 there were some 8,000 French immigrants in Westminster and 15,000 in Spitalfields, constituting roughly 5 per cent of the population. Gwynn (2001), p. 36.

[29] Generally, if a master died, the apprentice was maintained by the executor or administrator; but, the executor or administrator could bind the apprentice to another master for the remaining part of his time. The 'custom of London' ensured that a London apprentice would continue in the same trade but elsewhere a boy could finish with a master of a completely different trade or even be dismissed of no further use, and hence returned home to his father. Burn (1785), vol. I, p. 82.

a Strand optician (§ Chapter 2).[30] The other apprentice, John Margas, had nearly finished his apprenticeship and appears to have made his own arrangement to complete the final months under John Cox, a young optician working in Cow Lane (§ Chapter 2).

The arrangement for Francis to work at Scarlett's had solved Hannah's immediate difficulty: indeed, Hannah wasted little time herself before re-marrying (§ Chapter 3). For Francis, the arrangement meant that his apprenticeship continued without interruption although a change of circumstance must have soon occurred because, within seven months, Francis was turned over yet again. This second arrangement – under the elder Edward Scarlett himself – lasted until early in 1744 by which time he was dead. As executor of his father's estate, the younger Edward Scarlett retained the boy and returned him to Walder's supervision. Of course, turnovers were neither unusual nor unexpected; sometimes they were necessary when a master changed occupations or when an apprentice was unhappy with his binding or, as here, when a master died. Already, Watkins had experienced far more than his fair share of disturbances as shown by the record.[31]

The misfortunes that had befallen the Scarletts may well have caused the younger Scarlett to depend upon Francis, by now a young man of 20 or so years. But the misfortunes did not end at this point because Walder also died within a couple of years of the previous turnover. Since no further turnovers appear on Watkins's record, it is assumed that his final master's death occurred shortly after Francis had completed his apprenticeship.

That the elder Scarlett should have died in his mid-60s may well have been considered a reasonable lifespan for an eighteenth century tradesman; however, that Adams should have died at 34 years and Walder at only 28 years of age are baleful reminders that death was never far away at any age. Diseases such as tuberculosis, influenza, diphtheria and smallpox could hardly be avoided in the face of ignorance. Contaminated water and atrocious sanitation often led to death. Despite the complimentary remarks offered by travellers who proclaimed London water good when compared with other European towns, the quality of Thames water was, in fact, very poor and destined to become much worse as the century progressed. Life in early Georgian London – often mistakenly pictured as bright and cheerful – was more often gloomy; it may, therefore, be significant that the demise of Watkins's masters occurred during the three very worst years ever for smallpox and typhus which peaked respectively in 1740 and 1743.[32]

---

[30] Walder was apprenticed some seven years earlier than Watkins, on 10 September 1730. Webb (1998), vol. 14, *Spectaclemakers' Company 1666–1800, Loriners' Company 1722–1731, 1759–1800*.

[31] Indentures and Freedom Records. CLRO: ELJL/747/63.

[32] Inwood (1998), pp. 278–280.

Freeman's Oath of the Spectaclemakers' Company

You shall swear to be true to our Sovereign Lord the King's Majesty, his Heirs and Successors, and at all Times obedient to the Master and Wardens of this Fellowship and Society, and their Successors after them, in all honest and lawful Things touching the Affairs and Business of this Fellowship. You shall be ready at all manner of Summons, and bear Scot and Lot in all manner of reasonable Contributions of and to this Fellowship and Company of Spectacle Makers, of the City of London. You shall to the best of your Skill, Power and Ability, uphold and maintain the Weal of this Company. You shall not know or suspect any manner of Meetings, Conspiracies, Plot or Devices against the King's Majesty, his heirs or Successors, or the Government of this Fellowship, but you shall the same to the utmost of your Power let and hinder, and speedily disclose to the Master, or one of the Wardens of this Society. And this City of London, and Fellowship of Spectacle-Makers, you shall keep harmless, as much as in you lies. Also you shall be ready at all Times to be at the Quarter Days, and every other Assembly, Matter or Cause that you shall be warned or called unto for the Affairs of this Fellowship, unless you shall have lawful and reasonable Excuse in that Behalf. And the Ordinances of this Fellowship or Society, ratified according to the Laws of this Realm, or otherwise lawful for this Fellowship or Society to make and ordain, you shall to the uttermost of your Power well and truly submit yourself unto and keep.
So help you God.

James Wharton, Clerk.

Figure 1.4 'A copy of the oath incumbent upon all Freemen of the Spectaclemakers' Company decided on 28 June 1759'. Transcript courtesy of the City of London, London Metropolitan Archives

Beyond Apprenticeship

Since his arrival in Westminster, Francis would have been watchful and keen, and maybe even an ideal, obedient apprentice; probably a great deal of responsibility was placed on him during the crises just described. As a matter of course, he would have learned many skills – both manual and intellectual – and, viewed in retrospect, he appears to have been quick to comprehend and put into practice the latest ideas and techniques. In time, he may have been allowed contact with customers at Scarlett's shop, including persons from the Royal Society and eminent foreign visitors, thereby learning the art of how to oblige and, most importantly, how to present his master's business by offering thoughtful and tactful advice to those clients who were less aware than himself of the limitations of materials or the possibilities of design. In terms of his personal education and scientific understanding, he would have become familiar with the relevant texts, such as the optical works from Father Chérubin,[33] Isaac Newton and Robert Smith, or, at least, he would follow his masters' practical interpretations of these works. Of particular interest during the latter years of his training was the appearance of Henry Baker's *The Microscope Made Easy*, a work of immediate popularity which, inevitably, would have boosted Scarlett's trade (§ Chapter 3). Movements on the scientific front would certainly have greater appeal than the humdrum business of making up spectacles according to some standard procedure. He would also be conscious that the prices of certain instruments (microscopes in particular) had begun to fall in inverse proportion to their widening appeal. And so, despite the set-backs over the years – including the deaths of his mother and the three masters – he would have matured sufficiently to consider his own future when, one day, he would wish to trade on his own account. As a first step in this direction – presumably with Scarlett Jr's consent and encouragement – Watkins applied for his freedom from the apprenticeship bindings.

Becoming a Freeman of the Spectaclemakers' Company

Francis Watkins was granted freedom of the Spectaclemakers' Company shortly after Lady Day 1746, at a general court held on 27 March. The formality of this occasion – according to custom – was marked by swearing an oath, most probably read aloud before the master, the wardens and other assistants constituting the court. But, along with the freedom came a new set of obligations potentially more onerous than those in the apprenticeship indenture.

Sometimes masters were lax about their company obligations although there was still much value to be gained from its protection through control of the trade. Reminders of their responsibilities were issued from time to time, to wit 'to the End that no Members of the said Company may pretend Ignorance thereof for the future....'[34] On gaining his freedom Watkins would not only be made aware of the company by-laws and ordinances, but also he may have felt considerable pride in swearing allegiance to both king and company. Indeed, being a freeman of the Spectaclemakers' Company was just as much an honour in itself as it was a mark of standing within the trade.

---

[33] Michel Lasérré (1613–1697), a Carmelite known by his name in religion of Chérubin d'Orléans, was a well-known writer on optical matters. (Editors' addition.)

[34] For example, SMC Court minutes (28 June 1759). GL: Minute Book CLC/L/SH/B/001/2.

For the privilege of freedom, Watkins would have paid an initial fine (or fee) of £1 5s 10d and, thereafter, about sixpence (but rising to one shilling in 1759), as 'quarterage' on each of the Quarter Days (Lady Day, Midsummer Day, Michaelmas Day and Christmas Day) from the wage he received from Scarlett Jr.[35] In fact, being free did not mean that he would leave Soho immediately. He remained there for a further year in order to repay his debt of gratitude to this kindly family who had taken him in upon the death of his first master. During this period (possibly earlier) it appears that a plan was put in place for him to return to the vacant shop of his first master.

First Years at Charing Cross

The Golden Spectacles at Charing Cross, where Watkins had begun his apprenticeship under Adams, had been occupied – as a residence only – for a couple of years by John Margas immediately after Adams's death; thereafter, the premises remained empty. Whether as part of some general arrangement made at the outset between the Scarletts and Hannah Adams (and her new spouse) when Watkins had been taken in by the Scarletts, or, more likely, between Francis himself and Scarlett Jr following the death of Henry Walder, it seems that an amicable plan was evolved to ensure that Francis had a good start to his working life. Thus it was that he returned to Charing Cross – during March 1747 – to set in action a plan to capitalise on the broadening public interests in experimental philosophy (§ Chapter 3).

Not only was the young instrument maker about to enter into business on his own account but also a significant development took place in his personal life. During his time at Scarlett's – while working under Walder – Francis Watkins would have become acquainted with Walder's wife Clarinda and her children, Henry and Elizabeth. He may even have lodged with this family and, on Henry's death, he would surely have offered Clarinda comfort and consolation in her grief. In terms of age, Clarinda was six years older than Francis but, importantly, from Francis's standpoint, she knew the trade well and what was to be expected of an instrument-maker's wife. Whatever difficulties this may have caused within the Scarlett business, it seems that Francis may have offered Clarinda and her children the security that she needed and – in this close pattern of living and working – it is not surprising that they ultimately came together as husband and wife.

Exactly when and where Francis and Clarinda Walder (née Bontin or Bentin) married has, so far, evaded detection although – given the rarity of Clarinda's family

[35] SMC Court minutes. GL: Minute Book CLC/L/SH/B/001/2; see also Crawforth (1987), pp. 322–325.

name in England and the knowledge that her own parish was given as Spitalfields in the marriage allegation associated with her previous marriage to Henry Walder – it is possible that the event took place in one of the several Huguenot churches whose records are no longer extant. That they spent their very long married life together is absolutely in no doubt and the likelihood of Clarinda's French ancestry explains, albeit tentatively, how Watkins came to write a work on microscopy in the French language (§ Chapter 3). It is clear, however, that Watkins began his own enterprise with a ready-made family in tow. Once settled at Charing Cross, his own offspring – Frances (b. 1749) and Clarinda (b. 1753), duly baptised at St Martin-in-the-Fields – were added to the family life. At this time Elizabeth and Henry Walder were already eight and six years old, respectively. Francis Watkins did, indeed, fulfil his honourable intention of providing for Clarinda so that her children would have an upbringing every bit as solid and secure as his own had been at Newchurch.

Becoming a Freeman of the City of London

The registers of the Corporation of London record Francis Watkins being admitted a freeman of the City of London in October 1750.[36] This event was quite distinct from the freedom gained three-and-a-half years earlier from the Spectaclemakers' Company. There were, of course, several advantages to becoming a City freeman and for many it was an appropriate action to take because of the privileges that it conferred; it was not, however, the honour as might be construed today. In fact, thousands of ordinary company freemen went on to become City freemen simply by dint of their apprenticeship servitude.

The first step in the process actually began within 12 months of first registering an apprentice with a company when a master would enrol his apprentice at the chamberlain's court. It is therefore possible that young Francis was enrolled by Nathaniel Adams within the qualifying period although if Adams had been reluctant or unable to afford the fee then the decision for enrolment may have come from Francis himself at a later stage; in this case, a higher admission fee would have been be levied. In principle, admission had to be backed by a certificate from a former master although, in Watkins's case – with the decease of all three masters – it is possible that he was backed by the Spectaclemakers' Company or direct certification by Edward Scarlett Jr, his former employer. The formal proceedings of admission normally took place in the chamber of the court of aldermen whereupon the clerk of common council would administer the 'freeman's oath'. As with his oath as freeman of the Spectaclemakers' Company he would have to swear an allegiance to the sovereign (George-II) but rather

[36] CLRO: Alphabets of City Freedom Admissions.

than declare his intention to uphold the values of the company as already carried out in 1746, he would, instead, swear to be 'good and true' to the lord mayor and to the laws and customs of the City.[37]

Historically and uniquely, the City of London had preserved its individuality and distinctive jurisdiction over the original 'square mile' and its immediate extended boundaries, that is, encompassing the City's 26 wards together with 'the Bridge' (*London Bridge*) and 'Bridge Without' (*Southwark*). They did not include Westminster, where Watkins lived, nor was there any corresponding freedom to be sworn in the City of Westminster. Thus, along with scores of others, Watkins – at 27 years of age – applied and was admitted freeman of the City of London. Along with this freedom came several rights, including: the right to vote for members of parliament; exemption from naval impressment; exemption from market tolls; the right to trial in the City's own courts (which might have been useful had he lived within the City but traded outside); provisions for orphans (a serious matter to consider when death was never very far away); provision for old age (almshouses); the right to hunt in Middlesex; and, lastly, possibilities for election in the civic hierarchy (as alderman or even lord mayor). From what is known of Watkins's lifestyle, it would seem that few, if any, of these privileges would have made a difference to his daily life. For example, as a citizen of Westminster and *not* the City of London, he was already allowed to vote in a by-election held the year prior to attaining his City freedom. Yet, there were other matters to consider as a family man with wife and responsibilities of fatherhood. He may, for instance, have considered the prospect of working in the City of London where trade was strong and a City freedom could have been of advantage, but this is mere conjecture. These distinctions, often forgotten today, were still important in 1750 and it is therefore relevant to note that all legal documents relating to dealings in property were signed 'Francis Watkins, optician and citizen of Westminster' thereby emphasising that he was *not* considered a citizen of London, despite holding a freedom of the City of London.

Taking on an Apprentice

Becoming a freeman of the Spectaclemakers' Company not only meant that Watkins could trade as a master on his own account but also that he could take on an apprentice. Under ideal circumstances, a 'generation' – in terms of a succession from master to apprentice to master again – might be considered

37 The swearing of the freeman's oath on a bible probably differed little between 1750 (when Watkins took this oath) and 1830, by which time it could be sworn according to other religions. Aldous (1999), p. 34.

as the equivalent of two apprenticeship terms, that is, 14 years. Such a vague definition was sometimes used to explain why a 14-year duration was an appropriate term for the life of a patent because two apprenticeship terms were presumed an adequate period in which to hand on the vital know-how contained in an invention (§ Chapter 4). In fact, the period between the start of Francis Watkins's own apprenticeship in August 1737 and that of his first apprentice in October 1750 was a little over 13 years and so Watkins had completed the ideal 'master–apprentice–master' sequence just within one generation. Generally, a master was entitled to his first apprentice after about three years of trading, and then allowed a second apprentice when the first was about half way through the term, thus permitting a maximum of two at a time. Adhering to this simple and practical rule helped to control the trade although it was ignored by some masters. Nevertheless, this rule appears manifest in Watkins's case since he followed it by adopting a second apprentice in June 1753.

The first apprentice was Addison Smith, a young man from Macclesfield, Cheshire. The actual arrangement was somewhat unusual since he was already 18 years of age on joining Watkins, that is, well beyond the customary age for starting an apprenticeship.[38] He had been sent to the capital by his parents – William and Maria Smith – in order to seek out a life for himself. In Westminster, he lodged with (and possibly worked with) his uncle, Thomas Anderton, taking up an apprenticeship with Watkins on 4 October 1750. A less curious arrangement was that of Watkins's second apprentice, Henry Pyefinch, son of a Herefordshire gentleman, who was registered at the customary age on 28 June 1753. Both these apprenticeships were secured for a consideration of 40 guineas. Correspondingly, the court minutes of the Spectaclemakers' Company reveal Watkins's presence at the General Quarter Courts of 23 October 1750 (shortly after Michaelmas Day) and 28 June 1753 (shortly after Midsummer Day), that is, on the two occasions at which apprentice assignments were made to masters.[39]

In time, Addison became familiar with Clarinda's four children. Not surprisingly – given the daily routines with the Watkins family – Addison began to 'walk out' with Elizabeth Walder, presumably with the encouragement and blessing of both Clarinda and Francis. However, Addison did not take his freedom until 1763, that is, some six years beyond the minimum required for servitude, perhaps thereby intending to avoid payment of the quarterage but more probably influenced by his master who did not feel obliged to subscribe to a

---

[38] Addison Smith (bapt. 1732) was the son of William Smith and his wife Maria (née Anderton).

[39] SMC Court minutes. GL: Minute Book CLC/L/SH/B/001/2. Watkins made his first appearance as Assistant at a court meeting for the first Thursday after Lady Day 1750, probably the point at which he applied for his first apprentice. He did not then appear until 28 June 1753, that is, in connection with his application for a second apprentice.

City-based company whose rightful jurisdiction in Westminster was increasingly held in question. By this time, Addison was 31 years old and Elizabeth 22 years old. The couple married on 8 May 1763 at St Martin-in-the-Fields and, about the same time, set up shop in the Strand, towards Lancaster Court. But even if Francis Watkins had lost his apprentice and Clarinda her daughter by her first husband, the young couple were tied to their elders for some years through the partnership that Francis arranged for Addison at the start of his married life.

Meanwhile, Clarinda's son, Henry-II Walder – whose leanings were elsewhere than in the instrument trade – joined the navy to become a midshipman on board the fourth-class frigate HMS *Deptford*.[40] Whatever happy state may have existed at the time of Addison and Elizabeth's marriage, it was overturned when nearly six months later – possibly due to an accident at sea – Henry-II died. As requisite of all naval servicemen, he had prepared a Will, stipulating his sister Elizabeth as sole beneficiary of his meagre belongings. Also marring the daily life of the business – concurrently with the family grief at the loss of Henry – both partners (Francis and Addison) became engaged in a fearsome battle with the clever, self-absorbed and much opinionated optician, Peter Dollond who had taken out legal proceedings against the partners. This trouble had arisen after Peter Dollond succeeded to his father's patent for 'combined object glasses' for which, in 1758, there had been a formal agreement for Watkins to market these lenses in refracting telescopes (§ Chapters 4–6). It is sufficient here to note that Peter did not wish to continue his father's obligation with Watkins and, in an underhand way, he set in place measures to have the agreement annulled.

Launched in Business

The boy who left Radnorshire in 1737 bound for the capital had emerged, ten years later, fully fledged as an instrument maker and ready to embark upon a very demanding career. It was very fortunate that Francis Watkins had been accepted by the Scarletts soon after the death of his first master, Nathanial Adams. Whether this was an act of kindness or convenience on the Scarletts' part, it certainly was an opportunity for young Francis to gain a wider experience than he would have received under his original master at Charing Cross. Edward Scarlett Sr's reputation for inventiveness and quality of workmanship were known beyond the shores of the British Isles, as may be deduced from the wide range of goods appearing on his trade card.

---

[40] Will of Henry Walder [Jr]. TNA: PROB 11/893 f 536, pp. 376–377; proved 14 November 1763 by 'the oath of Elizabeth Walder (wife of Addison Smith) the sister of the deceased.'

During 1721 and 1722 Scarlett was Master of the Worshipful Company of Spectaclemakers and it is significant that he was one of the very few instrument makers among the 300 subscribers to Robert Smith's highly academic text, *A Compleat System of Opticks* (1738).[41] In his professional capacity, as optician, he is reputed to have made his mark by inventing the 'temple spectacle frames' but his distinction is probably better noted through his royal warrant as optician to King George-II. On a local basis, his parish duties as churchwarden at St Anne's, Soho, entailed administering aid to the poor. These were all marks of intelligence and respectability and so Watkins, it seems, had joined a well-ordered household and a top-class instrument house. Indeed, for five of his most formative years, he was guided through the remainder of his apprenticeship to emerge as one of the most competent opticians of the mid-eighteenth century.

Robert Boyle relates in *The Usefulness of Experimental Natural Philosophy* (1671) that: '... inventions of ingenious heads do, when once grown into request, set many mechanical hands a-work, and supply tradesmen with new means of getting a livelihood, or even enriching themselves.'[42] It was, in fact, the glass grinders and mirror polishers of the seventeenth century that became the bedrock for the newer optical trade. Boyle also adds: '... others have had profitable work laid out for them, by the newer directions of some English gentlemen, deeply skilled in dioptricks, and happy at mechanical contrivances; insomuch that now we have several shops that furnish not only our own virtuosi, but those of foreign countries...' It was these 'several shops' that took on a new lease of life following the foundation of the Royal Society in 1666, when microscopy was in the ascendant and the refracting telescope became of paramount importance to astronomers and seafarers alike. By the end of the seventeenth century, there were a dozen or so noteworthy optical instrument makers in the capital supplying the needs of scholars and gentlemen and those intent on naval or military careers.[43]

Knowledge of the optical trade before the foundation of the Royal Society is hampered by the loss of master–apprentice records in the Great Fire of 1666. We know of one important craftsman who was active before the trade found a secure footing. This was Richard Reeve, an educated man, possibly a turner by trade who, in the 1640s, quickly learned the techniques of optical figuring in response to the demands of Sir Charles Cavendish and the mathematician

[41] Robert Smith (1689–1768), Plumian professor of astronomy and experimental philosophy at Cambridge University. Another instrument-making subscriber was James Ayscough: Smith (1738), vol. 2, p. 407.

[42] Boyle in Boas-Hall (1965), p. 161.

[43] Clifton has revealed over 50 masters engaged in the making and selling of optical instruments between the early seventeenth century and 1720. Such a number would not be out of keeping with the dozen or so estimated here. Clifton (1993a), p. 342.

John Pell.[44] Reeve's name appears in letters and diaries of several eminent persons, including Samuel Pepys, Christopher Wren, John Evelyn, Robert Boyle and, not least, Robert Hooke who cited him in the *Micrographia* (1666).[45] Indeed, such was Reeve's eminence that he may now be considered founder of the Westminster dynasty of optical instrument makers. It was this distinguished line of 'master–apprentice–master' generations that Francis Watkins entered when he fell under the supervision of those who worked at Scarlett's instrument house.

These City of Westminster opticians were bound by the same rules and conventions of the Spectaclemakers' Company as their brethren in the City of London, although their tendency to remain in Westminster during their working lives creates an impression of a separate identity. To some degree, the cohesive forces that held this group together may have been due more to the sense of greater freedom in Westminster as provincial and immigrant workers flooded in to the rising Georgian capital. By contrast, those who worked within the City bounds were not only subject to the laws and by-laws of individual companies, but were also subject to the strict control of the mayor and his aldermen, whose jurisdiction and political import had changed little since the reign of the Tudors.

When Watkins's career began, Newtonian experimental philosophy was not unknown among those for whom opportunity for polite learning and the lecture courses offered by J.T. Desaguliers (d. 1739) were of considerable importance. The course of lectures offered locally at the King's Experiment Room in Duke's Court, near the Royal Mews, which followed the style and quantity of Desaguliers, may be seen as a continuation of the effort to maintain an appeal to Newton's legacy. Whether any of the trainees at Scarlett's workshop attended this series is not known although they might suggest a reason why Watkins directed his career towards experimental philosophy when he eventually opened up at Sir Isaac Newton's Head.

To introduce this period generally, the following chapter sets out to identify and locate some instrument-making contemporaries, in both London and Westminster, with whom Francis Watkins would have had an acquaintance (§ Chapter 2). That account straddles the mid-century, about the time at which Watkins entered the trade, although it is set against a backdrop of well-known history and topography in order that the reader should be orientated to a capital that was far less sophisticated and very much smaller than the one we know today.

---

[44] Simpson in: Hunter and Schaffer (1989), p. 37; see also Simpson (1985), passim.

[45] Simpson (1985); Richard Reeve (d. 1666) www.oxforddnb.com/view/article/48372 accessed 6 July 2011.

Chapter 2

The Optical Community in Eighteenth-century London

In 1700 metropolitan London was on the verge of becoming the largest city in Europe, challenged only by Paris. Some 600,000 persons lived within the City of London and the developed parts of Westminster and Southwark, with housing beginning to appear in the adjacent areas of Essex, Middlesex and Surrey.[1] After the devastation brought about by the Great Fire of 1666, commerce and trade had speedily been put back on an even keel. But this desire to return to normality meant that the grand plan of Christopher Wren to rebuild the capital with wide boulevards and piazzas was not taken up. Instead, much of the new building, now of brick rather than wood, went up along the same narrow roads and alleyways. According to Daniel Defoe, in 1710,[2] some post-Fire shop owners spent upwards of £200, not in stocking up but rather in the fitting-up with fine shelving, gilding, glass-paned doors and painting. Generally, most instrument-making workshops were to be found in the back streets, but a few fronted the main thoroughfares and so probably went to this sort of extravagance.

By the time that Francis Watkins was established at Charing Cross a visitor to London would have no difficulty in identifying the shops of opticians by their hanging signboards.[3] A few members of the Spectaclemakers' Company depicted their trade by the sign of 'Archimedes' accompanied by one or more pair of 'golden spectacles'. Others favoured eminent figures of more recent history: thus, Tycho Brahe's Head signified instruments of great accuracy (George Adams in Fleet Street); and Sir Isaac Newton's Head represented shops where the apparatus of experimental philosophy could be found (Matthew Richardson in the Strand, James Mann and James Ayscough in Ludgate Street, and Watkins at Charing Cross). Sometimes, the specialist trade of a shop was depicted by an individual instrument such as a globe (the elder Scarlett in Soho), or Hadley's quadrant and spectacles (Joseph Linnell).

---

1 For population estimates, see http://www.oldbaileyonline.org/static/London-life18th.jsp accessed 24 August 2011.

2 Defoe (1726), pp. 312–322, esp. p. 315.

3 Hanging signs were not banished generally until 1762 when street numbering was advocated. Heal (1947), p. 2.

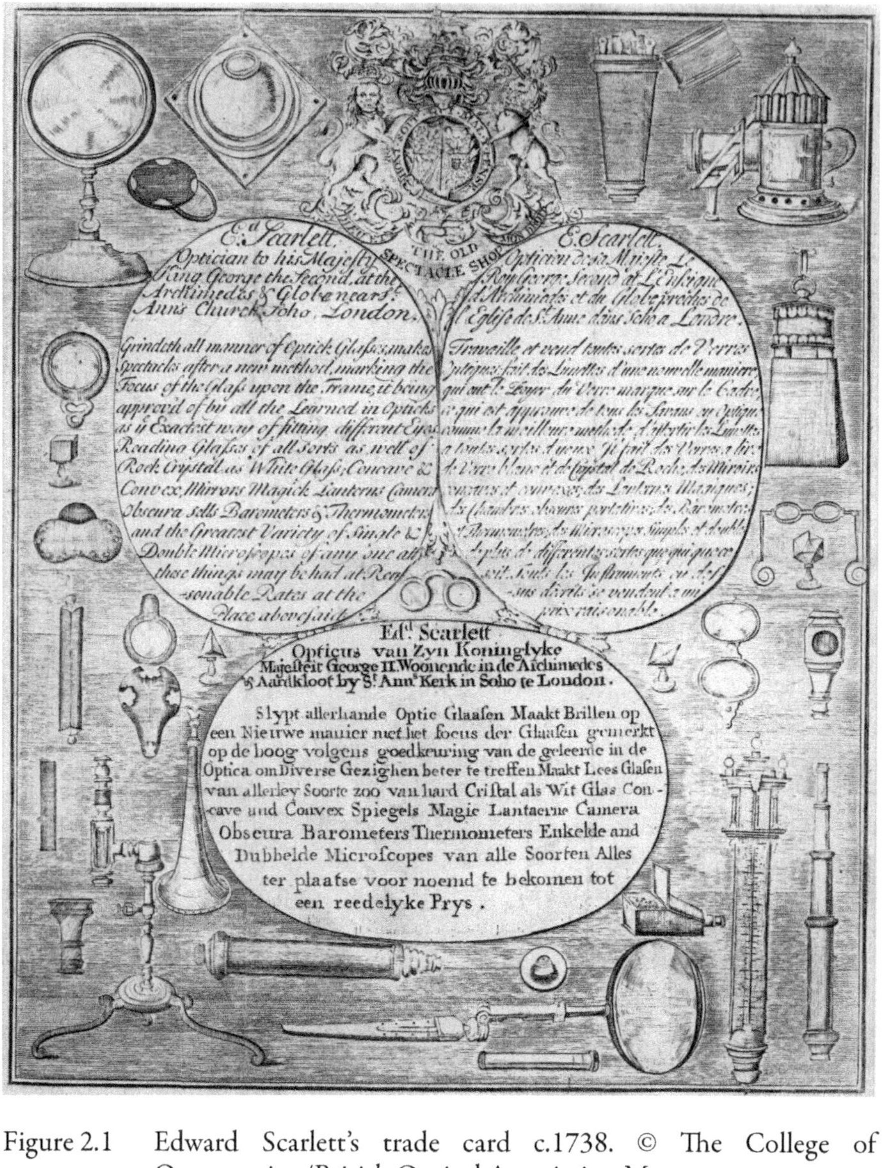

Figure 2.1 Edward Scarlett's trade card c.1738. © The College of Optometrists/British Optical Association Museum

Of course, window displays may have included other instruments to be found within the shop although less obvious to the prospective customer would be the price and quality. Recommendation by word of mouth would have been as important then as it is today; even more welcome, from the instrument-maker's standpoint, was to have a recommendation in a popular text. No doubt the optician John Cuff was forever grateful to Henry Baker, FRS, for twice mentioning his name and the exact location of his shop in *The Microscope Made Easy*.[4] A purchaser unfamiliar with the metropolis might consult one of the newly published street directories (such as that by Lowndes). Armed with such a guide, perhaps also with a copy of John Pote's popular compendium, *The Foreigner's Guide; or a necessary and instructive companion both for the foreigner and native, in their tour through the cities of London and Westminster*, a prospective customer from abroad or the provinces would quickly be able to locate the maker of his choice.[5]

Francis Watkins had no need of directory or guide. Over the years of training spent at Charing Cross and Soho running errands for his masters and obliging customers, he would be very familiar with the locations of instrument shops and suppliers. He would also glean much through daily gossip and soon become familiar with those masters who excelled as well as those of dubious reputation. Despite all that is now lost by way of record, it is still possible to catch a glimpse of the world of opticians into which this young man had entered from the remnants of extant notes to be found in a miscellany of sources. This chapter, therefore, presents some of this material in the context of those locations where commerce and scientific activity could be found, both within and without the City of London.

The City of London

London had spread well beyond its original Roman-walled 'square mile' bounded by Saxon times to include a gated area outside the walls. An even wider area was swept within the compass of the City by the thirteenth century with the newer boundaries being redefined according to the Bars. Echoes of former times can still be detected today by place names such as Aldgate, Ludgate, Temple Bar or Holborn Bar.

---

[4] Baker (1742) told his readers: 'Mr. Cuff, against Serjeants-Inn Gate, in Fleet-Street, has taken great Pains to improve and bring them [the microscopes] to Perfection....' (p. 21) and 'Mr Cuff, in Fleet-street, has already made some Boxes after this Pattern, which prove very convenient. He likewise makes and sells all the Microscopes with their Apparatus mentioned in this Treatise.' (p. 61n). See Turner (1974–1975), p. 61.

[5] Anon [John Pote] (1763). For other contemporary works of topographical interest, see Webb (1990).

Instrument shops were spread throughout this wider notion of the City: stretching from Limehouse in the east (well beyond the Tower) to Smithfield in the north, and to Holborn and Fleet Street in the west. The leading instrument makers could afford shops on the principal east-to-west thoroughfare which ran through the heart of the City from Aldgate to Newgate via the Royal Exchange and Cheapside, also round St Paul's Church Yard from the eastern entrance at Watling Street to the westward exit at Ludgate Street; and still within the extended area of the City, the continuation of the ancient roadway running from the City at Ludgate, leading westwards, and covering the region around Fleet Street, from Bridewell to Temple Bar. Lesser men, and many subcontractors inhabited an extended area north of the old Roman walls, from Moor Fields (to the northeast) to Hatton Garden (to the northwest), through West Smithfield and Clerkenwell Green (to the north); craftsmen serving the maritime trades settled on the old route closer to the Thames, from Wapping and Limehouse, via the Minories and Tower to Eastcheap and Cannon Street. The area behind the Royal Exchange, centred about the cross roads formed by Leadenhall Street and Cornhill (running east–west) and Gracechurch Street and Bishopsgate Street (running south–north) was established as a commercial centre in ancient times and this remained so.

The Exchange itself had been burned down in the Great Fire of 1666; although escaping the conflagration, poignantly perhaps, was a statue of Sir Thomas Gresham who had built the first Exchange in the 1560s. Gresham House, a considerable building in Bishopsgate and once the home of this venerated family, became Gresham College in 1596, with seven professors whose duty it was to lecture on astronomy, geometry, physics, law, divinity, rhetoric and music.[6] It was here that men of conversation and ideas, such as Pepys, Dryden and Evelyn, rubbed shoulders with others of a scientific persuasion, such as Boyle, Wren, Newton and Halley. It was after hearing one of Wren's lectures that the gathering decided to form a Society for promoting 'Physico-Mathematical Experimental Learning' which ultimately became the Royal Society of London and provided stimulus for the advancing natural philosophy. Robert Hooke was commissioned 'to furnish the Society every day they mett, with three or four considerable experiments.'[7] In return for his services, Hooke lodged at Gresham House until his death in 1703. Given Hooke's association with Gresham College and his position as operator for the Royal Society, it is not surprising that optical instrument makers could be found hereabouts, likewise the specialist glass grinder John Mellen.

In Little Tower Street, not far from the Tower and close to the Navy Office before its removal to Whitehall, was Watts' Academy where Benjamin Worster

---

6 'Gresham College' and its lecturerships continue, now housed by the Mercers' Company in Holborn, London; see http://www.gresham.ac.uk/about-us/about-gresham-college accessed 22 December 2011.

7 Cited in Andrade (1960), p. 5.

spread something of what he had learned by attending one of Desaguliers' inspirational courses.[8] But demands for instruments from schools would be minimal compared with the needs of shipping. Numerous mathematical, nautical and optical instrument makers could be found in the dockland region stretching as far as Limehouse and Poplar in the east.[9] One significant maker, located in the Minories – a district by the Tower called so after a Franciscan order of 'minor friars' (Grey Friars) – was owned by the Gilbert family, in business till about 1845. On the western side of the Tower on Tower Hill was the Royal Mint where Isaac Newton was Master from 1699 until his death in 1727. From here, the important roadway of Fish Street Hill led to the only Thames crossing at London Bridge.

Before the Great Fire the area round St Paul's Churchyard was occupied by booksellers; post-Fire, however, it was the newer interests in experimental philosophy that captured the interests of passers-by. It was hereabouts that several important optical shops flourished, particularly in the short stretch of Ludgate Street as it descended from the great west door of the cathedral to the Fleet Ditch. Beyond Ludgate, several optical instrument makers had shops in Fleet Street and workshops in its adjacent courts and alleyways. This road was paved even in Watkins's day. It had been a busy area in the seventeenth century when the Great Fire consumed half its length along the north side. However, despite this interruption, commerce was speedily restored. Child's Bank – the oldest bank in England – was set up here in 1673 and much used by the aristocracy and well-known figures such as Pepys, Dryden and Walpole. Other banks quickly followed, such as Gosling's at the sign of 'The Three Squirrels' and Hoare's at 'The Golden Bottle'. Taverns were plentiful in this region: 'The Apollo' was a good meeting place and at 'The Queen's Head' James Hodgson delivered his lectures on natural philosophy with the assistance of Francis-I Hauksbee.[10] Fleet Street already had a long-standing history of printing by the eighteenth century. William Griffith had a press in Shoe Lane in the 1560s and John Murray took over the long-established printing house of William Sandby, near the Temple Gates, in 1793. Of greater relevance here was the printing business of Francis Watkins's brother-in-law, William Owen who married Mary Watkins and moved to Fleet Street in the 1740s (§ Chapter 1). Owen published a wide variety of works, including most of Benjamin Martin's scientific output, and also Watkins's work on the electrical machine (§ Chapter 3).[11]

---

8 *A Course of Experimental Philosophy* to be Performed by Benj. Worster, A.M., and Tho. Watts at the Accomptant's Office in Little Tower Street (n.d.). See also Worster (1722).

9 James Legon has recently identified some 30 entries of mathematical, nautical and optical instrument makers from six of the principal dockland parishes registers (Rotherhithe, Wapping–Stepney, Shadwell, Limehouse and Poplar). Visit: http://www.parishregister.com/ accessed 24 August 2011.

10 Stewart (1999).

11 Millburn (2004).

The City of Westminster

The locations where optical instrument makers could be found west of Temple Bar are, to some degree, a consequence of the natural event of the Plague (1665), the accidental event of the Great Fire (1666), the political consequences arising from the effects of the Civil War (1642–1650) and the Restoration of the monarchy (1660–1661). Before the mid-seventeenth century there was very little development beyond West Church Lane (later called St Martin's Lane); indeed, the parish churches of St Martin-in-the-Fields and St Giles-in-the-Fields at each end of this lane were truly in a rural setting. Some spilling over of tradesmen and artisans from the City into the Lincoln's Inn district had already begun during the Plague and more changes were to follow. One of the main causes of the westward shift of commerce after the Great Fire was the high rental cost of the new brick-built properties in the City. According to a census of 1673, over 3,000 of these newer properties remained empty. Many tradesmen discovered that their businesses flourished better, taxes were lower and communal duties less onerous beyond the City boundaries, particularly in the district surrounding Covent Garden. This region, bounded by the Strand, Drury Lane, Long Acre and St Martin's Lane, was the oldest commercial region of Westminster and it was mostly to the Strand that the instrument trade came.

An electoral poll taken in 1749 – in which Watkins's name appears – shows that the voting population of Westminster (comprising some 395 occupations) was less than 9,500;[12] for contrast, another poll of Westminster (although, by now, combined with St Mary-le-Bone) taken in 1800 shows that the population had reached 225,000.[13] Although these counts cannot be fairly compared, like-for-like, they do give a sense of the dramatic increase in the number of inhabitants in the second half of the eighteenth century. There was a tremendous growth of new estates following the Restoration, when the landed gentry and ambitious gentlemen flocked back to London from the country. A few of those aristocrats, having great fortunes, built houses on a grand scale: surviving examples of which include Marlborough House, Dover House and Burlington House. Others preferred to live in spacious houses, set in squares, overlooking private gardens. Even before the Restoration, the Earl of Salisbury had caused Bloomsbury Square to be laid out and was well ahead in the sale of plots for good brick dwellings and shops. This was followed by Berkeley Square (built by the Duke of Devonshire) and Hanover Square (built by the Earl of Scarbrough).

---

[12] The totals within each occupation spanned from victuallers (846) and carpenters (395) to sedan chairmen (88) and watermen (41).

[13] Inwood (1998), particularly chapter 9 and p. 263.

Although the West End estates with their fine squares were intended for members of the highest class, they also provided residences for those of middle-class professions: lawyers, clergymen, teachers, physicians, surgeons and apothecaries. Encroaching on these, in the meaner streets around the squares, was a mixture of ordinary shopkeepers and skilled artisans: chandlers, peruke makers and sedan chairmen. Such was the diffusion of the lower classes into these regions that, by 1749 (according to the Westminster poll), it is clear that cheesemongers and cabinet makers were just as likely to be living in St James's as the lords, earls or other titled persons for whom the properties were built; a similar mixture could be found near St George's in Hanover Square where clockmakers and poulterers rubbed shoulders with the aristocracy. Moreover, the region around St Giles, only 300 yards from the very fashionable Bedford Square, had become an unsavoury district, tainted with beggars, thieves and prostitutes in the back-alleys. Despite these darker places, the new estates formed 'little towns' and so, at last, Georgian London, with its welcoming shops, was opened up. Indeed, to capture the custom of foreign visitors, the shopkeepers – instrument makers being no exception – sometimes printed their tradecards in the common foreign tongues: Spanish, Dutch, French.

On leaving West Strand at Northumberland House, the street turned towards White Hall. There had been no instrument maker here since the demise of Thomas Tuttell in 1702. Thirty-five years were to pass before Nathaniel Adams moved to a shop adjacent to the busy Golden Cross coaching tavern on the north side of the Cross. In fact, it was to this shop that young Francis Watkins came in 1737 although within the year Adams moved to more commodious premises on the east side of the Cross.[14] Adams traded under the sign of the 'Golden Spectacles' and claimed his privileged title as Optician to His Royal Highness, the Prince of Wales – Frederick, son of George-II. This appointment is engraved around the mirror of a Culpeper-type microscope.[15] Also arriving at Charing Cross with Adams was his first apprentice, John Margas. During the time that Francis Watkins lived here, the shops, taverns, coffee houses and residences that made up the local community had, as their principal focus, the famous bronze equestrian statue of Charles-I – the oldest statue in the capital still *in situ* today – erected by Christopher Wren in 1674 on the former site of an Eleanor Cross. Proclamations were read from here, even as late as 1901 when the death of Queen Victoria was announced. Congregations were common on these public occasions and a contemporary watercolour clearly shows that the upper stories of Watkins's premises provided a good view for spectators.

---

14 Poor Rates: Books & Ledgers (St Martin-in-the-Fields), WA: F499 (April 1737) shows Adams paid £1 13s. 4d (first rate) and £2 18s. 4d (second rate); the final entry on north side of Cross is given in WA: F502 (April 1738).

15 Turner (1989), pp. 34–35.

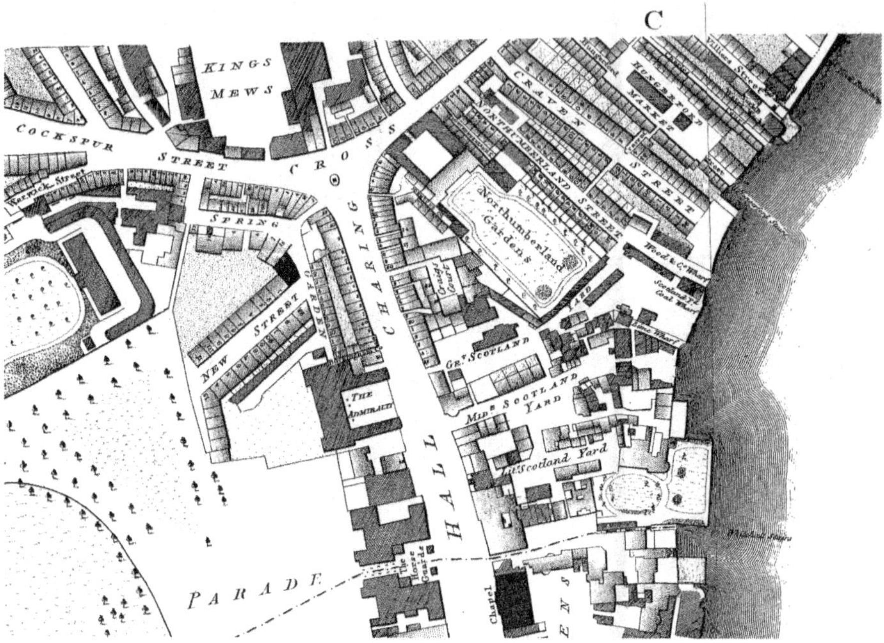

Figure 2.2 Charing Cross: detail from Richard Horwood, 'Plan of the Cities of London and Westminster ...', 1792–1799. City of London, London Metropolitan Archives

On the western side of the Charles-I statue was Drummond's – one of the oldest banks – where Watkins opened an account in 1750. Behind the bank was Spring Gardens; nearby was Cox's Museum. Taverns were popular places of exhibition and in the Long Room of 'The Mitre' tavern, in 1741, Henry Bridges's 'Microcosm; or World in Miniature' was displayed. This remarkable piece of clockwork invention, complete with integral paintings, music and moving figures, comprised two planetaria: one for the solar system (with acclaimed proper proportional eccentricities and capable of being driven to compress ten months into ten minutes) and the other was specifically for Jupiter and its satellites in orbit around the Sun.[16] Public shows (both scientific and mechanical) held here would no doubt have influenced Watkins and Margas in their formative years under Nathaniel Adams (§ Chapter 3).

Only a dozen or so optical instrument makers held any special importance during Watkins's first decade in the trade. There were, however, a great many more workshops hidden in the back streets and peripheral regions from where he

16 Millburn (1995b).

may have purchased materials or small items necessary for his craft. By the mid-century, a few of these more prominent opticians began to deal in philosophical instruments; five of these opticians – Richardson, Watkins, Lincoln, Johnson and Shuttleworth – incorporated 'Isaac Newton's Head' on their signboards, while others – such as Radford, Linnell and Scarlett – included symbols of instruments to attract attention. Among these leading instrument houses, as will become apparent in subsequent chapters, Francis Watkins emerged as an important maker and, moreover, one who could be trusted because he stood for principles of honesty and moderation, which, as we shall see, clashed with others who were more abrasive in their business tactics.

The metropolis and the optical trade changed in many ways after 1760. There was much ado about shedding the past and, under a Whiggish government, the pace of life seemed to quicken. There were visual changes too in the general topography when the old City gates were removed to improve the flow of traffic; New Gate being the last to be removed in 1776. Other effects resulting from the Paving Acts of the 1760s led to an entirely new appearance of the streets and so a new experience for pedestrians. Of the grand engineering schemes afoot before the mid-century, the new bridge across the river at Westminster – begun in 1745 and finished in 1750 – opened up the area to the south and eastwards of the Thames. As a direct consequence of this project, changes took place around Watkins's shop at Charing Cross when street widening was enforced to improve the approach to the new bridge. This entailed Watkins, jointly with the Earl of Dysart, in litigation with the Crown due to inadequate compensation for loss of some land (§ Chapter 10). New and very impressive buildings too were springing up. Just south of Watkins's premises, in Whitehall, was the Admiralty Office, where naval business had been conducted since 1720 – a fine example of modern architecture. Even Westminster Abbey took on a different appearance after 1745 with the erection of the western towers.

Politically, Britain was strong throughout the decades that straddled the mid-century. George II had succeeded in a battle at Dettingen in 1743 and success of a different kind occurred at Culloden in 1746. Even if not strongly of a political persuasion himself – although he did vote for Lord Trentham in a local poll of 1749 – it is conceivable that Watkins and his wife visited St James's Park in 1749 for the occasion of a 'grand pyrotechnic display' of philosophical fireworks. This extraordinary lavish event, mounted by the Royal Laboratory at Woolwich, and accompanied with a specially composed overture by Handel, marked the signing of the Treaty of Aix-la-Chapelle, ending the War of the Austrian Succession. But such an outward sign of pomp and circumstance also had another purpose, namely, it heralded the fact that the House of Hanover was fully settled in Britain. Also, as a matter of contemporary celebration, was the triumphal return of Robert Clive in 1753 after successfully leading troops of the East India Company to victory against the French. Indeed, events such as these marked out a completely new era when Britain began to enter a period of vast prowess and economic stability.

War and treaties brought opportunities for merchants and businessmen, and the instrument trade was among the first to benefit from exploration and the need to exercise control in the widening Empire. Indeed, trade blossomed on a scale not known before and the East India Company – from its headquarters in Leadenhall Street in the 1750s – effectively began to administer the Empire. By the end of the century, Watkins's successors – Jeremiah and Walter Watkins – benefited directly from overseas trade and the patronage of the East India Company (§ Chapter 10). At home, too, there was an upsurge in industrialisation when a wage economy was substituted for the paternalistic regimes of master, journeyman and apprentice. This shift in approach, although yet in the future, did not affect Francis Watkins's traditional style although it certainly affected his nephews, as successors. Sufficient has now been said about the man and the location of his business to move forwards to the opening years at Sir Isaac Newton's Head when trade depended as much upon the dilettante interests of the aristocracy and gentlemen of means as those rising entrepreneurs of industry who perceived profit and power through the application of science (§ Chapter 3).

Chapter 3
At the Sign of Sir Isaac Newton's Head

Francis Watkins left Scarlett's workshop in Soho at some point during March 1747, complete with his company freedom, a year's experience as a journeyman and comfortable in the knowledge that he was to return to the very same shop at Charing Cross where he had begun his apprenticeship under Nathaniel Adams; he was also secure in the knowledge that the widowed Clarinda Walder would join with him to share whatever future the business might bring. Francis was intelligent and competent and, most importantly, willing and able to take on all aspects of the trade. However, he would never have dreamt that the shop at Charing Cross where he began to trade on his own account in 1747 would be occupied continuously for a century by future generations of the Watkins family. At that early date, he would simply wish to make a living by using the skills and experience acquired under the Scarletts. As his own business progressed, he would naturally wish to provide for his own family, just as his father had done for his brothers and sister at Newchurch. The present chapter offers insight to some of the diverse activities that established the shop at Sir Isaac Newton's Head during the first decade or so of trading and, as a backdrop, begins with a short account of the shop itself, covering the interim period between Nathaniel Adams's passing and Watkins's return.

Inside the Shop at Charing Cross

Watkins had only been training under Nathaniel Adams about three years when tragedy struck, as we saw in Chapter 1. The cause of Adams's death at the early age of 35 is unknown but since he did not leave a Will it became incumbent upon his widow to organise the administration of her husband's estate. She was required to draw up an inventory of her late husband's moveable effects and to settle all outstanding business from the time of Nathaniel's death up to an agreed date upon which trading would officially terminate. In practice, this meant providing proper accounts, not simply for the business but also for the entire contents of the 'The Golden Spectacles' at Charing Cross where they had lived and worked. This was carried out on 27 March 1741 and shortly thereafter 'A true and perfect Inventory of all and singular the Goods and Chattels and Credit of Nathaniel Adams of the parish of St Martin-in-the-Fields in the County of Middlesex deceased ...'

was prepared.[1] The full inventory comprised a detailed valuation of every item in every room, beginning in the garret, descending the stairs, into the backrooms, dining room, parlour, the 'beaufet', and back kitchen before dealing with the shop itself. The total of the household goods and furnishings amounted to £143 1s 5d, a sum which is no more nor less than what might be expected for a young man, not long married and only a few years into his career.[2] Inside the shop itself:

> ... Two Wainscott Counters the Two Sides of the Shop with Glazed bases and nest of Drawers | each Twenty One and Twenty Four Squares of Glass with Twenty Six Panes of Glass | the outside Front Sashes and an Iron Gate | the Princes Armes gilt | Brass Branch for Three Candles in the Sign Post and Sign of the Spectacle | a Settee Bedstead ... Blankett | a Quilt and a Piece of Cover.[3]

The valuation of these items – including the privileged sign showing the patronage of Frederick, Prince of Wales – amounted to a mere £11 17s 0d which suggests a somewhat modest interior. The drawers with prized glass fronts would contain the many varieties of spectacles. To this we can add the knowledge that the outside appearance of the shop was double fronted and hence the need to arrange the inside with two counters, one on each side. The internal appearance would therefore be very similar to the instrument-maker's shop depicted by Benjamin Martin in his *General Magazine*.[4] This glimpse into the shop's interior, with its description of a settee bedstead, blanket, quilt and cover, also reminds us that Watkins, as junior apprentice, slept in the shop. This was the custom of many shop owners who gave the youngest apprentice the last duties at night and first duties in the morning, besides providing some security.

Preparing and costing the household items would have been a straightforward task for the professional valuers, but not for the specialist stock of an optician and instrument maker. As the inventory shows, two eminent members of the Spectaclemakers' Company – Edward Scarlett Jr (Soho) and William Radford (Strand) – were called upon for their expertise and assistance.

The bulk of Adams's stock lay in spectacles, comprising every kind to be seen in the streets of London. In total there were some 38 dozen pairs of spectacles, framed in a variety of materials (stained horn, leather, tortoiseshell, silver, metal, pearl and black skin), and several dozen covered spectacle cases in shagreen, fishskin and horn. Also there were some 64 dozen spectacle 'eyes' (kept aside for making-up the spectacles) and numerous reading glasses. Prices given at

[1] Administration (Nathaniel Adams). TNA: PROB 3/40/32 20 April 1741; probate granted April 1742. TNA: PROB 6/117.

[2] A fuller account of this inventory is given in Buchanan and Gee (2004).

[3] Note 1.

[4] Tallis (1847), Part 76–77, pp. 192–193 (frontispiece of § Chapter 10); Martin, *General Magazine* (1759–1764), part iv, figure 11 of plate iii, facing p. 181.

Figure 3.1 Interior perspective drawing of a Georgian instrument-maker's shop. From Benjamin Martin, *General Magazine* (1759–1764), pt iv, facing p. 181. Courtesy of the late John Millburn

Scioptick balls; pocket microscopes; double brass microscope; Wilson's microscopes; pocket microscopes; 20 inch, &c refracting telescopes (in brass and mahogany or vellum); 20 inch, 2 ft, &c. reflecting telescopes; camera obscuras; prospects; opera glasses; magic lanthorns and trumpets; prisms; air-pump and glasses; barometers; thermometers; loadstone.

Figure 3.2 A selection of the optical and philosophical instruments included in the inventory of goods for the administration of Nathaniel Adams's estate in 1741

this valuation generally reflected the different fashions and also the pockets of customers. Although not calling himself an optical and philosophical instrument maker, Adams had a small number of items of this category, as shown in Figure 3.2. There were also tools for making up telescopes and a lathe, though no sign of lens-grinding machinery, thus suggesting that he sourced his glasses from elsewhere, perhaps from Oliver Combs, Scarlett's glass grinder.

The inventory further records that Hannah Adams 'doth declare that after the said Deceaseds death and before the taking of this Inventory she did sell several Things of the said Deceaseds Stock in Trade and Received Money for the Same and kept a just and fair Accompt of this Amount thereof....'[5]

[5] Note 1.

Cash raised by these additional sales amounted to £6. 19s 8d. Hannah also received £19 0s 0d from a Mr Shipman and a further sum of £5 5s 0d from one Reverend Waring, both as part payments on their debts. In total, these sales, plus the valued goods in the shop, amounted to £256 12s 8d. The document summary showed there were 31 debtors, from the nobility and aristocracy (including the Duke of Bedford) and dignitaries (including the Secretary of the French Embassy) to ordinary working individuals (including a butcher in the St James Market). These outstanding bills amounted to a further £51 0s 0d.

None of the goods, chattels or ready money was available to Hannah personally until the final settlement during April 1742.[6] Interestingly, Hannah was not alone in the winding down of her late husband's affairs for – soon after completing the administration – one John Robertson had a presence at the shop and it was not long before this person became Hannah's new husband. In this interim period, the elder apprentice, John Margas, appears to have remained in residence at the shop (possibly to assist Hannah in completing orders and provide security of the premises); Margas also appears to have made an arrangement with the optician John Cox of St John's Court, Cow Lane, in order to complete the term of his apprenticeship.[7] Unlike Margas, the 17-year-old Francis Watkins was some considerable way from completing his apprenticeship and so was taken in by Scarlett and placed under the supervision of Henry Walder.[8] Both arrangements appear to have worked satisfactorily: Watkins went on to complete his apprenticeship in Soho while Margas – by mutual arrangement with Robertson – remained at Charing Cross until the end of 1744.[9] On 7 January 1745, however, the following announcement appeared in the press:

---

[6] That is, the inventory was signed in the presence of Robert Chapman (surrogate) and R. Wotton (notary) on 18 April 1741 and would have been considered 'proved' 12 months later.

[7] Clifton (1995), p. 178, states that Margas was made free of the SMC in 1741/2 after turning over to John Cox. Margas had only a matter of a few months to run before he was eligible for freedom but it seems that Cox may have tried to avoid the relevant turnover fee for he was ordered before the Court for taking Margas on without paying the relevant fee of 40s. SML: Court collection, Z/8, p. 57. In any event, Margas did earn his freedom on 30 December 1742. Cox had been apprenticed to Thomas Lincoln and was made free of the SMC on 29 June 1732.

[8] Hannah, in her declaration on the Inventory (note 1) referred to 'a boy and the Statute of Limitation' which appears to have alluded to Francis Watkins. The problem of what to do with the boy appears to have been sympathetically organised by Edward Scarlett Jr who absolved Hannah of all responsibility by taking Francis into his own workshop.

[9] That is, Margas's and Robertson's names are bracketed together in the St Martin-in-the-Fields parish registers (Poor Rate: Books & Ledgers) until April 1742. WA: microfilm F517 (May 1742). Thereafter, only Margas's name appears in the ledgers until May 1744. WA: microfilm F518 and 519 (May 1743 and May 1744, but not listed in May 1745).

> John Margas, Optician, is removed from Charing Cross to the Golden Spectacles in Long Acre, near James-street.[10]

About the same time as Margas made this move, he married and then remained in Long Acre with his wife (Ann) and son (Jacob) until 4 April 1758 when he was declared bankrupt.[11] He then moved to Dublin – presumably after clearing business with his creditors – where he traded in Capel Street from 1759 to 1767. Margas died in 1767.[12]

Francis Watkins, Optician

After Margas's departure from Charing Cross, the shop remained unoccupied for two years until Francis Watkins returned there. It is impossible to tell exactly from the local rate ledgers when Watkins actually arrived: he was not there in November 1746 when the rate collector called although his name was entered as resident when rates were collected in June 1747.[13] However, a more exact indicator of Watkins's arrival may be gained from the *Daily Advertiser* which, on 4 March 1747, ran the following announcement:

> Francis Watkins, Optician, late from Mr. Scarlet's has opened the shop kept by the late Mr. Nathaniel Adams, at Sir Isaac Newton's Head, Charing Cross.[14]

Whether this opportunity to acquire the shop occurred by chance or design is not known. Someone must have paid the reduced rates while the shop was unoccupied and, quite possibly, this may have been arranged by Scarlett Jr or Watkins himself, possibly in connection with the widow Clarinda Walder (§ Chapter 1). It was certainly time for Watkins to move on after serving 12 months as journeyman at the Soho workshop. Undoubtedly, a good deal of forethought preceded the move. For instance, the choice of 'Sir Isaac Newton's Head' for the shop sign would suggest a well-considered plan based on his ability to comprehend the rising demand for apparatus of experimental philosophy.

---

10 BL: (Burney). *Daily Advertiser*, 7 January 1745.

11 Margas's bankruptcy is noted in Benjamin Martin's *Miscellaneous Correspondence*, vol. ii, *General Magazine of Arts & Sciences*, 1755–1764, p. 771, and declared formally in the *London Gazette*, 4–8 April 1758 and *Lloyd's Evening Post and British Chronicle*, 7 April 1758.

12 *Freeman's Journal* (Dublin), 8 December 1767; Prerogative Wills of Ireland, 1536–1810 (Dublin, 1897).

13 WA: Poor rates, F521, November 1746 and F522, June 1747. The latter entry referred to 'Franc (sic) Watkins']. Another entry for April 1747 referred to the occupant as '?Wallis', with the interrogation mark possibly signifying a name volunteered by a neighbour.

14 Quoted in Goodison (1977), p. 269.

At least, Watkins veered away from the optician's traditional sign of 'Archimedes' and avoided the use of Adams's 'Golden Spectacles' which was already taken up by Margas.[15] By choosing an iconic symbol to represent the much lamented Newton, he was, in effect, positively joining those in the trade who were prepared to promote something of the legacy of Newton whose importance was just then dawning on a wider circle of the ordinary public. Alas, Watkins's sign was discontinued in 1762 when the Westminster Paving Act outlawed hanging signs in favour of street numbering. The shop then simply became '5 Charing Cross'.

What was needed by the 1740s was a place where enthusiasts and gentlemen amateurs could find instruments, machines and apparatus, not only at the right price but also with the assurance of technical back-up should it be required. Much of the interest that had been generated in the previous decades had come from the lectures to be found around town in coffee houses or taverns, or even in the homes of lecturers such as Desaguliers or on the premises of some instrument makers. With this widening appeal and the desire to partake in the pastime of experiment, Watkins became a leading purveyor of scientific instruments. Two popular starting points for home experimenters were the 'doctrine of colours' (following the emphasis given to Newtonian optics by Desaguliers) and microscopy (following the impetus given to the subject by Henry Baker). Thus, to cultivate customers and encourage an interest in experimental philosophy generally, Watkins advertised one-shilling demonstrations with an electrical machine.[16] Indeed, his instrument-making career had begun at a very favourable time and, as the reader will detect in the remainder of this chapter, he worked with confidence across the several domains of the instrument trade.[17]

> Machines for performing all manner of electrical experiments made and sold by Francis Watkins (from Mr Scarlet's) at Sir Isaac Newton's Head, Charing Cross; as also all kinds of optical and philosophical instruments, according to the latest improvements, executed in the best manner according to the designs of those who bespeak them, with the apparatus proper to accompany them.

Figure 3.3 Early advertisement of Francis Watkins, transcribed from his *Electrical Experiments* (1747) p. 75

[15] Lillywhite (1972), pp. 501–502, notes two precedents for the use of Sir Isaac Newton's Head: *viz.* a cabinet maker in Crane Court near to the Royal Society, and Matthew Richardson (from Scarlett's) at York Buildings in the Strand. To this we may add that Ayscough and Mann combined 'Sir Isaac Newton's Head' with 'Two Pair of Golden Spectacles'.

[16] *General Advertiser*, 25 May 1747.

[17] Advertisement within Watkins (1747), p. 75.

Spectacles

The making and selling of spectacles formed the base trade of all opticians and, in the mid-eighteenth century, any optician of merit was also a member of the Spectaclemakers' Company. So, in principle, a certain level of competence by the maker could be assured and the quality of product assumed. By and large, opticians did not need to advertise spectacles; by the age of 40 most people required spectacles and recommendations usually came by word of mouth. Of course, price was always an important factor, but so also was fashion. To this end, opticians stocked a wide range of spectacle frames, suiting every pocket and all manner of taste. By the mid-century, it also became a matter of choice for a discerning customer to select the type of glass that was to form the spectacle lens. For the most part, white crystal glass was recommended, although other materials – 'Brazil Pebbles' (natural quartz), 'Venetian Green' and superfine crown glasses – were also available. It therefore became a matter of some importance that the ordinary public should be instructed in how to make a suitable choice, based on a modicum of scientific knowledge. In this respect, James Ayscough took a lead by publishing a pamphlet with three aims in mind: firstly, to explain normal vision and how deviation from this could be corrected through the aid of spectacles; secondly, to provide tests sufficiently simple to allow a customer to judge the quality of the spectacle glass; and finally, to promote a newer glass – superfine crown glass – said to have advantages over other glasses commonly used. As we shall note, the advance brought about by Ayscough was not to Francis Watkins's liking and he appears to have been quick to organise other opticians in a campaign to diminish Ayscough's standing.

James Ayscough came to London in 1732 as an apprentice to James Mann at the Great Golden Spectacles in Ludgate Street, and ultimately finished up as master of this important City workshop.[18] Like Francis Watkins, his background appears to have been well ordered and, as the son of a parson whose own education was completed at Christ's College, Cambridge, he was literate and intelligent. Watkins and Ayscough began to trade on their own accounts in the same year, 1747, although Ayscough remained with his master much longer, having gained his freedom in 1740.[19] Under Mann he acquired knowledge of conventional optics, and contributed to the design of the universal microscope. In the present context, however, it is Ayscough's initiative in the use of a refined form of crown glass that is of prime interest. This he announced in 1750 through an advertising campaign.[20]

18 Millburn (2000), p. 225.

19 SMC Minutes CLC/L/SH/B/001/2, p. 253.

20 *London Evening Post*, 22 March 1750, and various other newspapers, until the *London Daily Advertiser and Literary Gazette*, 18 April 1751.

JAMES AYSCOUGH,
At the Golden Spectacles and Quadrant,

in

Ludgate-Street, London;
Takes the Liberty to acquaint the PUBLICK,

THAT he has finish'd a sufficient Sortment of Concaves, (for all Degrees of short-sighted Persons) made of *Crown-Glass*. Likewise *Spectacles* made of the same Glass; which for its several peculiar good Properties*, and also from the Experience of Numbers of People, he can recommend preferable to White Glass, or *Brazil Pebbles*; and, for the greater Exactness of fitting every Sight has encreased them to upwards of thirty different Degrees, set in all manner of neat and commodious Frames; some of which press neither on the Nose or Temples.

N.B. Very neat Playhouse Glasses, made in a particular manner.

**See a Treatise on the Nature and Use of Spectacles*

Figure 3.4 A new promotion in spectacle glass, by James Ayscough, repeated regularly in the *London Evening Post*, April till December, 1750

For the first time in print by any optician, Ayscough offered the public a printed account of the eye and its defects of vision. His pamphlet on the *Nature and Use of Spectacles* advised a customer on how to test a spectacle glass by checking the figuring – the grinding and polishing of the glass – which gave an indication on the uniformity of curvature. To do this test the customer was encouraged to look through the glass at a series of 'squares' (crossed parallel lines) drawn on paper. Any crookedness of these lines or variation in the magnification of the squares would indicate poor figuring. As for the defects within the glass, the customer was invited to discover the extent to which veins and specks appeared by viewing a distant lighted candle through the lens with the eye placed roughly in the near focal plane. These simple tests were fairly standard for checking objective lenses in telescopes and microscopes but Ayscough believed that customers should be aware that the same could be applied to spectacle glasses where distortion caused by veins and opaque spots caused by specks interrupting the incoming rays were important factors in making an educated choice.

Of course, such tests would have been explained to the more demanding customers by countless generations of tolerant opticians. Ayscough went one step further by urging customers to put the same tests to good effect in order to compare a newer form of crown glass 'of finer Composition than any commonly used'. From promotional advertising, we may appreciate that something new was being presented to the public:

> Superfine Crown Glass which has for many Years been approved of, (though not commonly made) by the best Judges in Opticks, to be much preferable for Spectacles than any other Sort, and far exceeding the Brazil Pebbles, are now made and sold by James Ayscough, Optician at the Great Golden Spectacles, in Ludgate-street, London.[21]

Another detail arising from this choice of glass concerned the colour:

> For the Colour of Glass, the whitest has been generally made use of for Spectacles, but I am inclin'd on many Accounts to prefer another Sort before it, or any other Material that has been made use of for this Purpose: For the common white Glass gives an offensive glaring Light, very painful and prejudicial to the Eyes, insomuch, that some advise green or blue Glass, tho' it tinge every Object with its own Colour, rather than this white Sort.[22]

Great play was made of the property of colour because the newer glass had the ability to moderate the reflective glare from paper, or, as Ayscough explained, it rendered vision 'easy and pleasant [to] the tenderest Eye'. Another benefit to be had from this glass was that the material was harder than common white glass and so, although taking longer to grind and polish, it was more scratch resistant.

Not all opticians in the trade were receptive to the notion that superfine crown glass could exceed any expectation of what might be achieved by the more expensive, but very fashionable, Brazil pebbles, or even by the cheaper, commonly used, white crystal glass. But such was the strength of Ayscough's argument that a good deal of consternation arose among those opticians with vested interests in fashion and profit rather than optical quality. In fact, it was not long before six opticians – headed by Francis Watkins – took space immediately below Ayscough's newspaper announcements to counter the notion that this new form of glass was an improvement.

Stopping short of mentioning Ayscough by name, they warned their affluent customers to be wary of 'specious Pretences' going the rounds of the classified newspaper advertisements.[23] Naturally, Ayscough retaliated and within a few weeks he published his defence against those same opticians for their 'fallacious Pretences'.[24]

---

21 *Whitehall Evening Post or London Intelligencer*, 27 March 1750.

22 Ayscough (1750), p. 13.

23 *Whitehall Evening Post or London Intelligencer*, 27 March 1750; *London Evening Post*, 3 April 1750; ibid., 10 April 1750.

24 *London Evening Post*, 3 and 10 April 1750.

THE undernamed Opticians take this Method of Acquainting the Nobility, Gentry, and Others, That they make and sell Spectacles ground in the most perfect Manner, as usual, of Brazil Pebble, Chrystal Glass, fine Venetian Green, or of the same Kind of Glass which has lately been reported as preferable to any other, though [in] reality such as is commonly used in Windows. And the Reason of this Address is, lest the Publick should be amused by specious Pretences, as if somewhat New and Curious were now invented, instead of a Thing which has been at all Times used, and that only in the most ordinary Goods, such as are sold by Wholesale into the Country, or hawk'd by peddling Jews about the Streets.

Francis Watkins, at Charing Cross
Peter Eglington, in the Strand
John Cuff, Fleet-Street
Joseph Hurt, Ludgate-Street
George Sterrop, in St Paul's Church-yard
Thomas Ribright, in the Poultry

Figure 3.5 Francis Watkins's organised group of opticians opposing Ayscough's superfine crown glass, reprinted from *Whitehall Evening Post or London Intelligencer* on 27 March 1750

In Answer to two anonymous Advertisements printed some time ago, and to a third, which has been publish'd for several Days past in the publick Papers, sign'd by six Opticians, *alias Spectacle-makers*, with an Intention to hinder the Sale of these Spectacles, Mr AYSCOUGH would beg Leave to observe, that if there is any Sort of Glass, which from its particular Properties is preferable to any other for Spectacles, it is neither the Name it is distinguish'd by, nor any other Uses that it may be apply'd to, that will make the least Alteration in those Properties, nor any Ways lessen its Value in the Esteem of candid and impartial Judges. That he is fully satisfied the Glass he recommends greatly exceeds (on many Accounts) any other Sort for Spectacles, and, when well chosen and truly wrought, will be found, by Experience, justly to deserve the Character he has given it in his Advertisement: And he doubts not but that whoever shall compare the mean Artificers by which the Advertisers would endeavour to hinder the Sale of these Spectacles from being made Use of, with the Manner in which they set off their own, will very easily distinguish between the real and pretended Intention of their Advertisements; but he hopes the Publick will never suffer themselves to be either amused or imposed upon by such weak and fallacious Prentences.

Figure 3.6 James Ayscough's response to opposing opticians, reprinted from *London Evening Post*, 10 April 1750

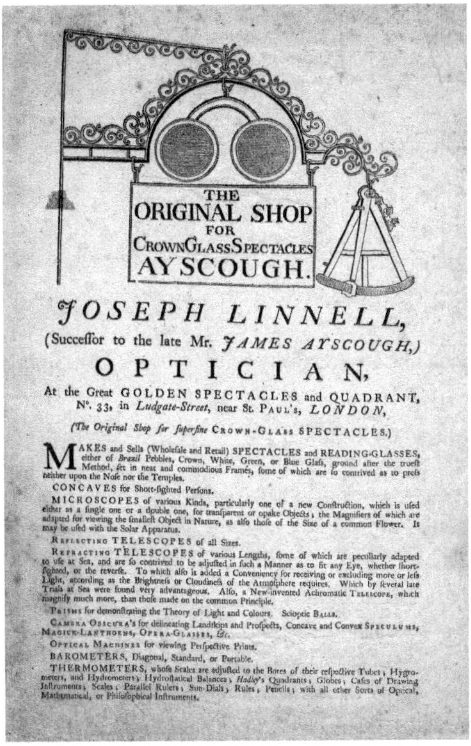

Figure 3.7 Joseph Linnell's trade card, showing the continued tradition for superfine crown glass. © The Trustees of the British Museum, Heal, 105.68

None of this squabble had any effect on Ayscough's determination for he repeatedly reminded the nobility and others (with particular vigour after October 1750) that 'he has finished a sufficient Sortment of Concaves, (for all Degrees of short-sighted Persons) made of *Crown-Glass*'. Indeed, even after Ayscough's death in 1759, Joseph Linnell – who had taken over the Ayscough establishment – continued to supply superfine glasses in the tradition established by Ayscough.

By the later editions of Ayscough's tract, dated 1754 and 1755, the arguments against the common white glass and Brazil pebble were mustered once more, better articulated, available also in French, and widely available from several outlets in London and Paris.[25] White glass, it was argued, did not allow for good

[25] Ayscough (1754); *Public Advertiser*, 1 April 1754; Ayscough (1755b); *Public Advertiser*, 9 December 1755.

figuring due to its softness, and the Brazil pebble produced double refraction (akin to Iceland Spar), although it could not be denied that it did allow a 'true figure' and 'good polish' due to its hardness.[26] By this time, Ayscough was able to respond fully to the various criticisms that had been levelled against him:

> (i) regarding cost
> That these Spectacles must be dearer, the greater Charge arising from the working, as well as from the higher Price of Materials; but, their being much preferable in every Shape to the common Glass, which Experience and the universal Approbation they have met with sufficiently confirms, makes them cheaper upon the whole, since the Difference of the Price is such a Trifle in comparison to their superior Qualifications.
>
> (ii) regarding opposition to superfine glass
> Notwithstanding the Opposition that was made to this Sort of Glass, some time since, by a few People of the Trade, yet some of the most violent of these Opposers are now so much convinced of the superior Excellency of it, that they themselves make Use of it for the same Purposes.
>
> (iii) regarding Rock Crystal and Brazil Pebbles
> Even at that Time, some of them acknowledged the Properties of this Glass, though they opposed it in Publick (like the Ephesian Goldsmiths, who raised a Clamour against Christianity, lest it should hurt the Craft, by destroying that lucrative Business of Shrine-making) giving it as a Reason, that it would spoil the Sale of Spectacles made of Rock Crystal and Brazil Pebbles, which for some Time had been much cried up for their extraordinary Virtues, while, in reality, there is nothing to recommend them but their exorbitant Price, and no small Gain which they brought to the Craftsmen.
>
> (iv) regarding reasons for introducing superfine glass
> Upon the whole, the Motive of recommending this Glass was no other than being fully convinced of its Properties from the above Reasons, nor was there any Pretence to an Invention; only, happening to be the lucky Discoverer, I thought it my Duty to communicate it to the Publick. Neither is it true, that this Glass was at all Times formerly used, and for the most ordinary Goods, as has been falsely insinuated, seeing the Price would not answer for the Glasses sold by Pedlars and common Hawkers.[27]

---

[26] Despite Ayscough's well-reasoned discourse, the double image of Brazil pebbles was probably due to internal reflections within the thick lenses rather than double refraction which, at that time, was only known to occur in Iceland Spar crystals.

[27] Version consulted: Ayscough (5th edition, 1755a), pp. 22–23.

It had been sufficient, Ayscough asserted, that all his findings – for which he sought no credit – had been put to good use. To emphasise this supposed altruism, he concluded by quoting from William Molyneux's *Dioptrica Nova* (1692) on the importance of dioptrics when applied to the benefit of human kind; that is, the importance ('inestimable Blessing') of the need to apply optical knowledge to the improvement of spectacles ('perfection of this Noble machine') for those with less fortunate eyesight ('blindness and darkness').[28] Such humanitarian views were very much in accord with John Locke's philosophy and may, in this case, be seen as an outcome of Molyneux's actual correspondence with Locke. But the adaption of Molyneux's writing here – so carefully selected by Ayscough – was surely meant to shame the six opticians (led by Watkins) who had objected so harshly to Ayscough's development in spectacle glasses merely out of self-interest, showing no regard to values of a higher order. Of course, moral justification had little place in the hard world of commerce although there was little harm in jolting the conscience of those who made a comfortable profit from the whims of their affluent customers.

Electrical Demonstrations

One of Watkins's principal concerns at the start of working for himself must have been how best to capitalise upon the flurry of excitement that had been generated by the appearance of the friction electric machine at several local exhibitions and public lectures over the previous 12 months. One such machine to appear at Charing Cross during June 1746, at the Mitre Tavern, was reported to be the very instrument that the French physicist Jean-Antoine Nollet had used to show spectacular effects before the French court.[29]

William Watson, FRS was less easily taken in by casual reports of this illustrious machine; he expressed doubts over its former ownership and scepticism about the Dutchman who demonstrated it.[30] While Francis-I Hauksbee had given demonstrations of electricity early in the century these were mostly intended to stimulate discussions among the Fellows of the Royal Society. Four decades on, however, these amazing improved continental machines were demonstrated before any who could afford it, with great aplomb. The paying public – always thirsty for the unusual or dramatic – was soon to be entertained by electrical shows throughout the capital.

---

[28] Ayscough made a particular point of stating that his tract was 'not a work for the instruction of philosophers, anatomists or opticians' (4th edition, 1755a), p. 7.

[29] *Daily Advertiser*, 2 June 1746, from Lysons's Collectanea, Vol. III, p. 140.

[30] A private note from Watson to Henry Baker cited in: Hackmann (1978), p. 173.

> We hear that the very same Machine with which the new Electrical Phaenomena was perform'd before the King and Queen of France, is now in the Hands of Mr. Marshall, and is to be seen at the Mitre, Charing-Cross. It causes the Electrical Commotion to be felt by every person in the Room, and the Fire to come out of every Point of the Body, and was communicated by Professor Muschenbroek, from Leyden, to the famous Academician Abbé de Nollet, of Paris. What is very surprising in this Apparatus, is, that it may be exhibited at all Times, and in all Sorts of Weather. It has been shewn before several Persons of Distinction, and has given Universal Satisfaction.

Figure 3.8 Announcement of the famed electric machine to be exhibited at Charing Cross during June 1746, reprinted from *Daily Advertiser*, 2 June 1746

Times had indeed moved on and the imagined 'effluvium' that was once supposed to underlie all electrical phenomena was, by the 1740s, understood to obey specific laws: in practice, this meant that the demonstrable phenomena could be controlled with greater confidence. Among the better known continental presentations of this epoch were those devised by Georg Bose who – at one celebratory dinner hosted by the Berlin Royal Academy in January 1744 – surprised his guests by contriving sparks to emanate from table cutlery.[31] Such electrical 'jokes' were always planned ahead, using hidden wires from the prime conductor of a good electric machine operated by an accomplice.[32] Other tricks included the 'beatification' of a guest, the 'electric kiss', and the extraction of sparks from water at the tip of a sword. A final *pièce de theatre* for the more confident showmen – such as Johann Heinrich Winckler, a linguist and colleague of the celebrated J. S. Bach at the famous Thomas Schule in Leipzig – was to substitute themselves for the prime conductor, thereby enabling sparks to be drawn from their fingers and so cause the vapour from a dish of pre-warmed alcohol to ignite.

In London, too, Stephen Gray – whose experiments on electrical conduction were well known among Fellows of the Royal Society – performed experiments in 1730 using a small boy from a charity school, suspended horizontally from a roof beam by bands of silk. The boy was 'made attractive' (that is, charged) by frictionally rubbing the glass drum of the electric machine against his boot, and thus he could then amaze an audience by luring pieces of leaf brass from a dish placed near to his hand. The business of explaining precisely what was happening to the jumping pieces of leaf brass usually came from the ground rules laid down by Charles François Dufay concerning 'vitreous' and 'resinous' (later

---

31 Heilbron (1982), p. 182.

32 Hackmann (1978), pp. 174–175, expands a little on some of these demonstrations.

termed positive and negative) electricity; such rules, however, once thought settled, were then only at a beginning.[33]

William Watson – son of a London corn chandler and trained by apprenticeship as an apothecary – came to this subject through his attendance at the public meetings of the Royal Society. In the late 1730s, these events took place at the The Mitre in Fleet Street, but, from 1743, when Watson joined the Society's private dining club, meetings were held at the Crown and Anchor in the Strand. Although Watson's early interests lay in botany, he quickly turned to electrical investigations after hearing of Winckler's experiments. In 1744, he set out to lay a scientific foundation to the visible effects seen in the common demonstrations and, from these observations, he was led to the concepts of 'power', 'virtue' and 'aether'. Although these notions were applied ambiguously and inconsistently, he envisaged a link between 'electric effluvia' and 'active fire' through the belief that electricity was particulate although a subtle fluid moving by impacts and behaving so as to restore any imbalance in distribution. Theories of electricity of this kind were highly imaginative and experiments to test the consequences of these ideas were difficult to conceive although, to his credit, Watson appears to have accepted a concept of 'circulation' and the need to 'earth' the electric machine. Furthermore, contrary to the views of some who believed that the 'electric shock' was instantaneous, Watson suggested that its speed, though fast, was simply 'immeasurable'. For all his electrical studies, he was awarded the Royal Society's Copley Medal on 30 November 1745, 'On account of the surprising discoveries in the phenomena of Electricity, exhibited in his late Experiments.'[34] Views on the nature of electricity were to change yet again in the late 1740s when Benjamin Franklin – an ally of Watson's – took up the subject and introduced a one-fluid concept to model the various observed phenomena.[35]

Despite the deficiency of theory, sufficient was known from practice to enable the construction of moderately reliable machines capable of demonstrating a variety of spectacular effects. As an adjunct to the electrical machine came 'Musschenborek's [*sic*] wonderful bottle', the Leiden jar, an invention of the mid-1740s which had the peculiar ability to deliver dangerous shocks by storing the 'charge' derived from an electric machine. With it, for example, a gentleman (holding the jar) and a lady (holding the prime conductor) would, indeed, suffer a violent shock on clasping hands. Inevitably, the question of how many individuals could feel the shock when joined in a train followed. The academician Louis-Guillaume Le Monnier shocked 140 courtiers before the French king and, likewise, Nollet shocked 180 Parisian guards

---

33 Gray (1731–1732), pp. 18–44.

34 *RS: Journal Book* XIX (1742–1745), pp. 489–490.

35 Heilbron (1982), pp. 187–195.

and 200 Carthusians at a nearby monastery.[36] These spectacular discharges or 'commotions' (as they were called) were not without their intrinsic scientific value because it was also noted that the effect occurred only along a closed chain, with nothing apparent along any side chains of those holding hands. Gradually, 'know-how' was accrued and it was generally understood that the effect was feeble when those in a chain stood on damp ground and that the jar itself might hold its punch for hours or even days in good dry weather.

What had first amazed the intelligentsia and social élite in the salons of Amsterdam, Paris and Berlin could, by 1746, be displayed to the paying public around London. In addition to demonstrations at the Charing Cross tavern, other popular venues included 'The One-Tun' and 'The Punch House' (both in the Strand), and 'The Swan' (in Exchange Alley).[37] Promoters of these grand spectacles were usually opportunists who capitalised on sensationalism: for example, Benjamin Rackstraw, later renowned for a museum of extraordinary objects in Fleet Street; and a Mr Pearce, proprietor of a coffee house in Little Britain near West Smithfield; both advertised, the latter stating that 'there is lately brought over [from the Continent] the wonderful and so much celebrated Machine, or Instrument of ELECTRICITY ....'[38] From the description of this machine, it would appear identical with that employed by Marshall at The Mitre.

Given the popularity of these events, it seems very likely that Francis Watkins attended these epic performances. By the summer of 1747 – only three months into his independence as optician – he too began to reap an income from electrical demonstrations. From a run of advertisements in the *General Advertiser*, beginning 25 May 1747, it was announced that Francis Watkins: 'has finished a most compleat Electrical Apparatus for performing, at any Time of the day, all the known Experiments ... Each Person to pay 1 s[hilling] ... Electrical Apparatus's [*sic*] of all Sizes sold by the Operator.'[39] However, Watkins's demonstrations would not be before large audiences but only to those customers likely to purchase the machine described in his text.

Not everyone in the Royal Society approved of those who demonstrated electrical phenomena in this popular way. Henry Baker – perhaps reflecting upon a time only a few years previously when electrical demonstrations were the privileged domain of William Watson, rather than as mere entertainment of pleasure seekers – commented derisively on those who earned 'a great deal of

[36] Heilbron (1982), p. 186.

[37] Lysons, Collectanea, Vol. III, p. 140.

[38] *General Advertiser*, 6 September 1746. Electrical demonstrations also took place in the provinces at this time. Millburn (1986a), pp. 10–13.

[39] Thereafter (without reference to the one-shilling demonstrations) announcements of the publication of Watkins's book appeared in: *London Evening Post*, 6 August 1747 and *Whitehall Evening Post or London Intelligencer*, 8 August 1747.

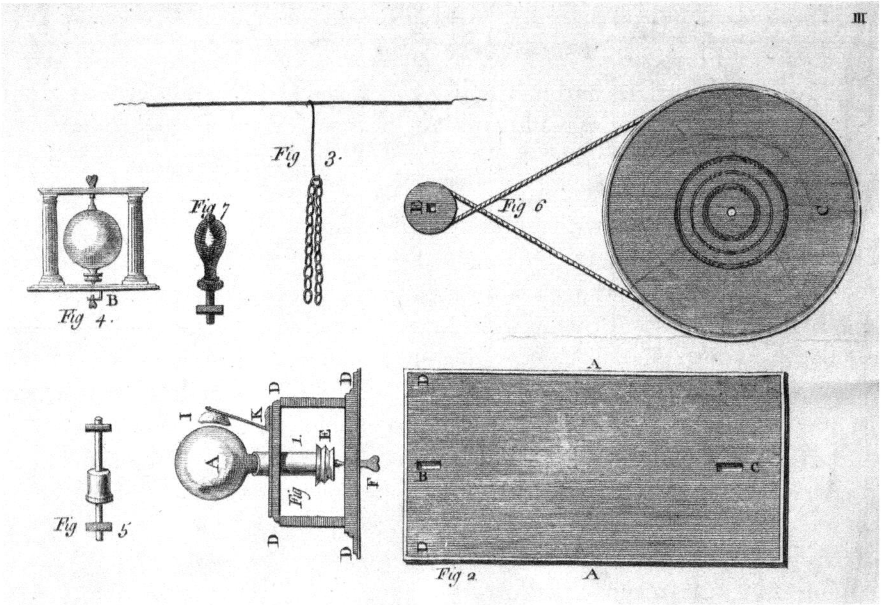

Figure 3.9 Francis Watkins's electrical machine, plate from his *Electrical Experiments* (1747). By permission of University of Glasgow Library, Department of Special Collections

money shewing a course of Electrical Experiments at a Shilling for each Person.'[40] More constructively, however, John Smeaton, in a letter to Benjamin Wilson dated 24 September 1746 remarking on the monetary gains possible from showing 'ye wonders of Electricity', wrote:

> ... if 200ℓ could be got by a 'worthy employment' y^t^ way I don't see where is ye harm as there is no Fraud or Dishonesty in it.[41]

At least Baker concurred with Smeaton when it came to the need for caution over 'the common Herd of conjuring Philosophers about Town'.[42] And yet, according to the Duchess of Bedford, Baker himself was not averse to performing a little light after-dinner entertainment with an electric machine for, in May 1746, the Duchess wrote: 'I supped at the Duchess of Montagu's on Tuesday night,

40 Turner (1974–1975), p. 64.

41 RS Wilson papers, fol. 22.

42 Smeaton to Benjamin Wilson, 24 August 1747. BL: Add. MSS 30094, fol. 41, quoted in Stewart and Weindling (1995), p. 39.

where was Mr Baker of the Royal Society, who electrified; it really is the most extraordinary thing one can imagine.'[43] Such showings at aristocratic dinner parties were very private affairs and hardly likely to touch the general public. Indeed, neither Baker nor Smeaton could avoid the plain and simple fact that it was only by public exhibition that ordinary people could gain an acquaintance with the various philosophical developments. There were, however, other, more serious, routes forward for those with ambition to learn and understand the mysteries of electricity. Listed among the purchases of the Spitalfields Mathematical Society for 1747 was 'a proper Apparatus for Electrical Experiments' and, from the date of purchase, it is possible that Watkins was the supplier (§ Chapter 5).[44]

For those who purchased one of Watkins's machines there was help at hand. Watkins appreciated that it was not easy for a novice or would-be showman to succeed in performance without a little help. Electrical machines were temperamental, required maintenance, and special care was needed in the preparation of the parts to overcome atmospheric dampness. It was to difficulties such as these that he addressed himself during his first months at Sir Isaac Newton's Head. He made a machine to his own design, advertised as 'a most compleat Electrical Apparatus for performing ... all known Experiments, with several new ones, and a very singular one, wherein a Horizontal iron bar is forced several feet out of its Place, only by the Electrical Attraction of a Bit of Down weighing less than Half a Grain ....'[45] He also supplied a handbook – *A Particular Account of the Electrical Experiments hitherto made publick* – explaining how his machine should be managed and experiments conducted 'as shewn in his course'. The text was advertised during early August. A note on the title page informs the reader that the book could be purchased from 'W. Owen, Bookseller and Publisher, at Homer's Head, Temple-Bar' who, as noted in Chapter 1, was Watkins's brother-in-law.[46]

Just before the appearance of Watkins's handbook, another instruction manual on the same subject had appeared with identical text and pagination, but a different author and publisher on the title page. This was John Neale's *Directions for Gentlemen who have Electrical Machines, how to proceed in making their Experiments* ...[47] Neale was a watch- and globe-maker, who patented a form

---

43 Thomson (1940), p. 129.

44 Cawthorne (1929), p. 158. For a listing of apparatus bought by the Society, see Spitalfields Mathematical Society (1804).

45 *General Advertiser*, 25 May 1747.

46 Watkins (1747). Announcements of Watkins's text are noted in *London Evening Post*, 6 August 1747, and the *Whitehall Evening Post or London Intelligencer*, 8 August 1747.

47 Neale's book was announced in July 1747 in the *Scots Magazine*, 9 (1747), p. 352. He gave electrical demonstrations 'at Mr Horne's', according to an advertisement in the *St James's Evening Post*, 12 February 1747.

> *To the* PUBLICK.
>
> Whereas from several late influences of the success of electrical applications to persons afflicted with nervous disorders, especially of the paralytic kind, it appears highly probable, that great benefit may accrue from thence to the community; and that it will be well worth while to determine further by experience in what particular cases success may be expected, or despaired of; it is earnestly requested of all those who have made or undergone any such trails, that they would be pleased to transmit circumstantial and well attested accounts thereof, either to *Francis Watkins*, at Sir *Isaac Newton's Head, Charing-cross,* or to Mr *John Neale*, in *Leaden-hall-street*, to be inserted in a narrative in due time to be published, of the success of electrifying in the aforesaid disorders.

Figure 3.10 Advertisement, transcribed from *Electrical Experiments*, suggesting a connection between John Neale and Francis Watkins

of planetarium in 1744, but went bankrupt in 1750, and again in 1758.[48] Neale, too, had given a course of 40 demonstrations at the Swan tavern in Exchange Alley.[49] From an advertisement by Watkins, in the endpapers to his own text, there is a hint that he had a working arrangement with Neale.[50]

Nor was this subject a passing fad or fashion. One year later, Watkins informed the public of a newly designed machine, convenient for its 'portableness' and said to be 'of singular service in many disorders' following intelligence from abroad.[51] Indeed, the Royal Society had received such communications from France. More importantly, from Watkins's standpoint, this new machine – then extolled for its mechanical 'Qualities' and ability to deliver electrical 'Virtue' and 'To be had of me only' – was clearly something in which he had invested much time and effort, although precisely how exclusive Watkins was at capturing this market is unknown.

48 Clifton (1995), p. 197; King and Millburn (1978), pp. 196–197. There is an example of Neale's planetarium in Hauch's collection at Søro Academy, Denmark, for which see Andersen (1995), p. 256.

49 Lysons, Collectanea, Vol. III, p. 141.

50 Watkins (1747), endpapers.

51 *The Remembrancer*, 14 May 1748.

Francis Watkins

AT

Sir ISAAC NEWTON'S HEAD, Charing Cross,

Has lately finished a most compleat ELECTRICAL MACHINE which, with its whole Apparatus, is contained in a Box scarce exceeding the Size of an ordinary Tea-chest, may be taken out fixed up fit for Use by one Person in a Minute; and at the same Time the Person turns the Machine with one Hand, has the other at Liberty to do what is necessary; besides the above Qualities, this Machine furnishes the Electrical Virtue as powerful as the largest, and is never liable to be put out of Order as most others are: To be had of me only, where its Operations may be seen, or any Gentleman may be waited on with it, on proper Notice.

And as there is great Reason to believe, by the Account received lately from Abroad, and communicated to the Royal Society, that Electricity may be of singular Service in many Disorders, I presume as soon as any Gentleman sees my Machine, he will prefer it to any yet made, by Reason of its Portableness, as it may be easily carried from one place to another, without misplacing any-thing.

N.B. At the same Place may be had, Price 1s. 6d A particular Account of all the Electrical Experiments hitherto made public, and at Mr Owen's, Temple Bar; and at Mr Birt's in Ave Mary-Lane.

Figure 3.11 Advertisement of a new electrical machine, transcribed from the *Remembrancer*, 14 May 1748

Trial of Jack's Patent Telescope

Besides the electric machine – which introduced Watkins to a wider public – his mainstay rested in the field of optical instruments, for which it was paramount to have earned the freedom of the Spectaclemakers' Company. There were, however, instrument makers elsewhere in other livery companies who also made instruments with optical components, thereby causing a degree of irritation among the opticians who considered this as encroachment. One such maker, long before Watkins, was John Marshall, a freeman of the Turners' Company, whose move to make microscopes was met with disapproval by a court of the Spectaclemakers' Company. Yet, by the 1740s, this sort of development was inevitable, especially with the increasing application of telescopic lenses to navigational instruments. The issue about to be described below is not about the breadth of the optical domain but rather it portrays an overconfident young man, edging his way upward in the trade. Francis Watkins – then only four years into his livelihood – had decided to tackle one of the most respected instrument makers of the age. That person – not an optician in the strictest sense but a mathematical instrument maker – was none other than George Adams Senior, a freeman of the Grocers' Company.

The conflict under scrutiny here began towards the end of 1751 when Adams had just announced the completion of a new telescope for use at sea in conjunction with a quadrant, an arrangement designed by the Scottish military engineer Richard Jack.[52] Regarding the telescopic part of the combination, Adams had claimed:

> The Magnifying Power of this TELESCOPE greatly exceeds that of any other of the common Sort, although they are considerably longer than this; also its particular Property, of various magnifying Powers without changing the Glasses, (which no other Construction hath) renders it extremely serviceable for viewing distant Objects.[53]

Several announcements of this sort appeared in the *General Advertiser* but, even before the end of December 1751, Watkins – who had purchased one of these instruments – could no longer restrain himself at the imputation that telescopes 'of the common Sort' (which he sold) were not comparable with those of his Fleet Street competitor.

On 28 December, the following protestation appeared in the columns of the same newspaper:

> Having seen repeated Advertisements in public Papers by Mess. Jack and Adams, setting forth a new invented Refracting Telescope, made and sold by PATENT, I think myself obliged, in Justice to the Publick, and to myself, to declare, that I have carefully compared one of the said *Patent* Telescopes, with others of the common Construction, and have really found all its boasted Excellencies to be intirely false and groundless .... [54]

Of course, scrutinising the design of a competitor's product was one way to learn how a similar instrument in his own range might be improved, and by ridiculing the 'puffs' – which makers were prone to make in their advertisements – it was sometimes possible to gain a commercial edge. In this case, Watkins appears sceptical over whether Adams's claim to patent had any justification in fact – it may be Patent 656 of 1750 for a sea quadrant to which this argument probably

[52] Millburn (2000), pp. 60–68; Richard Jack (d.1759), www.oxforddnb.com/view/article/46517 accessed 30 June 2011.

[53] *General Advertiser*, 12 December 1751, quoted in Millburn (2000), pp. 61–62.

[54] Watkins's response was printed in the *General Advertiser*, 28 December 1751, p. 4. The patent referred to is probably number 656, dated 1750, for a sea quadrant, see *Abridgements* (1875), No. 656.

refers, though no patent for a telescope has come to light. More blatantly, Watkins insinuated that Jack's optical claims were unsubstantiated:

> ... And whereas Mr. J. has affirmed, in the Manner in such Cases required, that *his* Invention overcomes that hitherto insuperable Difficulty arising from the different Refrangibility of Rays, despaired of by Sir ISAAC NEWTON; I also farther declare, that this *his* new Telescope colours the Image of the Object more than any other of the ordinary Construction, and that Charity therefore obliges me to believe this Gentleman to have been totally ignorant of the true Nature of Refractions, at the time of his making the said Affirmation: He appears to me also equally mistaken in affirming this Telescope to be properest for Sea Observations, since its Field is not above one-Third of that of others of the ordinary Construction....[55]

Since none of the Adams-Jack advertisements assert that the 'insuperable Difficulty' was overcome, as Watkins claims here, then it must be assumed that Jack may have stated this view elsewhere.[56] What is important here, and of considerable relevance in later chapters, is that Watkins was conversant with the problem of chromatic aberration and of Newton's flat denial of the possibility of there ever being a corrective solution for the coloured edges around optical images. However, having this knowledge did not help him and, at this stage, he appears ignorant of concurrent investigations elsewhere to reduce the chromatic effects of optical images.

If the chromatic effects due to differing 'refrangibilities' (refractivities) of white light passing through the objective lens were not the most troublesome issue, then the 'correction' applied must have been otherwise. As Watkins himself suggests:

> ... The Difference between Mr. *Jack*'s new Telescope and those of the common Construction, is in the *Eye Glasses* only; and there is not the least Contrivance to remove or lessen the Errors arising from the Object Glass. I have also carefully examined the Proportions of the Eye Glasses in this *his new-invented Patent* Telescope, and find the same to have been before described, or mentioned, by *P. Cherabim* [*sic*] *D'Orleans*, in his *Dioptrique Oculaire*, Page 187, but rejected by him as not so good as those of other Proportions; and indeed all Compositions of three Eye Glasses, unless of equal Convexities, are rejected by the best Optical Writers....[57]

---

[55] *General Advertiser*, 28 December 1751, p. 4.

[56] It is possible that Jack had private intelligence of Chester Moor Hall's achievement (§ Chapter 4). However, given the later revelations (§ Chapters 6 and 7), this seems unlikely.

[57] Note 54.

To Watkins's credit, he had uncovered an old problem and the fact that he had consulted Chérubin d'Orléans *La dioptrique oculaire*, a text published in 1671 though still used for reference, suggests that he had seen a copy at Scarlett's or possibly had access to the Royal Society's library. Watkins delivers a sophisticated analysis, especially in his evaluation of the telescope:

> ... But since this *Inventor* [Jack] seems to value himself most on account of his great Increase in the *magnifying Power*, by means of his Eye Glasses being all of *unequal* focal Lengths and *Plano convex*; I have taken the Pains (for the Satisfaction of the Public, the preserving of our Credit abroad, and in Justice to myself) to construct two Telescopes, of the same Length with the *Patentees*, with Eye Glasses all of *equal* Focus's, and *double* Convexities, one of which magnifies equal with his, and the other more, takes in a *larger* Field, and is at least as *distinct* as his; for I have got Object Glasses ground in the same Manner with those in the *Royal Observatory* at *Greenwich*; and which were found to exceed all others....

With the comparisons done, the assertions made and the suggestion given that the reputation of the trade itself was at stake, he concluded with the following challenge:

> ... Being myself satisfied of the Truth of the above related Facts, I beg the Favour of the Curious to satisfy themselves, which I am at any time ready to give them the Opportunity of doing; and farther, do hereby challenge the said Mr. *Jack*, or Mr. *Adams*, to a fair Comparison before impartial Judges.
>
> FRAN. WATKINS, Optician, *at Charing Cross*.[58]

The challenge may have appeared impertinent, having come from one so fresh to the trade although, from Watkins's standpoint, it was an opportunity to raise his profile among those who mattered. Having decided to pit his own telescope against that of Jack's design was a very daring act. Watkins may have been proficient in his practical optics and abundantly confident but, besides having a military engineering background, Jack was also a competent teacher of mathematical subjects, and not so easily put off by the upstart at Charing Cross.

Irked by Watkins's repeated challenges, which began in mid-December 1751, Jack finally responded. On the evening of 20 February 1752, he sent two persons to Watkins's house with the following note:

---

[58] Note 54.

For Mr. Francis Watkins *att Charring* Cross, London.
S I R,
I expect you'l meet Mr. Adams and me, *tomorrow*, at half an Hour after 12 o'Clock, at Mr. Bridgmans, the Sign of the White Hart, the corner of Welbeck Street next the *Fields* near Oxford Chapel, where you will meet with Impartial Judges to prove whither our Telescopes, or those of the Common Construction are the best,
Golden Square, Thursday
the 20th: of Feb: 1752 From Richd: Jack.[59]

Watkins could hardly refuse to accept what was now a matter of honour. Nonetheless, the unexpected arrival of the letter – with its arrangement at so 'short notice at so obscure a place' – caused consternation for he conveyed it immediately to the *London Daily Advertiser* who, the next day, published the letter from Jack to Watkins, together with an appeal by Watkins to an 'impartial public' that his several earlier challenges had never been acknowledged, and so he doubted that this proposed meeting could be fair.

Fearful of what size telescope Adams and Jack might produce at the challenge, Watkins added the following:

> ... If they [Adams and Jack] will abide by the Comparison of a particular two Foot Telescope of theirs, (which they were told when it was purchased, was to be compared with those of the common Construction) or another of the same Dimensions, let them publickly signify this, and I will meet them the next Day. But if they propose one or more of other Dimensions, let them fix those Dimensions in the News Papers, and allow me a reasonable time, as a Week, to provide the like, and on such Conditions I do now repeat my former Challenge.
> FRANCIS WATKINS.[60]

A further announcement in the *London Daily Advertiser* from Watkins's pen, dated 28 February 1752, indicates that Adams and Jack allowed a further week's grace for Watkins to prepare for the challenge. This ran:

> MESS. JACK and ADAMS having appointed this Day, at Twelve o'Clock, for comparing their Patent Telescops [sic] with mine, in Tubes of the same Length: I hereby give notice to all proper and impartial Judges of such Matters, that the Place appointed by them for this Comparison, is Mr. Bridgman's House, the Sign of the White Hart, near Oxford Chapel, where they shall be sure to find
> FRANCIS WATKINS.[61]

---

59 *London Daily Advertiser*, 21 February 1752. *Note*. In publishing Jack's letter, Watkins described it as both 'exact and literal', that is, in repeating the spelling errors.

60 Ibid.

61 *London Daily Advertiser*, 28 February 1752.

In the interim, Watkins inquired of Adams how many impartial judges he and his partner intended to bring to the meeting. To this, Adams answered:

> The impartial Judges we mean are Gentlemen of Rank, Character, and Knowledge; but we cannot limit the Number which will be at the Place of the Meeting: You may bring as many such Judges as you think proper.[62]

There are no references to members of the Royal Greenwich Observatory or Board of Ordnance or Lords of the Admiralty being present at the 'trial'. In all likelihood, both parties had probably drawn on trusted friends and influential customers to judge whatever tests were put in place:

> We are authorized and concerned to disabuse the Public, in the Relation which was published of the Trial of the Patent Telescope, and those of the common Construction; which, in most of the material Circumstances is either false or absurd, impertinent to the Question, defective in the Particulars, and injurious to the Honour of Personages of the most distinguished Rank and Abilities. And that such Gentlemen as choose to take unquestionable Authority for the Measure of their Judgement in this Affair, may be fully satisfied of the Merits of the Competition, and of the superior Properties and Excellencies of the Patent Telescope, by condescending to call at Mr ADAMS's House, in Fleet Street, they will see the same determined and authenticated under the Hands of the most Noble and Honourable Personages and Gentlemen, who were present at the Trial; whom they will find to be of so superior Rank, Honour, and Abilities, as to be above the mean and base Suspicion insinuated, of being partial in their Decision; and to be incapable of mis-applying their Authority, to give Sanction to any Design to mislead or impose on the Public; and where such Gentlemen, as in scientifical and experimental Researches, would choose to receive Conviction from the Evidence of their own Senses, may have an Opportunity of making the Experiment. And if the great Demand for this Instrument can be of any Weight to determine the Preference, we can assure, and prove to the Public, that the Sale thereof, within the Compass of a few Days after the Trial, to Gentlemen who were present, or who waited for the Issue and Report thereof from Persons, on whose Judgement and Veracity they could depend, amounted to above the Sum of £100. We are satisfied to rest the Merits of our Pretensions to the future Favour and Patronage of the Public, as these Facts shall appear to them to be established, or conclusive.
>
> The Patentees. RICHARD JACK
>
> London, March 16, 1752 GEORGE ADAMS.[63]

---

62 Ibid.

63 From the *General Advertiser*, 18 March 1752 (and repeated 20 and 23 March), and quoted in Millburn (2000), p. 63. From the formality and style, it seems probable that this

The above declaration of 16 March would suggest that notable personages and successful sales rather than technical superiority were the mark of the telescope's success. As Adams and Jack report of this telescope:

> It hath more than thirty different magnifying Powers, without changing the Glasses; a Property which no other Telescope hath.[64]

No doubt this instrument – 'By his Majesty's Royal Patent' – was well suited to its task in terms of robustness, finish in mahogany tubes and superior magnification and so, it must be accepted that it did have an edge over those of the 'common sort' sold by Watkins. Customers could choose the same in a variety of lengths from pocket size to 10 feet which, of course, provided increasing power in proportion to the length. Thus, this new range of telescopes prevailed, that is, until the Dollonds made their entry into this field later in that same year with a development of even greater scientific interest.

Model Steam Engines

The slow development of the steam engine is usually recounted in terms of the contributions of famous British engineers such as Thomas Savery, Thomas Newcomen, John Smeaton and James Watt but not forgetting ground-breaking experiments and machines by others such as Otto von Guericke and Denis Papin in the previous century. Savery's patent 'fire engine' (1698), Newcomen's 'atmospheric engine' (*c.*1714), Smeaton's extensive work on engine efficiency (from 1769) and Watt's major contributions of the separate condenser and double acting piston and, not least, application of an understanding of 'latent heat' (after 1769), all led to improvements in efficiency and a widening application of steam technology. In essence, it was this new form of motive power that fired imaginations and created that very peculiar 'English Revolution' which drove the nation inexorably towards industrialisation. These successful engineers with their various investigations into the power of steam often depended upon ideas and designs tried and tested by others, but whose names and incremental contributions are now submerged or entirely forgotten.

In the mid-eighteenth century, there were many minor engineers – sometimes instrument makers – who endeavoured to meet the challenges brought about by the new pumping engines. The widespread application of the Newcomen engine was accompanied by a persistent demand for greater power and

---

letter had passed through the hands of a notary or solicitor. Millburn states that no example of the Adams/Jack patent telescope has been identified.

64 *London Gazette*, 24 March 1752.

increased efficiency. The history of this development is rarely passed by without some account of the work of John Smeaton and James Watt, but their seminal contributions often overshadow others, mostly forgotten, who also rose to meet the challenges of the new age. Such a forgotten individual could be Francis Watkins, who was well versed in mechanics and the use of materials. Long before Smeaton's efforts to understand the factors governing power and efficiency of steam engines and long before Watt's radical re-think of the Newcomen engine, Watkins was among the first to promote an understanding of what he termed 'the distinguishing Invention of the last Century,' namely, the 'Steam Engine'.[65]

It is not difficult to see how, by the mid-century, Watkins entered this particular arena. By 1745 – with the significant take-up of the Newcomen engine – Desaguliers began to give currency to the subject in his lecture courses, although he had nothing to show by way of a working model. However, as a special effort to illustrate the subject, he chose to add to the second (corrected) edition of *A Course of Experimental Philosophy* (1745) an explanation and illustration of a fire-engine built by the late Richard Newham, formerly an engine-maker at Cloth-Fair in Smithfield. Bearing in mind this stage in the development of the engine and the likelihood for makers with a mechanical flair to closely follow Desaguliers's teaching, it became appropriate for philosopher mechanics, such as Francis Watkins and Jeremiah Sisson, to develop small working engines by which prospective customers could ascertain for themselves just what might be achieved by consideration of their ability to raise water.

The widespread promotion of the Newcomen engine had led some opportunistic engineers to make inaccurate claims but, noting this tendency, it was singularly due to Desaguliers that 'Persons who have Money' were cautioned to be wary of the 'boasting Engineers' whose claims often exceeded their expectations. Indeed, some landowners, with aspirations to develop mines but ignorant of the nature of engines, were easily misled and it was Desaguliers who educated these would-be owners in the science of pumping engines. In fact, the whole matter of the new engines had been a concern of the Government who, only a couple of years earlier, in 1743, nearly passed bad legislation which – had it not been for Desaguliers's warning – might have led many to financial loss. So, those with an interest in this field heeded the following warning:

> No Man must hope that with any Engine in the World, a Man shall raise more than one Hogshead of Water ten feet high in a Minute, with moderate Work, or such as he may hold all Day; nor an Horse above one Hogshead 50 feet high in the same time.[66]

---

65 *London Evening Post*, 6 June 1751.

66 Desaguliers (1745), vol. 2, pp. vii–viii.

The unwitting landowners had been warned, and so had those engineers who were guilty of excessive claims.

Watkins was certainly aware of how he should represent the performance of an engine after the rule introduced by Desaguliers. In the absence of any concept of *work* – that is, the product of weight and height through which a load is raised – it was simplest and best to present the rate at which the engine was working in terms of the *speed* with which it could raise a specified quantity of water.[67] For the same reason, it was, as yet, impossible to present the *efficiency* of an engine as the ratio of work output to work input since that concept only followed after Smeaton's results using model steam engines in the late 1760s. However, as a measurable quantity, the model steam engine developed by Watkins in 1750 was said to raise 'seven Hogsheads an Hour'. This *volumetric rate* was very comprehensible to the layman who understood that a hogshead was a barrel holding 54 gallons. Thus, Watkins's modest claim of seven hogsheads per hour was well within Desaguliers's upper limit of 60 hogsheads per hour.

Watkins's achievement – as stated in the *London Evening Post* for 16 August 1750 – might well have caught the eye of educated landowners, although what was offered there was more suited to pumping water through modest heights, not deep mining. The announcement ran:

> We hear there is now finished, by F. Watkins, Optician to his Royal Highness, the Prince of Wales, at Charing-Cross, a curious working Model of, The Steam Engine. 'Tis, perhaps, needless to observe, that this is look'd upon as the greatest Improvement of Mechanical Performances of the last Century; for, by a Power utterly unknown, produces the greatest Force of all Mechanical Performances. It is very neatly finish'd, in Brass, and contain'd within a Frame of four Feet by two, and six Feet high, and raises near seven Hogsheads an Hour, with a small Quantity of Fire. It is purchased by a Foreign Prince, and will be sent Abroad in a few Days.[68]

Thus, it was possible to visit at least one instrument shop where an engine could be seen in action. Without any quantifiable or comparable statement of what was meant by the 'improved mechanical performances', a visitor to Sir Isaac Newton's Head would have been hard put to appraise this engine, even if it was of acceptable workmanship sufficient for a foreign prince. Nor was this engine the only one attempted; another was completed in the following year.

---

[67] Watt did *not* introduce the concept of *horsepower* (32 400 ft lb/min) – that is, the power exerted by a dray horse pulling a capstan with a force of 180 pounds around a circle 24 ft across at a rate of two-and-a-half times a minute – until 1782.

[68] *London Evening Post*, 16 August 1750, p. 2, col. 2.

The purpose in placing an advertisement in the *London Evening Post* was not only to attract the 'Nobility &c.' but also it was directed at experts who might care to cast their judgement.[69]

At an entry fee of half-a-crown a head, this 1751 exhibition was certainly for the serious-minded, not the entertainment-seeking public. This second engine may have had some fair success because the advertisement ran well into the following year. By January 1752, the design must have been thoroughly tried and tested because Watkins then published a drawing and explanation '... Price 2 s[hillings] ... AN Accurate Print of the STEAM ENGINE, with a Mechanical and Philosophical Description of the Parts and Properties of that most curious and useful Machine'.[70] This expensive tract must have contained valuable 'know-how'; indeed, should a copy come to light, it may fill a significant gap in our knowledge of this phase in the development of the steam engine.

During those years, in the mid-1750s, there was in London one young man not yet known to the engineering world, but destined to have his name immortalised with the steam engine. He was, of course, James Watt (b. 1736). Accounts of his contributions to the development of the steam engine tend to start in 1763 when – as technician-cum-instrument maker at Glasgow University – he was asked to repair a defective model Newcomen engine.[71] Confronting this situation gave him ample cause for thought, particularly on the problem of heat losses; ultimately, he was led – as is well-known – to the idea of a separate condenser. However, his first insights to this kind of engineering had most likely begun during his working year 1755–1756 in London, that is, just before his move to Glasgow in 1757. During this spell, he worked for John Morgan, a mathematical instrument maker in Cornhill and also spent time with John Neale, who was known to Watkins.[72]

The Neale connection is particularly interesting because – as described earlier in this chapter – Watkins had links through their common interests in the electric machine. As a keen engineer, it is likely that Watt would have met many others with kindred interests. For example, it is known that he met Christopher Stedman of the Stationers' Company, James Tomlinson of the Spectaclemakers' Company and Nathaniel Hill of the Merchant Taylors' Company, a diverse group

[69] *London Evening Post*, 6 June 1751; the second machine was also advertised in the *London Daily Advertiser*, 16 January 1752.

[70] *London Daily Advertiser*, 16 January 1752. No trace of this work has been found. It is possible that the print was a diagrammatic engraving, together with a description of the working of the engine.

[71] Holbrook (1992), p. 134, entry G5.1. The illustrated model steam engine (Fig. 93, p. 60) in the Hunterian Museum, Glasgow University, is attributed to Jeremiah Sisson: Clifton (1995), p. 253.

[72] Hills (2002), pp. 55–58.

of instrument makers, each of whom supplied Watt with instruments for re-sale.[73] It is, therefore, not unreasonable to suppose that he also visited Watkins's shop for the same reason and, perhaps, while there witnessed the recently built model steam engine in action; he may also have visited the shop of Jeremiah Sisson in the Strand, another maker of pumps and engines, including – as it happened – the very engine that famously finished up in his Glasgow workshop for repair some eight years later.[74] But even long after Watt's mechanical improvements, the struggle to understand the power of steam remained a mystery as may be seen from a pamphlet by Cuthbert Clarke in which, as with Watkins, the subject was approached philosophically as a search for 'the origin, vicissitude and power of steam employed in fire-engines'. Indeed, the quest to devise the most efficient 'parts, powers, and true construction of that ingenious machine ...' ran on ceaselessly.[75] It would, of course, be of considerable interest to compare Clarke's thoughts with those advanced, but alas lost, by Francis Watkins some 20 years previously.

The Microscope

By the mid-1750s Francis Watkins's attention was focussing on the microscope. To appreciate this instrument in terms of its popularity, it is useful to reflect upon the sort of publicity given to it over the preceding decades. Interest in the microscope had begun in the previous century when the Fellows of the Royal Society were inspired by Robert Hooke's weekly displays of microscopical wonders using a large compound microscope. The sheer cost of owning a good instrument of this kind was a limiting factor and so some of the initial enthusiasm inevitably waned. Addressing the problem of cost in 1665, Hooke suggested that those wishing to pursue the subject should engage in making their own lenses; this, however, was a very tedious process – involving countless hours of grinding and polishing – hardly likely to attract more than a handful of enthusiasts. Despite this drawback, the subject itself was intrinsically fascinating and so never actually became moribund. Eventually, the appearance of the simple Wilson screw-barrel microscope at the beginning of the eighteenth century ensured that interest was popularly sustained.

When Ralph Sterrop joined with John Yarwell (c.1707), their joint trade card proclaimed: 'a new double Microscope invented by the said Yarwell, fitted

---

73 Clifton (1995), articles on Watt, p. 292; Stedman, p. 264; Tomlinson, pp. 279–280; and Hill, p. 136; Morrison-Low (2007), pp. 112–113.

74 Clifton (1995), article on Jeremiah Sisson, p. 253; Morton and Wess (1993), pp. 168–170; Hills (2002), pp. 312–319.

75 Clarke (1773).

for all Uses, particularly that admirable Curiosity of seeing the Circulation of the Blood in small Fish, and Thousands of Living Creatures in a drop of Pepper water.'[76] Shortly afterwards, about 1710, Culpeper advertised 'Several sorts of new contrived microscopes ... so well adapted to the Purposes both of Use & Entertainment, in discovering the curious Works of Nature, in those minute Objects that the bare Eye cannot reach ....'[77]

Three decades later still, Joseph Hurt (successor to Sterrop), also advertised the importance of the microscope in confirming Harvey's circulation of the blood hypothesis, adding that his microscopes not only detected 'Animalculæ in Fluids' but also 'many other surprizing Phænomena of Nature.'[78] By the 1760s, Samuel Scatliff announced that he 'Makes & Sells ... Microscopes Single or Double, which amazingly Magnifie ye Minutest Objects discovering the circulation of the Blood in Animals, & many other surprizing Phænomena'.[79] Descriptions such as these – extolling the virtues of the microscope according to its ability to reveal the circulation of blood or the wriggling 'animalculæ' – were simply crude indications that their instruments could resolve and magnify sufficiently to see exactly what one was supposed to see. It was, in effect, a basic test of the instrument, akin to that of an astronomical telescope through which one expected to pick out the moons of Jupiter or Saturn's rings. Evidence of this sort, drawn from trade cards and newspapers over five decades, clearly shows that the microscope had been in continuous demand.

By the start of the 1740s, these instruments were affordable to a wider audience. Henry Baker, then newly elected to the Royal Society and fresh from intensive studies in microscopy, argued:

> ... we are now so fortunate as to have this Instrument greatly improved amongst ourselves, the Apparatus made much easier as well as more useful, and the Price considerably reduced ... Nothing therefore is now wanting, but a general Inclination to employ these Instruments, for a farther Discovery of the Minute Wonders of the Creation ...[80]

---

[76] Calvert (1971), Plate 60.

[77] Good (1707), p. viii; Collins (1710), p. 52, and cited in Bryden (1992), p. 325 and footnote 114.

[78] Heal Collection, BM: (Prints and Drawings), item 105.58, reproduced in Crawforth (1985).

[79] For Hurt, see Crawforth (1985), p. 512 (Figure 28) and for Scatliff, see Calvert (1971), Plate 45.

[80] Baker (1742), Dedication to Martin Folkes, PRS, p. iv. For Baker, see Turner (1974–1975), and www.oxforddnb.com/view/article/1120 accessed 25 August 2011.

Indeed, the optical instrument makers had gone a long way towards designing and making microscopes that were well fitted to their task; moreover, they were made in sufficient numbers (in response to demand) to ensure that they retailed at competitive prices. What was lacking however was a suitable text on how best to apply the instrument to the study of microscopic objects. This became available in 1742 when Baker published *The Microscope Made Easy*. As Baker himself explained:

> In this inquisitive Age, when the Desire for Knowledge had spread itself far and wide, we sit not down contented, as heretofore, with the Opinions of ancient Times, but resolve to examine for Ourselves, and judge from our own Experience; it may not, perhaps, prove unacceptable to point out some proper Subjects of Enquiry.[81]

What Baker had perceived was that the microscope, perhaps more than any other instrument of science, was complete in itself and catered perfectly as a means for training an individual in the experimental processes of observation and rational judgement. In place of Hooke's advice on how to construct a microscope Baker assured readers:

> As my Desire is, to make People sensible of the Pleasure and Information the Microscope can afford, and instruct them how to manage and understand it, rather than how to make it, I shall take up none of their Time with the Manner of melting, grinding, polishing, or setting of Glasses; a Work very few of my Readers will ever trouble themselves about. But, instead thereof, I shall endeavour, with all the Brevity and Clearness I am able, to explain the Effects of Glasses on the Sight, and lead them gradually into the Nature, Uses, and magnifying Powers of Microscopes.[82]

Individuals, hitherto wary of constructing a microscope (a skill-based process that was bound to end in disaster in the hands of an untrained eye and ham-fisted worker), now had a ready-made solution to both the instrument and what was to be done with it by way of study. *The Microscope Made Easy* became an overnight success, for no sooner had the print run of 1,000 copies appeared in bookshops than it was sold out, thus necessitating re-runs, twice in quick succession, in March and April 1743.[83]

---

81 Baker (1742), p. xi.

82 Baker (1742), p. 3.

83 Early editions (1742 and 1743) were published by R. Dodsley. No further English editions appeared until 1754 (R. & J. Dodsley); the final English edition being 1769 (J. Dodsley). The second edition and its reprint are essentially the same as the first, although a copper plate of a solar microscope with explanatory footnotes was added.

As Francis Watkins neared the end of his apprenticeship with Scarlett he would certainly have been aware of the rising demand for microscopes and so he familiarised himself with the contents of Baker's text. He would also have been aware of those in the trade who rose to meet the challenges of this urge for self-education using the instruments of science. John Cuff, for example, was highly recommended by Baker but he alone could not cope with the stepped-up volume in demand. Others keen to capitalise in this field were Scarlett in Soho, George Sterrop at St Paul's Churchyard, James Mann in Ludgate Street, and George Adams in Fleet Street. Each produced their own version of the universal microscope.

A second boost came in March 1745 when Baker's second project – to re-issue the illustrations in Hooke's *Micrographia* – came to fruition. These amazing drawings by Wren had not been reprinted since the second edition of *Micrographia* in 1667. Fortunately, all but seven of the original plates had survived and these, with the missing plates redrawn, were published as *Micrographia Restaurata: or, the Copper-Plates of Dr. Hooke's Wonderful Discoveries by the Microscope* along with some explanatory (though anonymous) notes. For Baker, this work substituted for the lack of illustrations of microscopic objects in his own text *The Microscope Made Easy*. During 1744–1745 George Adams studied the works of Louis Joblot (on animalculæ) and Abraham Trembley (on the polyp), and incorporated these topics in his *Micrographia Illustrata* (January 1746).[84] According to Adams himself, this work was 'rather a faithful collection of everything that has hitherto been mention'd by the best writers upon microscopes, than matters of my own invention'.[85] Unfortunately, Adams did not defer to Baker and, in consequence, Baker grumbled that this Fleet Street instrument maker had plagiarised his work.[86] Regardless of this, Adams had promoted the subject and along with it the new universal microscope.[87] In turn, an interest in the microscope was developed by James Ayscough who, during the time that he partnered with James Mann, *c.*1745–1747, produced a universal microscope (comprising single, compound and solar microscopes, for use with opaque and transparent objects) which he sold for six guineas along with an explanatory pamphlet.[88] A few years later, in 1751, George Sterrop announced that '... by about the Middle of next Month, he shall have finished a curious new invented single and double Microscope, which may be applied to the solar Apparatus'.[89]

---

[84] Adams (1747); first published 1746, second edition 1747, fourth edition 1771.

[85] Quoted in Millburn (2000), p. 36.

[86] As Millburn (2000), p. 34 pointed out, Adams's text cost 16 shillings whereas that of Baker's was only five shillings, thus suggesting that more was to be obtained in the book of higher price.

[87] See Millburn (2000), p. 32.

[88] Classified advertisement in the *London Evening Post*, 10 March 1747.

[89] *General Advertiser*, 15 May 1751.

Each of these developments would have made an impression on Watkins. It is even possible that he had some inside information on the strength of the demand in the early 1750s through his friend John Margas, who at this time was working for Cuff. He would note that profit was to be made from this trend. He would also be conscious that each of the instrument makers cited here had developed simpler forms of pocket microscope which were widely advertised. Thus, by April 1751, it appears that his interests in the electric machine had diminished while at the same time he repeatedly advertised that he had developed his own pocket microscope which had overcome the usual problem of screwing and unscrewing, in order to focus – a common objection to the Wilson version – by the use of draw-tubes.[90]

However, an interesting insertion running below the Watkins's advertisements contained the following cautionary (anonymous) remark:

> An Old Optical Instrument-maker humbly begs Leave to give this Hint. A Fine adjusting Screw is the old and truest Method in Optical Instrument, (especially) where an extraordinary magnifying Power is employed.[91]

The identity of this 'old optical instrument maker' was perhaps John Cuff, because this insertion was inserted below the Watkins announcement and Cuff's list of products above it.

Baker's bestselling text *The Microscope Made Easy* became Watkins's bible and, ultimately, the basis for his own publication in 1754. Baker had prefaced his text with the promise:

> I shall avoid as much as possible all Affectation of Learning, or Expressions that are not in common Speech, being desirous that every body may understand me.[92]

Despite this laudable aim, it was still too demanding a work, particularly for the widening audience that had begun to emerge by the mid-century. A novice needed a guide offering sufficient handling instructions, together with a simple indication as to the kind of discoveries that might be achieved with easily available specimens. Baker's text – beautifully written and moderately definitive – was far too authoritarian, although well suited to educated persons; simplification and reduction was therefore essential if the subject was to be diffused to a wider readership. This was what Watkins had in mind when he wrote a 97-page primer,

90 *London Evening Post*, 25 April 1751. This insertion was repeated on no less than 18 occasions. *Note:* All advertisements of this period (*General Advertiser*, the *Whitehall Evening Post* and the *London Evening Post*) were preceded with the proud reminder that Watkins hailed from the Scarlett workshop.

91 *Whitehall Evening Post or London Intelligencer*, 30 April 1751.

92 Baker (1742), p. xvi.

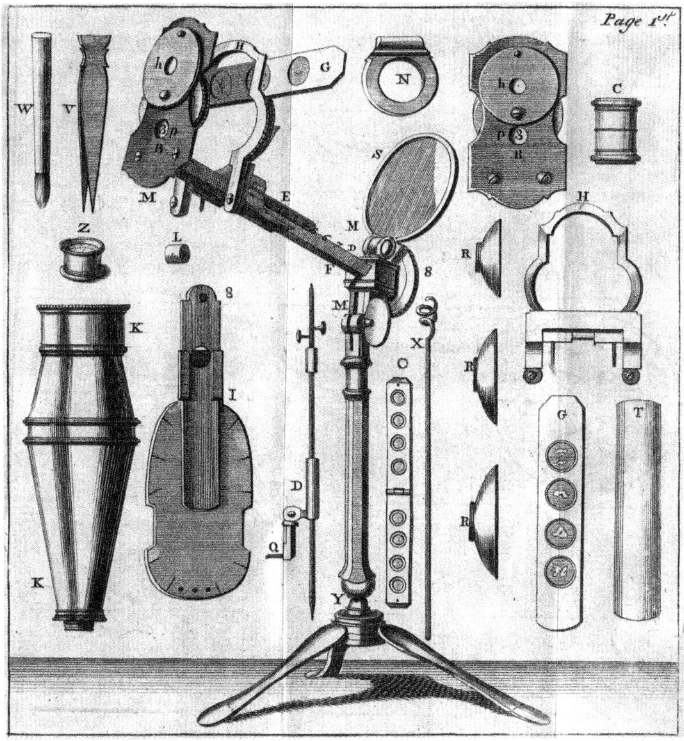

Figure 3.12 Plate showing his microscope, from Watkins's *L'Exercise du Microscope* (1754). Museum of the History of Science, University of Oxford

L'Exercise du Microscope (1754), as an elementary textbook largely for the numerous French-speaking immigrants around the metropolis.[93] As Watkins himself explained:

> I could not do better than to give the explanation in the most universally understood language: having been much encouraged to do so, by the realisation that there is nothing substantive in the French language on this subject.[94]

[93] It is noteworthy that James Ayscough also published his tract on the microscope in French around this time (1753) but, unlike Watkins's comprehensive work, he did not elaborate on what could be observed by prescriptive experiments. Watkins duly acknowledges Baker as his inspiration.

[94] Watkins (1754), 'Au Lecteur': '... j'ai cru ne pouvoir mieux faire que d'en donner l'Explication dans la Langue la plus universellement entendüe: ayant été d'autant plus incité à faire par la Consideration qu'on ne trouve encore rien de fort considerable dans la Langue

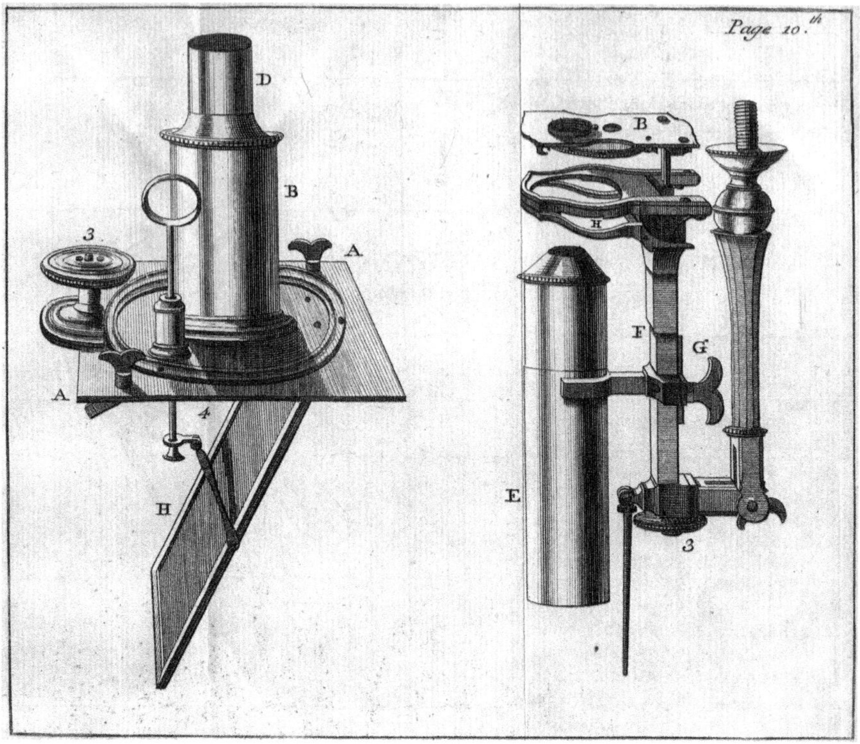

Figure 3.13 Plate showing 'le microscope solaire' from Watkins's *L'Exercise du Microscope* (1754). Museum of the History of Science, University of Oxford

The text was an invaluable contribution to the diffusion of the subject because it set out a number of observational exercises in a graded fashion. It was also a practical guide to handling and adjustment of the instrument. This double microscope represents the culmination of Watkins's microscope work.

Baker's texts, all superbly presented by the publishers R. & J. Dodsley and sold at Cooper's in Paternoster Row as well as at Cuff's in Fleet-street, were in a class of their own. The works of Ayscough, Adams and Watkins were each privately printed, and were intended to be sold at each maker's shop rather than through a bookseller. Watkins's text lacks perfection and from the numerous typographical mistakes it seems that it was cheaply printed. Nevertheless, it was good value in educational terms and so may have inspired many juveniles and

Francoise sur ce Sujet.' It seems very likely that Watkins's wife, Clarinda, herself of Huguenot stock, assisted him in this work.

young adults lucky enough to own a Watkins boxed-up microscope. In the same year that Watkins's text appeared, the Dodsleys reissued Baker's *The Microscope Made Easy* along with his *Employment for the Microscope*, to be sold together as a compendium.[95] How widely or seriously Baker's texts were studied is not known. Perhaps, as with a few extant Cuff microscopes whose specimen objects remain intact to this day, many copies may have finished up, unread, on the shelves of a family library. One set of Baker's works – apparently actively used – resided in the library of the Reverend George Austen, father of the celebrated Jane Austen, who reputedly introduced his daughters to the 'wonders of the world in miniature'.[96] By contrast, Watkins's universal microscope, together with his plain-speaking French text would have held greater appeal for those tyros who truly wished to explore the microscopic world in a spirit of investigative science; this alone may explain why only a few well-worn copies can be found today.

From this outline of the first decade of business at Sir Isaac Newton's Head, it is plain that Francis Watkins worked across the full range of practical activities expected of an optical and philosophical instrument maker. His reputation in spectacle making ensured that he was patronised by Prince Frederick, son of George-II, and so held a royal warrant as 'Optician to His Royal Highness the Prince of Wales'. More importantly, perhaps, Sir Isaac Newton's Head became one of those focal places where visitors called on their visits to town. Customers came from far and wide to purchase spectacles or have them repaired, but also they would call to examine instruments of an optical or philosophical nature such as those listed on the billhead.[97]

From a miscellany of newspaper advertisements, we also know that Watkins sold prints, books and maps; took subscriptions for John Newbery's lectures; and may well have offered tuition himself in mathematical subjects.[98] A similar array

---

95 *Whitehall Evening Post or London Intelligencer*, 23 May 1754.

96 Tomalin (1997), p. 28.

97 Taken from a bill of sale to Mr Elder, dated 19 April 1755. WSRO: Petworth House Archives 6643.

98 For example, during January 1751, Watkins advised those with an artistic or classical persuasion that they could purchase a copy of Piranesi's views of Rome (by the print seller Sayer), along with a zograscope – 'most delightful for the Concave and Diagonal Mirrours' – by which it was possible to make sense of the exaggerated perspective, thereby enabling an effect of three-dimensionality in these curious drawings: *London Evening Post*, 22 January 1751; during 1754, an announcement 'To be had at Mr. Watkins's, Optician to the Prince of Wales, at Charing-Cross', stated that James Hardy's text *A Complete System of Interest and Annuities* was available: *London Evening Post*, 7 May 1754; during November 1754, Watkins took bookings for John Newbery's lecture course, probably given at his shop in St Paul's Church Yard where he sold a range of books and medicines: *Daily Advertiser*, 13 November 1754; during 1755, a joint enterprise existed between Watkins and Robert Sayer (opposite Fetter Lane), Mrs Bakewell (Cornhill), Mr Eynon (behind the Royal Exchange) and William

of extraneous reasons for visiting the Dollonds was also built up in the following decade where, for example, subscriptions were taken for the 'lying in charity.'[99] The optician's shop was also an important place for travellers to call should they require globes. Both Watkins and Peter Dollond retailed those made by Nathaniel Hill, as did a host of other metropolitan instrument makers, including Heath & Wing (Strand), John Bennett (Soho), Johnson (Ludgate Street), Sterrop (St Paul's Church Yard), Ribright (Poultry) and Lincoln (Cornhill).[100]

As for personal interactions that Watkins may have had before 1760, very little is known, although from later glimpses into his life it is clear that he was recognised and respected within his trade. This is apparent through his connection with John Dollond and the litigation into which he was forced with Peter Dollond (§ Chapters 5 and 6) but also, it is implicit, that he was held in some regard as Master of the Worshipful Company of Spectaclemakers. Therefore, to prepare the reader for the issues that arose through Watkins's connection with the Dollonds, the following chapter concentrates on the subject of spherical and chromatic lens aberrations, and chromatism generally in order to understand the optical problem that beset the development of the refracting telescope. These issues had been a part of Isaac Newton's research on 'light and colours' and ultimately formed an important part of his seminal text, *Opticks* (1704). It was upon this work that opticians based their understanding of chromatism. However, as we shall see, Newton was not always correct and, in this case, he drew an erroneous conclusion with regard to spectral dispersion. It took time for this error to percolate through and even more time to produce a new direction in optical theory. Ultimately, the path was cleared to develop the achromatic lens and, hence, a very much improved refracting telescope.

Harbert (London Bridge), in the sale of a new map produced by the Society of Anti-Gallicans (who opposed everything French) showing forts and harbours of the English Empire in North America: *Whitehall Evening Post or London Intelligencer*, 23 December 1755; during March 1757, the press announced that those requiring instruction in 'arithmetic, algebra, geometry, surveying, fortification &c.' should inquire of 'Mr. Watkins's, Optician, Charing Cross': *Daily Advertiser*, 28 March 1757.

[99] *Gazetteer and New Daily Advertiser*, 28 August and 1 September 1764.

[100] *Gazetteer and New Daily Advertiser*, 11 February 1765.

Chapter 4

The Chromatic Problem: From Newton's Error to Hall's Solution

There are few episodes more complex, yet more revealing, about the optical instrument trade, than that surrounding the early history of the achromatic lens. Here, at last, was an invention destined to rid refracting telescopes of the coloured fringes that so annoyingly appeared around the edge of an image. Of course, such an innovation should have been greeted with acclaim but, in fact, its origin and early development were cloaked in secrecy and beset by confusion and acrimony.

Before we can consider these matters of litigation, it is vital that the reader should appreciate a little of the scientific background to the optical problem of chromatic aberration and attempts made at its correction. We begin with Isaac Newton's study of this problem. As shown in the present chapter, he abandoned all hope of ridding the refracting telescope of this particular nuisance.

During the half-century between Newton's death and Dollond's successful experiments, two lesser-known individuals – Chester Moor Hall and Caleb Smith – each realised that the problem of chromatic aberration was not as insoluble as Newton had supposed and so modest headway was made. The presence of these two individuals in the chronology of events – particularly that of Hall – is of considerable importance when later considering the arguments proposed by Francis Watkins and the Spectaclemakers' Company who wished to have the Dollond patent annulled. Therefore, the positions of Hall and Smith are introduced in the present chapter, at the very outset of this intriguing early phase of the development of the achromatic telescope. However, John Dollond's entry into this field was relatively late and since his position as originator of the patent is quintessential to what follows by way of litigation, an understanding of his background – leading to the partnership with Francis Watkins – is reserved for a separate examination in Chapter 5.

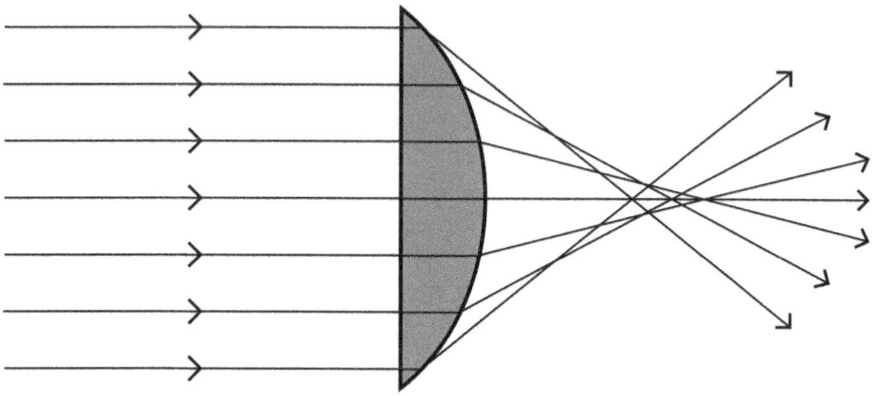

Figure 4.1 Diagram showing longitudinal spherical aberration

The Problem of Chromatic Aberration

In the eighteenth century there was no material that could be claimed as 'optical glass', at least in the sense that this term is understood today.[1] The quality of an optical image could be spoilt by the presence of bubbles, specks and striations within the glass. Of the several suppliers in London, the Whitefriars glasshouse – on the south side of Fleet Street near St Bride's, between Bridewell and the Temple – was popular with optical instrument makers as their source of glass blanks.[2] After obtaining a blank – which was rarely more than four inches in diameter – the next step in production was to grind and polish it to a required specification. Finally, a finishing stage remained, to correct *spherical aberration* arising from the uneven curvature of the lens surfaces. Three areas of concern here included: (i) the differing focussing error due to the extremes of the curvature from outer to inner zones; (ii) the distortion in an image such that what should appear, say, perfectly round might instead appear in the shape of a pin cushion or barrel; and (iii) the degree to which 'stopping down' the aperture of the object glass could help in image formation. Unfortunately, (iii) introduced the double effect of reducing the light intensity while at the same time extending the focal length due to use of the flatter central zone of the lens.

[1] A succinct account of the types of glass available for optical work is provided in Turner (2000), pp. 399–404.

[2] Museum of London: Whitefriars Glasshouse, trading ledgers, 1778–1792. For a backdrop to the Whitefriars glasshouse, see Evans, Ross and Werner (1995) ch. 1. King (1955) p. 159 refers to other crown glass works used by optical instrument makers in Lambeth and Ratcliffe. Also see Lysons (1795), 3, p. 473.

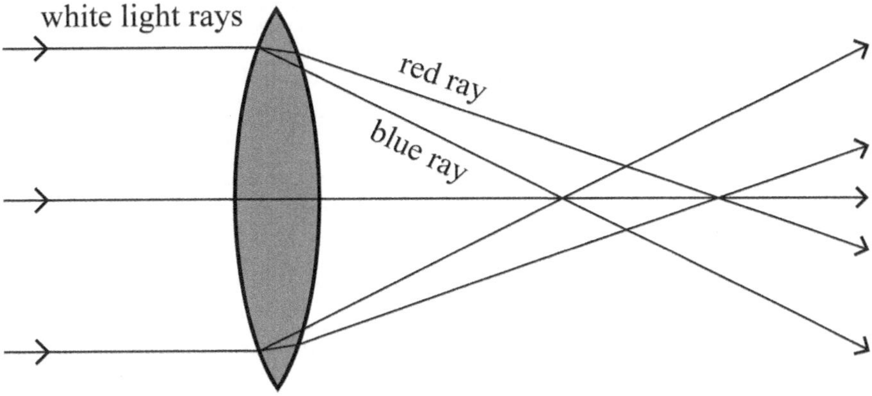

Figure 4.2 Diagram showing longitudinal chromatic aberration

Thus, although the magnification was potentially greater, the brightness of the image was much reduced. The principal difficulty is illustrated in Figure 4.1.

To address such problems it was important to seek out a good lensmaker whose competence at reducing spherical aberration was known. Such a person was George Bass who worked in the Bridewell Precinct of Fleet Street, not far from the Whitefriars glasshouse. We shall meet him in this and again in later chapters.

Another optical hindrance – the principal one relevant here – was that due to the prismatic effect close to the perimeter of a lens (*chromatism*). This was manifest as a coloured fringe around a bright image (*chromatic aberration*). It was well known that a piece of glass with non-parallel sides, such as a prism, could not only bend (*refract*) a beam of white light but it could also split (*disperse*) it into an array of coloured light (its *spectrum*). Much of the understanding of prismatic behaviour is credited to Isaac Newton, who established: (i) that the spectral coloured lights are not created by the prism but are intrinsic components of ordinary white light; (ii) that white light, once dispersed into its spectrum, could not be further dispersed by a second prism;[3] and (iii) that a pair of identical prisms – with refracting angles inverted with respect to each other – could recombine a dispersed beam into its original beam of white light.

Although the sum total of Newton's studies on light and colours did not appear in print until his famous text *Opticks* in 1704, he had actually begun to experiment systematically in 1666 with a pair of prisms, reputedly purchased at the annual fair held at the glassmaking town of Stourbridge, but his earliest

[3] The experience that light cannot be further dispersed using a second prism was recorded long before Newton.

confrontation with the problem of coloured fringes probably arose a few years earlier, during his first attempts to grind lenses for a telescope in 1663.[4] The overall problem of chromatic aberration is illustrated in Figure 4.2.

Taken together, it is not difficult to understand why the mixture of spherical and chromatic aberrations of a single lens forms an image that is both indistinct and tinged with colour. However, Newton's studies revealed two important facts: (i) that it was not the spherical aberration that limited telescopic performance but rather the chromatic aberration; and (ii) that a narrow beam of sunlight from a circular hole in a window shutter of a darkened room, on passing through a prism, emerged to form an elongated spectrum, some five times as long as its breadth. The latter has been illustrated many times previously but is so important that it is worthwhile repeating here.

Of course, this result was surprising because it was generally held that all rays were equally refrangible and so – for a prism arranged for *minimum deviation* – the image should have remained circular.[5] Clearly, then, sunlight consists of rays of different *refractive indices*. Here was the essential clue to how colour dispersion might be measured, analysed and ultimately expressed in terms of a mathematical model. Studies on this phenomenon engaged Newton's attention from the time of his appointment as Lucasian professor of mathematics at Cambridge in October 1669, and culminated in a seminal paper on 'light and colours', read before the Royal Society on 8 February 1672.[6]

A Missed Opportunity

A new understanding of the spectrum and, therefore, the function of the prism as a device for splitting white light into its component colours could easily be applied to account for the coloured fringes around an image. The practical problem of removing the unwanted coloured fringes from the optical image was, however, quite another matter. Early experiments had shown that since white light could be achieved by recombining the dispersed colours from a prism with

[4] Whittaker, introduction to: Newton (1704, Dover edition, 1952), p. lxvi.

[5] Early on in his experiments with prisms, while investigating the deviation caused by varying the angle of incidence of a beam of sunlight on a prism, Newton discovered a minimum angle of deviation when incident and mean emergent rays were symmetrical. Thus, increasing or decreasing the angle of incidence about this position of symmetry caused the angle of deviation to increase. *Lectiones Opticae* [Optical Lectures] of 1669 (Sec. III, Prop. 25).

[6] Newton (1671–1672). Newton's mathematical work on light and colours was recorded in his *Lectiones Opticae* of which there were two versions (the second becoming available only after Newton's death). A discussion on the intricacies of Newton's theory of colours (with its simple and compounded colours and whether such are immutable by reflection, refraction and transmission), is set out by Alan E. Shapiro in Cohen and Westfall (1995), pp. 191–202.

the aid of a second inverted prism then it ought to be possible to obtain a colour-free image by combining lenses of opposite powers (*convex* and *concave*), the one chromatically compensating for the other, while still sufficiently refracting the beam as it passed through the optical system to form an image. Newton understood that to achieve such 'refraction without dispersion' required two optically different materials, that is, glasses of different 'dispersive powers', as he wrote to Henry Oldenburg, Secretary of the Royal Society:

> ... it seem'd not impossible for contrary refractions so to correct each others inequalities ... [because] by a Complication of divers successive Mediums, as by two or more Glasses or Crystals, with Water or some other fluid between them; all which together may perform the office of one Glass, especially of the Object-glass, on whose construction the perfection of the instrument chiefly depends. But what the results in Theory or by Tryals have been, I may possibly find a more proper occasion to declare.[7]

Unfortunately Newton was then led astray by his experiment with a water prism.

We shall consider, albeit briefly, just what Newton did understand at the outset of his optical studies regarding the passage of a beam of light through a prism. During the early experiments, he employed six prisms with different refracting angles in the range from 45° to 64°. He noted that a prism with a larger refracting angle would not only deviate a beam more than a prism with a smaller refracting angle, but also the prism with the larger refracting angle would create a longer spectrum than a prism with a smaller angle. So, for example, if the 64° prism deviated a beam four times as far as one of 45°, then Newton expected the respective spectra would have a length ratio of 4:1. In the course of these experiments he also considered other likely variables: namely, the *size* of the hole in the window shutter through which the beam of sunlight came; the *thickness* of the prism at the point where the beam was incident; and the *inclination* of the beam to the horizontal. Of these, however, he concluded there was 'no sensible changes in the length of the Image'.

Finally in this sequence, Newton investigated the proportionality between the remaining factors (*mean angle of deviation* and *length of spectrum*) for other materials using a specially prepared water prism. Of this, he concluded:

> Neither did the different matter of the Prisms make any [sensible changes to the length of the Image]: for in a Vessel made of polished Plates of Glass cemented together in the shape of a Prism and filled with Water, there is the *like Success* of the Experiment *according to the quantity of the Refraction*.[8] [Author's italics]

---

7 Newton (1672), pp. 5084–5085.

8 Newton (1704, Dover edition, 1952), p. 31.

In this statement lies a weakness because he should not have found a 'like success'. More precisely observed, the lesser mean deviation of the water prism would have produced a shorter spectrum than that of glass. The reason for this blunder or oversight has sometimes been explained by the suggestion that he added 'sugar of lead' (lead acetate) in order to increase the refractive index of water.[9] Unwittingly, this addition would also have increased the angular dispersion to the point where the dispersive power of the water prism may well have cancelled that of a glass prism for an inverted pair. However, had Newton increased the refractive angle of his water prism until it produced the same deviation of the mean ray of the spectrum, then he would have found that one prism would not have corrected the colours of the other. Newton's 'like success' might imply that equal and opposite dispersions necessarily entail equal and opposite refractions and so, of course, 'refraction without dispersion' would not be possible. Whatever the true situation with regard to these prism experiments, these results pointed to the futility of exploring further the possibility of lenses in combination to eliminate chromatic aberration.[10] Newton had correctly understood that that white light was a heterogeneous mixture of differently refrangible rays but, peculiarly, he believed that all transparent bodies had the same dispersive power. He had, therefore, missed completely the possibility of achromatism.[11]

Newton Abandons his 'Glass-Works'

The results of Newton's water prism experiment did not bode well for chromatic correction. Perhaps a more overriding reason for the abandonment of the hope of employing a compound object glass was the straightforward fact that a heterogeneous beam of sunlight could not be brought to a single focus due to each colour having its own refrangibility. Certainly, the strength of this argument may also have caused Newton to veer from the task of ever achieving an achromatic combination.[12] Explaining these thoughts to Henry Oldenburg by letter dated 6 February 1671/2, that is, just prior to submitting the paper on 'light and colours':

---

[9] Brewster (1855), vol. 1, pp. 110–111.

[10] Newton did make an argument for the compound lens but only in connection with correcting for spherical aberration while emphasising that it was impossible to correct for chromatic aberration by this method. Bechler (1975), pp. 119–122.

[11] So intriguing is Newton's water glass prism experiment (because of the inaccurate conclusion) that it has been the subject of several scholarly accounts.

[12] The argument showing the perfection of telescopes was impeded by different refrangibilities is far more complex than intimated here. Newton (1704), cf. book I, prop. VII, theor. VI: (Dover edition, 1952), p. 82.

> When I understood this [*viz.* that white light comprises rays of different refrangibilities] I left off my aforesaid Glass-works; for I saw, that the perfection of Telescopes was hitherto limited, not so much for want of glasses truly figured according to the prescriptions of Optick Authors, (which all men have hitherto imagined,) as because that Light it self is a *Heterogeneous mixture of differently refrangible Rays* ... This made me take *Reflections* into consideration, and finding them regular, so that the Angle of Reflection of all sorts of Rays was equal to their Angle of Incidence; I understood, that by their mediation Optick instruments might be brought to any degree of perfection imaginable ...[13]

So, Newton clearly understood that even if lenses could be ground to their ideal shapes (paraboloidal rather than spheroidal, a proposal stemming from Descartes and much elaborated in the *Philosophical Transactions* with schemes for grinding and polishing lenses of this shape), the greater problem of image distortion was due to chromatic aberration and this would always persist, no matter how perfect or ideal the shape of the lens.[14] His course of action, therefore, was to switch emphasis from the refracting to the reflecting telescope, an idea not entirely his own but one derived from reading the *Optica Promota* (1663) of James Gregory, professor of mathematics at the University of St Andrews, a copy of which had fortuitously fallen into his hands.[15]

Apart from the overriding advantage of a reflecting telescope having no chromatic aberration, it has, as its optical principle, the immediate possibility of shortening the tube length by one-half of that of an equivalent refracting telescope. Newton had already produced such an instrument by his own hands

---

13 Turnbull (1959), vol. 1, p. 95.

14 Astronomers did accept Newton's decision to move towards a reflecting telescope but for the wrong reasons, mainly believing that the errors in refracting telescopes were due to the spherical aberration arising from not having satisfactorily ground their lenses into cartesian ovals (paraboloids), despite Newton consistently referring to the 'impediment' caused by refrangibility being the greater problem. He also refers to the use of compound lenses of glass and water for the correction of sphericity rather than attempting hyperbolic-shaped lenses. Newton, *Principia* (1687), pp. 234–235 (See Hall (1993), p. 104). Proposition VII of *Opticks* provides an argument for the greater confusion and hence difficulty in chromatic, rather than spherical, aberration.

15 In a letter to Henry Oldenburg of 4 May 1672, Newton revealed: '... for when I first applyed my selfe to try the effects of reflexions, Mr. Gregory's Optica Promota (printed in ye yeare 1663) being faln into my hands, where there is an Instrument described (pag. 94) like that of Monsieur Cassegrain's, wth a hole in ye midst of ye object-Metall to transmit the light to an eye glasse placed behind it; I had thence an occasion of considering that sort of constructions, & found their disadvantages so great yt I saw it necessary before I attempted any thing in ye Practique to alter the designe of them, & place ye eye glasse at the side of the Tube rather then at the middle.' Turnbull (1959), vol. 1, p. 153. Gregory did make some attempt himself but, even using the best makers, his attempts failed. See Simpson (1992).

in 1668–1669. By the autumn of 1671, Oldenburg had learned of a second telescope by Newton and was sufficiently intrigued to invite him to present it to the Royal Society. Although this instrument was a little less than seven inches in length it yielded a magnification of about ×38 and enabled the observer to discern the satellites of Jupiter and the horns of Venus. These features and accomplishments were of sufficient merit for the instrument to be shown to the king by a select group comprising Sir Robert Moray, Robert Hooke and Christopher Wren. And yet, despite all this attention, it does not appear that those who examined it had noticed its ability to produce colour-free images.[16] The Royal Society then commissioned Christopher Cock (or Cox) to make a four-foot version of the same although this instrument failed to discern celestial phenomena as well as the refractors of Johannes Hevelius and Christiaan Huygens. Nevertheless, Newton's contribution was approved, and so led to his acceptance and election on 11 January 1672.

Experiment 8: Towards a Dispersion Theorem

Two complementary actions seem to have arisen in the wake of Newton's recommendation of the reflecting telescope: firstly, there was a need to justify the preference of the reflecting telescope by improving on its design and manufacture, and secondly, there was an obligation to prove, mathematically, that refracting telescopes were inescapably flawed due to the inherent and irremovable defects of chromatism. Concerning the first action, sufficient interest was generated for Francis-I Hauksbee, the Royal Society demonstrator, to make and sell reflecting telescopes; in time, his know-how was passed on to Edward Scarlett through John Hadley and Samuel Molyneux.[17] James Short, too, made vast improvements in the shaping of paraboloidal specula and, as a measure of the excellence of one of his reflectors, it was possible to read the print in the *Philosophical Transactions* at a distance of 500 feet.[18] Indeed, so successful was Short in obtaining prestigious commissions that he was soon well on track towards earning a considerable fortune.[19] Concerning the second, it was unfortunate that in presenting a dispersion theorem, Newton had chosen to compare the seven prominent spectral colours analogously with the seven intervals in an octave of a musical scale; that is, by dividing the indices of refraction between the extreme prominent colours, from red to violet, in

16 Bechler (1975), p. 102; Mills and Turvey (1979); Hall and Simpson (1996).

17 Brewster (1855), vol. 1, pp. 56–57.

18 Brewster (1855), vol. 1, p. 57. See also Simpson (2009).

19 Two examples of Short's expensive instruments are: (i) a 12 feet reflector sold to Lord Thomas Spencer for £600, and (ii) another sold to the King of Spain for £1,200. Brewster (1855), vol. 1, pp. 50–54.

the same mathematical proportions as for the musical intervals within a single octave. These two spectra – the musical and the optical – became the subject matter of experiment 7 in book I, part II, prop. 3 of *Opticks*.[20]

Sequentially, experiment 8 then proposed a dispersion theorem with supporting experimental results. In dealing with the theorem, Newton carried forward the earlier formed notion that the dispersion of a spectrum was independent of the nature of the refracting medium and dependent only upon the deviation of the mean ray.[21] Put otherwise, the analogy for the separation of the seven prominent colours with the seven intervals in an octave of a musical scale (not elaborated here) was influential upon the form of the theorem and therefore had hindered his realising the physical dependency of dispersion on the nature of the medium through which the light was refracted.[22] His experiences led him to propose a theorem of the following form: '*the Excesses* [or differences] *between the Sines of refraction of several sorts of Rays above their common Sine of Incidence* [for various colours] *when the Refractions are made out of divers denser Mediums ... suppose of Air, are to one another in a given Proportion*', which was, however, *erroneous* [Author's italics].[23]

The purpose of the experimental values presented along with the dispersion theorem in experiment 8 appears illustrative in intent. Briefly, they relate to the passage of a beam of white light through successive media with the incident and emergent beam both in air (*air* → *water* → *glass* → *water* → *air*). No practical description of this appears in *Opticks* although it is generally accepted that the arrangement was the same as that used in 1672 at the outset of his investigations. Thus, a glass prism was placed within a water prism such that their refracting angles were opposing and their bases parallel.

According to Newton in 1704:

> I found ... that Light as often as by contrary Refractions 'tis so corrected, that it emergeth in Lines parallel to those in which it was incident, continues ever after to be white. But *if the emergent Rays be inclined to the incident, the Whiteness of the emerging Light will by degrees in passing from the Place of Emergence, become tinged in its Edges with Colours*. This I try'd by refracting Light with Prisms of Glass placed within a Prismatick Vessel of Water.[24] [Author's italics]

---

[20] Newton, *Opticks*, book i, pt ii: (1704, Dover edition, 1952), pp. 125–129 (exper. 7) and pp. 129–132 (exper. 8). See Hall (1993), pp. 112–113 for comments on the musical analogy.

[21] Newton, *Opticks*, book i, pt ii, exper. 8: (1704, Dover edition, 1952), pp. 129–132.

[22] Although Newton often spoke of seven prominent colours for convenience, he did not believe the spectrum to comprise these colours alone, but rather a continuity of colours.

[23] Shapiro indicates that, at one point in his studies, Newton suggested a 'quadratic model' for spectral dispersion but this also was erroneous. See Shapiro in: Cohen and Westfall (1995), pp. 196–199.

[24] Newton, book I, pt II, prop III: (1704, Dover edition, 1952), p. 129.

Although no explanatory details were offered in *Opticks* it is worth repeating the argument of 1672 here. In that experiment, initially with the incident and emergent beams parallel to the base of the outer water prism (that is, passing symmetrically through the arrangement) with refractions 'as near as may be equal to one another', colours will appear in the emergent beam. Afterwards, by rotating the inner (glass) prism within the outer (water) prism, so that 'the refractions of the interior Prism be something lesse than those of the exterior [prism]' a position is reached when the emergent beam is white.[25] But, in this instance, the incident and emergent beam would then no longer be parallel. By this argument Newton had demonstrated that parallel incident and emergent beams may show colour and yet a refracted beam could conceivably be white. But, if this particular argument was correct then it surely would have supported the notion that an achromatic arrangement ought to be possible.

When *Opticks* appeared in 1704 there was little reason not to accept Newton's authority on the subject. Few then knew that his published views on dispersion in experiment 8 ran contrary to those held in 1672 and this would not become apparent until his *Optical Lectures* were published some time later. Moreover, the results of the water–glass prism experiment, as stated in 1704, did not truly meet the consequences of the dispersion theorem although – as will be noted in the following section – this lack of concord was not apparent at the time because no one bothered to check what Newton had laid down. Many years were to pass before these discrepancies became apparent and, even after these had been exposed, there were forgiving voices willing to re-accommodate experiment 8. Peter Dollond, for instance, in 1789, believed that the 'awkward' experimental results might be accounted for by Newton having used Venetian flint glass rather than English crown glass: an interesting proposal, respectful of Newton, but of little assistance in achieving a true account.[26] Of more relevance here was the simple remark in experiment 8 concerning 'tinged edges', thereby implying the impossibility of 'refraction without dispersion' and, in consequence, the impossibility of an achromatic telescope objective. This alone was sufficient to cause optical practitioners to take Newton's lead and shift their attention to the reflecting telescope.

---

[25] This was a clever argument used in 1672, possibly intended to confuse Hooke who held his own views on the matter. Bechler (1975), pp. 113–116; a simpler overview is provided in Hall (1993), pp. 107–108.

[26] P. Dollond in RS: MS L&P decade IX, item 146. Because of the lack of concord between experimental results and dispersion theorem, some scholars have denied that Newton ever tried this experiment: for example, Bechler (1975), p. 115 and Hutchinson (1991), p. 140. However, the present author draws attention to Newton's use of an amanuensis of whom he wrote: '... whose Eyes for distinguishing Colours were more critical than mine.' Newton (1704, Dover edition, 1952), p. 126.

Blind Acceptance of *Opticks*

This unquestioning attitude to Newton's views on optics is particularly noticeable over the years of his presidency of the Royal Society, that is, from his election in 1703 until his death in 1727. Indeed, the passing of Robert Hooke in 1703 – whose antagonistic views in optics Newton found so loathsome although with whom he had corresponded in 1672 on the subject of compound glasses at the outset of his optical experiments – eased the way for Newton's forceful opinions. But, if Hooke was gone, Gottfried Wilhelm Leibniz remained. He had been a critic of Newton's gravitational theory and, by the start of the new century, his attention had turned to optics. As if to damn his English rival, Leibniz urged experimenters to challenge Newton's theory of colours by replicating the experiments of Edmé Mariotte, whose experimental work in 1673 had cast doubt on the immutability of individual coloured rays passing through a second prism.[27] Newton – from his presidential position – dealt with this continental opposition simply by ordering the Royal Society's demonstrator, Francis-I Hauksbee, to reinforce the prism experiments by repeating them; likewise, Hauksbee's successor, John Theophilus Desaguliers, was instructed to disprove Mariotte's observations[28] Henceforward, the famous prism experiments became part of the standard lecture repertoire, to be repeated over and over again. On one such occasion the run of prism experiments were performed before visiting savants from France who in 1715 were in England to view an eclipse.[29] Doubters then were instructed to follow to the letter the procedures laid down in *Opticks* if they truly wished to understand the experiments.

Thereafter, the 'doctrine of optics' – as it was termed – was taught in London, first by James Hodgson, diffused further by Desaguliers and spread generally around the country by itinerant lecturers such as Benjamin Martin. Texts, such as Desaguliers's published course of *Mechanical and Experimental Philosophy* (1725 and 1734). Francesco Algarotti's *Newtonianism for Ladies* (1737), Voltaire's *Elements of Sir Isaac Newton's Philosophy* (1738) and Martin's texts, *A New and Compendious System of Optics* (1740) and *Lectures in Natural and Experimental Philosophy* (1743), each helped to promote the new science generally and optics particularly.[30] Of all the philosophical instruments that came out of these developments, it was the simple prism

27 According to Schaffer, Mariotte claimed that a single violet ray from a prism, on refraction through a second prism, had its images tinged with red and yellow, thus violating Newton's claim to the immutability of coloured light: Schaffer (1995), pp. 214–215.

28 Schaffer (1995), pp. 213–217.

29 Westfall (1980), p. 795.

30 Further popularisation of Newton's optics in the latter half of the eighteenth century may be noted in Martin (1759–1763), Newbery (1761), Ferguson (1768).

that provided a convincing experience of the capabilities of experimental philosophy. Nature was very much susceptible to experiment and, hence, an understanding of God's hidden universe could be revealed by the process of interrogation and demonstration. Whether this was the motive behind those who took up experimentation, or whether it was simply to satisfy the curious and bookish gentry in their desire for polite learning and conversation, the prism itself became a 'must have' for followers of Newtonian philosophy. There is no clearer sign of the speedy take-up of the prism experiments than the detail to be found on the trade cards of contemporary instrument makers. In line with the texts of this period, James Mann in the City advertised 'Prisms for demonstrating the surprizing Theory of Light and Colours' and Matthew Richardson in Westminster advertised 'Prisms for Sr Isaac Newton's Experiments of Light & Colours'.[31] One person caught up in this fervour – who obtained prisms for his private investigations – was Chester Moor Hall, a youngish law student at the Inner Temple, to whom we shall return shortly.

It now seems strange to think that the endless repetitive testing of the simpler aspects of the 'doctrine of optics' helped to add to Newton's authority after his death while the more difficult part (the dispersion theorem) was accepted axiomatically, that is, there was a singular lack of checking of theory against experiment. Indeed, optical practitioners generally accepted the implications of experiment 8 and, without further question, abandoned all hope of improving the refracting telescope.

Out of sheer frustration, quite unable to rid the common sort of telescope of its chromatic aberration, Newton had changed tack, thereby avoiding all the problems associated with refraction and dispersion. Explaining this in *Opticks*, he wrote:

> ... Seeing therefore the Improvement of Telescopes of given lengths by Refractions is *desperate*; I contrived heretofore a Perspective by Reflexion, using instead of an Object-glass a concave Metal.[32] [Author's italics]

Benjamin Martin – always an ardent propagandist for Newton – was among those who did not feel a need to check whether the experimental values given in experiment 8 matched the corresponding dispersion theorem. But, in 1759, he apologised for the omission of such a check:

---

[31] Calvert (1971), Plate 33 (Mann) and Plate 42 (Richardson).

[32] Newton (1704), bk I, pt II, prop VII, theor. VI (Dover, 1952), p. 102. One solution practised for the retention of refracting telescopes was to use objectives of long focal length but this was dismissed by Newton as being too cumbersome and introducing trembling and bending.

Figure 4.3 James Mann's trade card advertising 'Prisms for demonstrating the surprizing Theory of Light and Colours', undated. Science Museum/Science and Society Picture Library – all rights reserved

> I took it for granted, that there was nothing more to be known of it [refraction] than what Kepler, Des Cartes, Hugenius [sic], Halley, and Sir Isaac Newton had delivered ... And this seems to have been the general Case of other Opticians, till about the Year 1740. *They all acquiescing in what Sir Isaac Newton had said in his Optics* ...[33] [Author's italics]

Thus, Newton's use of the word 'desperate' (*despaired of*) was accepted without further ado. Interestingly, Francis Watkins (1751) and John Dollond (1752) supposed it to mean that chromatic correction would be *impossible*, whereas Caleb Smith (1739) and Benjamin Martin (1759) took it to mean *impractical.*

Generally, therefore, the refracting telescope was thought to have an irremovable problem although Caleb Smith held an alternative, more hopeful, view. In assessing the improved capabilities of optical instrument makers of the late 1730s, he wrote:

> ...we now see, that most of these Artificers are capable of making them [refracting telescopes] to such a Degree of Perfection as was formerly despaired of.[34]

This opinion was echoed later by Martin who realised the cause of Newton's desperation:

> Experience often convinces us that the wisest Men may be mistaken, and that the Annihilation of this Sort of Aberration or Error [of the chromatic kind] is so far from being impossible or desperate, that it may be effected [in] more Ways than one, as has been discover'd more than 20 Years since.[35]

By contrast, Francis Watkins had accepted Newton's word without question. As we have already seen in Chapter 3, Watkins – at only 27 years of age – criticised Richard Jack for his claim to have overcome 'that hitherto insuperable Difficulty arising from the different Refrangibility of Rays, *despaired* of by Sir ISAAC NEWTON' [author's italics]. Astounded that Jack should dare to propose a new refracting telescope that ran into grounds of supposed impossibilities, Watkins retorted: 'Charity therefore obliges me to believe this Gentleman to have been totally ignorant of the true Nature of Refractions.'[36] In fact, Jack's optical design may well have been cleverer than he imagined but ignorance and confusion about 'the true nature of refractions' abounded. As we shall see in the following chapter, neither was Dollond correct in his understanding of

---

33 Martin (1759), p. iii.
34 Smith (1740), pp. 339–340.
35 Martin (1759), p. 54.
36 Watkins, *London Daily Advertiser*, 25 December 1751.

the nature of refractions and, at first, he too accepted the pronouncement in experiment 8 without experimental checks.

Both Watkins and Dollond were unaware of other attempts at chromatic correction before 1751, that is, both were ignorant of the ingenious experiments of Chester Moor Hall and Caleb Smith. For the time being, therefore, both Dollond and Watkins remained blissfully ignorant and completely entrenched in a paradigm of Newtonian complacency.

Newton's desperation regarding the chromatic problem of telescope objectives was not the view of David Gregory (nephew of James Gregory) Savilian professor of astronomy at the University of Oxford. About the same time that Newton concluded it was unproductive to expect anything further by way of chromatic correction using a combination of water and glass lenses, Gregory proposed a solution using an analogy with the human eye. As expressed in his *Catoptricae et Dioptricae Sphericae Elementa*, first published in English in 1715, this ran:

> ... perhaps it wou'd be of Service to make the Object Lens of a different Medium, as we see done in the Fabrick of the Eye, where the Chrystalline Humour ... is, by Nature ... join'd with the Aqueous and Vitreous Humours, ... in order that the Image may be painted as distinct as possible upon the Bottom of the Eye.[37]

At the time, it was not understood that the images formed upon the retina by the eye lens were not free of coloured edges due to the surrounding vitreous and aqueous humours but rather due to a function of the brain in removing the chromatic effects.[38] Desaguliers, who presented his lecture courses to all and sundry in the capital, would certainly have been aware of Gregory's view, and so would those attending his courses.

Achromatism Achieved – Introducing Chester Moor Hall

One person who may have considered a two-guinea course of Desaguliers's physico-mechanical lectures worthwhile was Chester Moor Hall, an Inner Temple law student then in his mid-20s. Certainly, attendance at such a course would have provided just the sort of social and intellectual milieu that would have great appeal. As a student of law, Hall would surely have been intrigued by

---

[37] Gregory's text on catoptrics appeared in three English editions, 1715, 1726 and 1735. The 1715 edition, edited by William Browne, MD, is quoted here, pp. 110–111. See also Brewster (1855), vol. 1, p. 111, note 1. James Gregory of St Andrews should not be confused with James Gregory, the Oxford Savilian professor of astronomy.

[38] The false analogy with the human eye entailed a consideration of curvature of the surfaces separating the vitreous and aqueous humours, which was also erroneous.

the rigorous style in which these lectures were delivered. As Desaguliers himself explained: 'the Experiments made at the First Lecture prove the Precepts given at the Second, and so on ....'[39] This was altogether a new mode of delivering knowledge and quite unlike the more familiar pattern delivered in courses of Euclidean geometry where knowledge was derived logically from definitions and axioms. One quarter of Desaguliers's course (comprising some 22 lectures) was devoted to optics and, if Chester Moor Hall had been in attendance, he would have witnessed some 30 demonstrations. If, then, he had direct contact with the man who once had worked closely with Newton on the very same experiments, he may well have been alerted to those parts of *Opticks* over which Newton was known to have struggled. Moreover, as a confident young student of law, Hall may well have held a healthy degree of irreverence for great authority, especially when that authority had flatly denied any likelihood of improvement to the refracting telescope.

Whether or not Desaguliers was the motivating force behind Hall's interests is mere conjecture. More certainly, we know that Hall did engage in some aspects of the testing of Newton's *Opticks*. According to evidence revealed in the diary of the Danish astronomer Thomas Bugge, we have the following record of a visit paid by him to the Golden Quadrant, the instrument shop of Addison Smith, on 15 September 1777. Bugge records:

> He [Smith] told me that long before Dollond, Mr Hall, a gentleman and amateur, had begun to combine two pieces of glass in order to cancel out the colours and that he had taken the idea from Newton's *Opticks* 1st Book, VIIth Proposition.[40]

Proposition VII deals with colours arising from refraction and this leads to proposition VIII, specifically dealing with colours arising from prisms. Also, from a distinct account, also by Addison Smith, we know that Hall did succeed in making an achromatic combination using crown and flint glasses as early as 1733 (§ Chapter 6). Thus, at this stage, it is sufficient to note that Addison Smith held the key to vital evidence regarding the contribution of Chester Moor Hall although such attribution today is often afforded to Jesse Ramsden, whose support of Hall many years later carried considerable weight.[41]

---

[39] Desaguliers (1725), p. 2.

[40] I am indebted to Anita McConnell for directing me to this important detail from Bugge's Diary. Pedersen and de Clercq (2009), pp. 130–131 and to Peter de Clercq for allowing me access to the translation in the provisional copy of the Diary.

[41] Elsewhere in Bugge's journal (note 40) we note that he purchased from Martin a copy of his book on achromatic optics, costing 6 shillings, and crown and flint glass from Smith, costing £8 9 shillings. He also writes: 'I ordered a triplet prism in order to prove experimentally the nature of Dollond's telescopes.'

Ramsden became one of London's major instrument makers, but at this period he was apprenticed to the instrument maker Mark Burton during which time he spent many evenings in the nearby Dollond house, learning about optics. He subsequently married John Dollond's daughter – Peter Dollond's sister. In his later years – provoked by Peter Dollond's reluctance to concede any formal credit to Hall and fearful that all recognition and memory of that worthy gentleman would be lost forever – Ramsden set about compiling the little evidence that could be mustered in support of the deceased Hall's achievement. This is discussed in considerable detail in Chapter 8, although it is useful here to bear in mind his conclusion:

> I have not the smallest doubt that Mr. Hall discovered the fallacy of Sir I. Newton's experiments, and under many difficulties had actually caused to be constructed Achromatic Telescopes and of considerable aperture too, *long before Mr Dollond applied to the subject.*[42] [Author's italics]

Most of Ramsden's evidence in favour of Hall as inventor of the achromatic telescope objective came from hearsay and, for this reason, he sought written evidence from Francis Watkins and Addison Smith whose intimacy with Hall had arisen in 1763 in litigation at King's Bench (§ Chapter 6). For the moment, however, we shall simply note that Hall – although not making his findings public as might be expected today of an important technological leap – did, nevertheless, successfully combine glasses of different optical materials to reduce the effects of chromatic aberration and in doing this he was ahead of John Dollond by over 20 years.

So, in the absence of any account by Hall himself, we must begin from the simple fact, as told to Addison Smith, that he was motivated to repeat some of Newton's experiments. Even if he did not actually repeat the preliminary water–glass prism experiments, we do know that he appreciated the necessity to employ two lenses made of glasses of different refractive indices and opposite powers (*convex* and *concave*) in order to cancel out the effects of chromicity due to the extremes of the visible spectrum. At some stage he would have substituted flint glass for water (in repeating Newton's experiments) and, thereby, arrive at a combination capable of producing 'refraction without dispersion'. This process would have entailed calculation on a basis of experiment, but how Hall came to choose the less readily available flint glass is not known. He may (i) have noted the vastly different dispersions arising from a beam of white light on passing through prisms of crown glass and flint glass, the latter having the greater spreading effect (*dispersive power*); or, otherwise, (ii) have derived his preliminary knowledge by actually visiting George Bass's glass workshop since Bridewell was very close to

42 Ramsden (1789). RS: MS. L&P, IX, item 138.

the Temple where he studied.[43] But, whatever his source of knowledge about glasses, observations made along these lines would have allowed him insight as to how he might proceed in the design of an achromatic pair.[44]

Hall proceeded along the lines recorded by Addison Smith in the latter's testimony to Ramsden.[45] That is, he commissioned different opticians to grind the lenses for the intended combination: one glass blank was given to Edward Scarlett senior and the other to James Mann junior. From their widely separate geographical locations – the one in Soho near St Anne's and the other in the City opposite St Mary Ludgate – Hall probably imagined the secret of his invention would remain secure, at least until the trials were finished. What he did not anticipate, however, was that both these makers would subcontract their commissions to the same third party, namely, George Bass. For Bass, it may have seemed somewhat curious that two independent instrument makers should coincidentally request lenses with identical specifications (*curvatures* and *diameters*), albeit with opposite powers and different materials. Of course, each optician – Scarlett Sr and Mann Jr – may have divulged to Bass that their commission was for a Mr Hall.[46] But, even if they had not given Hall's name, it would still have been possible for Bass himself to make an association and so try the glasses together and to observe their combined effect. Whether this was the case is unknown although it is assumed here that Bass would have abided by any secrecy that may have been asked of him. It is also assumed that when the two commissions were completed and returned via the respective instrument makers, Hall would have been encouraged with the outcome. Simply by placing the two new glasses in contact – even without cementing together – he would have noted an immediate improvement in image definition.

Caleb Smith – Achromatism by Catadioptrics

Caleb Smith was a person known within the Royal Society for his skills in optics.[47] He was also conscious of Chester Moor Hall's efforts, as is apparent in the tract *The Description and Use of a New Instrument or Sea Quadrant* (c.1735):

---

43 Crown glass comprised 75 per cent sand (silica), 15 per cent potash or soda and 10 per cent lime (percentages varied): this was commonly used for many purposes such as window glass. Flint glass was similarly composed but with ground flint as the siliceous ingredient: this was already used at the beginning of the century for chandeliers. Blanks were cut from circular sheets made by blowing and whirling.

44 Willach (1996), p. 197.

45 Letter 'Addison Smith to Jesse Ramsden' (4 June 1789). RS: MS. L&P IX 139.

46 According to Ramsden (1789), this first achromatic lens was 2½ inch diameter and 20 inch focal length.

47 It appears that Smith may have been involved in the insurance business: his pamphlet (c.1735) states that it can 'be had of the INVENTOR, at his OFFICE for Insuring Ships and Merchandize, in Castle-Alley, over against the West Passage of the Royal Exchange ...'.

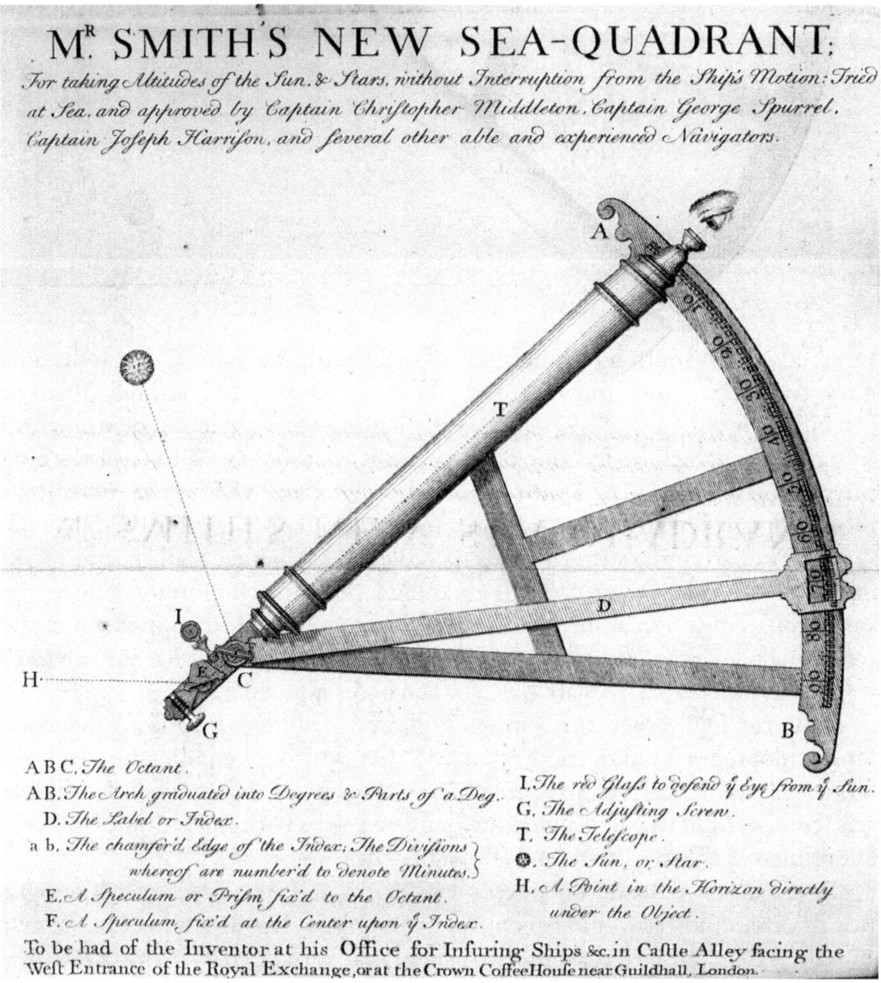

Figure 4.4 Caleb Smith's New Sea Quadrant, c.1735. Whipple Library, University of Cambridge

> About the same Time, or before this [that is, before Hadley's Quadrant was described in *Philosophical Transactions* in 1731], (if I am rightly informed) *Mr Hall, an ingenious Gentleman of the Temple*, invented an Instrument of the *like Kind*: But as it was never made publick, that I know of, I am not able to give any particular Description of it.[48] [Author's italics]

[48] Extracted from Caleb Smith (c.1735), and quoted in Talbot (2002a).

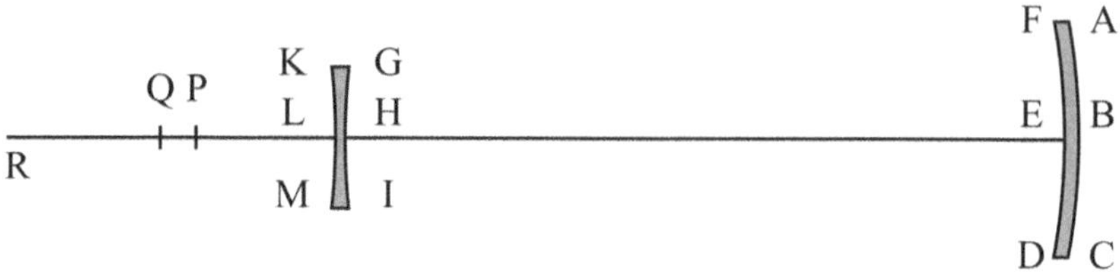

Figure 4.5 Optical arrangement for a telescope by Caleb Smith (1740)

Indirectly then, Smith was aware of Hall's efforts although he, too, was set on a quest to design an instrument of the 'like kind' but based on entirely different principles. By 5 April 1739, he was sufficiently confident in his own method to have dispatched an account of it to the Royal Society. With still nothing further from Hall's pen, Smith reported: '... no one, that we know of, has proposed any Remedy [for chromatic aberration]; apprehending, perhaps, the Difficulty of attaining such to be insuperable; ....'[49] Whether Hall became aware of Smith's Royal Society paper is unknown but more important here is the simple observation that both he and Hall were motivated to improve upon the refracting telescopes of the common sort, after having realised that the situation was not as desperate or negative as Newton had supposed.

Whereas Hall relied on combined glasses of different optical properties, Smith adopted a catadioptrical approach, that is, using a combination of both refraction and reflection. The principle was ingenious and entailed the double refraction of light through a meniscus lens having its back surface quicksilvered, as outlined in the caption to Smith's original drawing.

In the absence of the bi-concave lens (KLM-GHI) a beam of white light from a distant object would encounter the meniscus lens and refract at the first surface (FED) but reflect from the silvered rear surface (ABC) and, therefore, refract again on leaving the surface in a dispersed state (FED). These dispersed rays would then focus along the principle axis with the blue extremity of the spectrum at P and red at Q. On interposing the bi-concave lens as shown, the rays passing through, formerly appearing between P and Q, would instead focus together at the single point R, thus reducing the chromatic effect. Smith provides very accurate positioning for the bi-concave lens and equally accurate dimensions to which the four lens surfaces should be ground. A further extension shows how the arrangement could be made into a Newtonian-type for viewing by use of a reflecting prism.

Its actual construction, however, would have made heavy demands upon the makers' ability:

[49] Smith (1740), p. 327.

> Whoever shall think fit to put the Method here proposed in Execution, we dare venture (from a Trial that has been made) to assure him of Success; provided the same Diligence, Care, and Accuracy, be applied, in choosing, figuring, polishing and foiling the Glass, that has of late been employed for the forming Speculums of Metal; and let none be discouraged, though the first and second Attempt should fail; for that must be expected, if the ordinary way of grinding and polishing be used: Greater Exactness is here required, than is usually thought sufficient for the Object-glasses of refracting Telescopes; Let it also be considered how many Essays, for a long Term of Years, were made by Mr. *Gregory*, Sir *Isaac Newton*, and others, to reduce their Constructions of the reflecting Telescope into Practice, without answering, in any tolerable Degree, what their Theories promised: The Workmen they employed were chiefly Optical Instrument-makers, and had it been left to such Persons only to Perform by themselves, we have every reason to think, that it would have been pronounced impracticable to this Day, to make a reflecting Telescope that should equal or excel refracting ones of Ten times its Length; though we now see, that most of these Artificers are capable of making them to such a Degree of Perfection as was formerly despaired of.[50]

Thus, by 1740, with higher standards expected of the optical practitioners, ideas once thought impossible were then within reach. However, it is not thought that Smith's catadioptrical method was put on trial although, as we will see towards the end of the present chapter, it had a strong adherent in Benjamin Martin.[51]

Handing on the Achromatic Formula

Chester Moor Hall's ideas interested one or two makers in the trade, although, by his own admission, his efforts had been spasmodic, owing to the need to concentrate on his legal studies (§ Chapter 8). At one point, he appears to have sought assistance from John Bird, a newly-established mathematical instrument maker.[52] This contact could have been made as late as 1747 (after Bird opened his Strand shop), but before 1748 (when he became heavily engaged in work for

[50] Smith (1740), pp. 339–340.

[51] Martin (1759), pp. 107–114 devoted an entire chapter to Smith's catadioptrical achromatic telescope. However, King (1955), p. 261 does not think that there was much take-up of this instrument due to difficulties in its manufacture. Talbot (2002a) discusses survivals.

[52] Evidence for this comes from Ramsden (1789), who related that Admiral Campbell, 'perfectly remembered hearing Mr. Bird declare that Achromatic Telescopes had been made for Mr. Hall long before the date of Mr. Dollond's patent.' Further evidence of Bird's involvement comes from Watkins (§ Chapter 6), because Bird accompanied Watkins and Smith on their visit to the home of Chester Moor Hall in 1764. Bird's involvement is also noted in Peter Dollond's account of 1789 (§ Chapter 7).

the Royal Observatory); more likely, it could have been made much earlier since Bird had been in London since about 1740. But, in any case, the project was not dropped entirely because Hall also appears to have turned to James Ayscough, Mann's successor.

Ayscough must have taken Hall's ideas seriously because, in 1752, he engaged the Clerkenwell optician William Eastland to make up several telescopes according to a specification that stipulated a combination of concave flint and convex crown glasses in a ratio according to their focal lengths. Of this formula, the Leadenhall Street instrument maker Christopher Stedman who bore witness in court to Ayscough's written instruction to William Eastland which, as he remembered it, ran along the following lines:

> *Quere.* Suppose a Crown Glass flat on one side and Convex on the other sufficient to make it 2 foot focus then with a flint Glass flat on one side and Concave on the other sufficient by adding together make it 3 foot.[53]

At first sight, this vital detail might suggest that the combination was to provide a total focal length of 3 feet. However, this value would not work for an achromatic combination. If, instead, we suppose that the detail actually 'remembered' by Stedman in his testimony of 1765 in response to a 'bill of complaint' from Peter Dollond (§ Chapter 7) was wrongly written by the attorney who may have placed the figures in the wrong order (or simply misunderstood by Stedman), then we have $f_1 = 2$ feet (crown convex) and $f_2 = -3$ feet (flint concave), then, by the customary expression for lenses in combination, $1/F = 1/f_1 + 1/f_2$, the overall focal length would become $F = +6$ feet. This is approximately achromatic and such a combination would offer a considerable reduction in spherical aberration; in practice, it is little different from that eventually achieved by Dollond. After 1753 Eastland claimed to have sold many such refracting telescopes with compound (flint–crown) object glasses. Moreover, from this time, Ayscough advertised a refracting telescope of considerable importance that had undergone 'several late Tryals at Sea by the most experienced and competent Judges' which appears to have employed the new achromatic principle.[54]

We have already noted that Ayscough was no mere jobbing optician. He had subscribed to Robert Smith's *A Compleat System of Optics* (1738), where his name appeared alongside other high-class instrument makers. Although this was a highly mathematical text, book iii (of the four books comprising the set) was essentially a mechanical treatise, detailing, for example, Christiaan Huygens's methods for casting, grinding and polishing lenses for refracting telescopes.

---

53 From Stedman's evidence in response to Peter Dollond's bill of complaint of 1765 (§ Chapter 7).

54 Ayscough (1754), advertisement following p. 26.

As a leading practitioner, Ayscough would have been capable of understanding any theory posed by Hall and certainly competent enough to fulfil his practical needs. Of course, the intricate process of correcting for spherical and chromatic aberrations at the same time – so that the correction for colour compensation would not be destroyed by extra compensatory grinding to overcome spherical aberration – would have been time-consuming and so, most probably, considered costly, even prohibitively expensive.[55] Thus, although Ayscough was quite aware of the advance of using crown and flint glasses, he preferred to subcontract the optical work to Eastland, most likely for an agreed price.

Even had Ayscough taken extra effort in the improvement of the achromatic objective himself, none of the investment of time and money would have been secure without a patent. As shown by an advertisement for his sea telescope, Ayscough preferred an older style of trading, relying on the 'Honour and Generosity' of his customers for whom he willingly traded his ingenuity and labour for their loyalty. Patents were limited monopolies and, in Ayscough's opinion, it was much fairer to submit improvements and designs for 'publick Trial', as was the tradition.[56] By the mid-1750s (according to Ramsden), Ayscough's affairs became 'a little derang'd', although he seems to have avoided actual bankruptcy.[57] Taking this fact into consideration, it may be seen that several reasons militated against the development of the combined object glass in this particular workshop. Another factor could have been Ayscough's ill health, for in 1759, at only 40 years of age, he died.[58] There was, however, every reason and sufficient motivation for the compound crown–flint compound telescope objective to be developed elsewhere. Indeed, this is precisely what came to pass at the Dollonds' workshop (§ Chapter 5).

Caleb Smith's Solution Revisited

Although the crown–flint compound (achromatic) telescope was made and sold by the mid-1750s, it does not appear that there was any corresponding take-up of Caleb Smith's idea to achieve achromatism by catadioptrics although the idea – first published in the *Philosophical Transactions* in 1740 – was not entirely forgotten. Achromatism and optical aberrations were subjects about which very little was properly understood in the mid-1750s and these topics

---

55 Willach (1996) offers this cautionary reminder to historians who often ignore the nature of commercial work: namely, that for traders like Ayscough, 'time meant money'.

56 *London Evening Post*, 12 August 1755.

57 Ramsden (1789).

58 James Ayscough, of the parish of St Martin, Ludgate. TNA: PROB 11/850, Will proved 6 November 1759, f. 346, p. 84. To satisfy probate, Samuel Nicole Edlyne, Ayscough's apprentice, was brought in to verify the deceased's signature.

certainly required explaining to the serious user of telescopes. Indeed, here was subject matter ripe for the pen of Benjamin Martin whose prodigious efforts to diffuse Newtonian experimental philosophy were well known. After a long career, first as a teacher in a school at Chichester and later as an itinerant lecturer, Martin had finally settled in London in 1756 as an optical and philosophical instrument maker. As a prolific author of numerous texts with a proven talent for communicating complex ideas, the lack of any suitable text must have appeared as an opportunity upon which to establish and capture a share in this particular market. Nearly 20 years earlier, he had published *A New and Compendious System of Optics* (1740) and although not yet out of date, this text was certainly lacking any explanation of the common aberrations to be found with lenses and mirrors.[59] But now, with a new trade developing through Ayscough and Eastland, it was time for Martin to bring forward a project already in hand, namely, to publish an explanation of the problems of spherical and chromatic aberrations.

Some of this explanation was offered by Martin in the preface to his new text, *New Elements of Optics* (1759), in which the term 'new elements' referred specifically to the hitherto unelaborated problems of the principal aberrations.[60]

> Many Years have laps'd since this Treatise was first began, in all which Time I have seen nothing on these Subjects [i.e., the theorems relating to aberrations due to specula and lenses]; and as I was obliged to proceed in a trackless Field, and many and difficult Experiments necessarily occur'd, my Progress was but slow; and it has been with much more Labour than any one would imagine, that I have at length finished this small Treatise, by which the Reader will be fully certified, that all the former and latter Improvements in Optics are owing (not to *Strangers* and *Foreigners*) but to *Englishmen* only, whose peculiar Glory it is to have an acknowledged Superiority, in regard to Invention, *over all the World besides*.[61] [Martin's italics]

The new work comprised three areas: telescope types, spherical aberrations and chromatic aberrations. Apart from explaining the subject of optical aberrations, the intention was to lay bare the earlier practical attempts at achromatism. Here, at last, was a full explanation of how the crown–flint combination worked.[62] The preface was probably written last of all and so, after John Dollond's

[59] Martin was by no means the first to write on optics in a textbook fashion. He explained that his first optical text was based largely on Humphrey Ditton (for catoptrics) and Edmond Halley (for dioptrics) but avoiding David Gregory (which contained too much geometry), likewise avoiding Dr William Browne (which was a labyrinth of analytical investigation), and skirting Robert Smith (because it was too expensive). Millburn (1976), p. 29.

[60] Millburn (1976), p. 128.

[61] Martin (1759), Preface, pp. viii–ix.

[62] Martin (1759), pp. 53–83.

patenting of the achromatic telescope objective in 1758, he was able to add unashamedly his distinction between 'strangers' and 'foreigners', on the one hand, and 'Englishmen', on the other hand (as italicised). Here, already, are overtones of denying any credit to Dollond, the descendant of an immigrant Huguenot. Indeed, in Martin's estimation, only Caleb Smith – an Englishman whose solution of the chromatic problem was deemed 'quite perfect, and the only one that is so' – was deserving of any claim to this invention.

Martin's decision to proclaim Caleb Smith inventor on the grounds of his nationality as an Englishman was hardly a scientific rationale upon which to decide the appropriateness or success of the telescope for its claims to achromatism. Nor was it particularly helpful, although undoubtedly appropriate in terms of religious context, to remind his readers never to expect absolute perfection in the Works of Art, that being solely 'the Prerogative of the divine Being, and to be fought for in the Works of Nature only'. However, his stance was made plain by a stronger argument, mainly based on negative opinions concerning the optical deficiencies of combining crown and flint glasses. Ten objections were raised in his 'Memento', among which the first and second related to the imperfections of the optical medium and polishing of the surface; the third dealt with the confusion of refractions from internal and external reflections; the fourth cited the loss of light due to increased number of glasses; the fifth concerned imperfections from figuring when combined with imperfection of vision; the sixth stated that attempts to correct the imperfections of combined glasses were palliative only; the seventh told that mending one fault creates another (that is, reducing the chromatic nuisance only leads further problems of sphericity); the eighth a disadvantage due to the form (plano-concave) and positioning (concave side outwards) being the best arrangement but forbidden by theory; while the ninth and tenth mentioned that difficulties arise from additional lenses in the eyepieces.[63]

The path taken in Martin's text – after steering readers away from all thoughts of purchasing a telescope with a crown–flint combination – led to the method of chromatic correction introduced 18 years previously by Caleb Smith. Martin argued that the catadioptric method was, in fact, the only route by which to achieve a true achromatic telescope. Finally, he devoted an entire chapter to this telescope, reinforcing Smith's earlier recommendations.[64] Otherwise, his general recommendation was that a reflecting telescope would suffice all practical needs, and these he was prepared to supply.

Undoubtedly, this resurrection of an old idea would have annoyed John Dollond, particularly since Martin had outlandishly damned 'the other method' (meaning achromatism by compounded flint and crown glasses). This alternative

---

63 Martin (1759), pp. 84–87; Appendix 6 in this volume.

64 Smith (1740). Paper dated 5 April 1739.

was stated to be 'extremely imperfect, as it makes the Errors from the Figure [shape of the lens] vastly greater than it is by the single Object Lens'. All this, of course, was worded to exaggerate his preference for Caleb Smith's method:

> And it is no Wonder if Mr. Smith, who consider'd this Affair principally in Theory, should give the Preference to the catadioptric Method; for he expressly mentions other Methods, by which the Errors of Refraction in common Telescopes may be remedied, and there is Reason to think he means the different refractive Powers of Glass. In short, *the two Methods are essentially the same*, only the Mode or Way of applying the same Principle in Practice, is more perfect in one than in the other.[65] [Author's italics]

Of course, when Martin told his readers that the *two* methods towards achromatism were 'essentially the same', he simply meant that the aim of both methods was to bring the dispersed coloured rays to a common focus. As we shall note later, Dollond was to succeed in perfecting the crown–flint combined objective (§ Chapter 5) and Martin was to support Francis Watkins in an attempt to have the Dollond patent annulled on the grounds that it was obtained unfairly and that he was not the true inventor (§ Chapter 6).

In the mid-1750s achromatic telescope development had barely begun, although three new telescopes had been constructed, each with claims to having reduced the chromatic effects. These were a catadioptric telescope by Caleb Smith, a refracting telescope with flint–crown object glass by Chester Moor Hall, and a refracting telescope by George Adams based upon a design by Richard Jack. Caleb Smith's telescope, though ingenious in design, appears to have been developmentally a blind alley, although this catadioptrical approach was later championed by Benjamin Martin in 1759, but on the unsatisfactory grounds that the inventor Caleb Smith was an Englishman. Chester Moor Hall's instrument was less mysterious in principle, and probably was made and tested eventually by John Bird; while the Adams–Jack refracting telescope, which so incensed Watkins, was not truly achromatic, despite Jack having allegedly made claims to that effect. Each of these instruments had been intended for use at sea and none would have been suitable for astrometry (positional stellar astronomy).

Commercial interests in Chester Moor Hall's version picked up in the early 1750s when Ayscough supplied a formula to William Eastland, a detail to which we shall return (§ Chapter 6). Eastland was not alone in applying this formulation for an achromatic telescope. James Champneys took up the idea later, over 1765–1766, and in doing so attributed the origin of the idea explicitly to Chester Moor Hall during an advertising campaign (§ Chapter 7). What was needed in the interim was that someone with good theoretical and practical

---

65 Martin (1759), Preface, p. vii.

understanding and sufficiently inclined to take the development forward, that is, to make it commercially viable. Indeed, it was also a task demanding of an individual with the courage to show that Newton had erred in the formulation of a theorem of colour dispersion. Of course, such an individual would need time and a good measure of dextrous ability and, not least of all, personal motivation when spurred on by others familiar with the field. John Dollond was such a man and, as Jesse Ramsden later admitted, 'he [Dollond] was the only man in London who either had the knowledge or wish to improve Optical & Mathematical Instruments.'[66] Before Dollond could proceed, however, he had first to acquire vital evidence and that now becomes the focus of the following chapter.

66 Ramsden (1789).

Chapter 5

The Rise of John Dollond and his Patent

The serious problem posed by chromatism encouraged the belief that an achromatic telescope could never be produced, a conviction based on the acceptance of experiment 8 in book I of Newton's *Opticks*. In Chapter 4 we noted that a hint of the way forward did arise in the 1730s in the wake of Chester Moor Hall's experiments. Ingeniously, he had devised a way of combining flint and crown glasses to eliminate the worst effects of chromatic aberration; however, few then were privy to his ideas. It is not known how much development may have taken place before 1750 although some minor commercial headway was made in the early 1750s when James Ayscough supplied William Eastland with a basic formula for crown–flint achromatic combinations. Apparently ignorant of these tentative steps, both Francis Watkins and John Dollond, each in their separate ways, unquestioningly adhered to Newton's forbidding proposition. This much is clear in Watkins's case from his reaction to Richard Jack's claim to have reduced the effects of coloured fringes in his improved refracting telescope (§ Chapter 3); it is also clear from Dollond's astonishment at Leonhard Euler's proposal for an achromatic combination of optical media based on a non-Newtonian dispersion law (discussed in this chapter). Dollond's conviction wavered in the mid-1750s when he learned of a fault in Newton's dispersion theorem. Jolted from his original position, he then began a series of refraction experiments which not only confirmed Newton's error, but also led him to a method of preparing achromatic telescope objectives. Watkins, too, was persuaded to change his views when Dollond told him of *his* invention which would solve the problem of chromatic aberration in ordinary telescope objectives.

The present chapter identifies those factors which led to Dollond's success in the manufacture of achromatic doublets. Not only did he formulate the proper combinations of crown and flint glasses, but he also understood how spherical aberration could be reduced in the process at the same time. The technique was considered novel and, with Watkins's financial assistance, the process was patented. To compensate Watkins for his part in the plan, a profit-sharing partnership was drawn up for the sale of Dollond's 'New-improved refracting telescope'. Unfortunately, Dollond's dealings with Watkins and other third parties were far from transparent and, all too late, Watkins realised that his partner lacked integrity. This crucial issue had arisen when Watkins learned of Chester Moor Hall and his success in prior experiments and, therefore, it was considered that he was more deserving than his partner of the title 'inventor';

more important in the present chapter are the insights provided to Dollond's character. In fact, there were repercussions within the Royal Society when it was realised that Dollond was motivated more by jealous self-interest than any desire to uphold the unwritten ethic of integrity so demanded by the process of science. Indeed, John Dollond's lack of regard for protocol was to cause much unease among the practising opticians and experimental philosophers.

A Huguenot Origin

Biographical remarks about the life and work of John Dollond usually begin with a sympathetic reference to the Huguenot origin of his ancestors and the hardships they endured. The 'Dolon' family was French Protestant and, as such, they suffered religious persecution as had countless earlier generations of the self-proclaimed 'reformed religion'. The troubles had begun in the sixteenth century, when the 'Catholic league' rose up in opposition to teachings of Jean Calvin of Picardy. Thereafter, there were sporadic outbursts of harassment, not the least of which was the horrific massacre of 1572 in Paris (politically motivated by Catherine de' Medici and orchestrated by the Duc de Guise) which generated considerable anti-Catholic feeling in England. A 'Return of Strangers', ordered at the time of the Massacre, showed that the number of aliens living in London peaked at around 700 although this number was to increase with further waves of incomers.[1] Under Henry-IV of Navarre (himself a Protestant) this maltreatment lessened on the proclamation, in 1598, of the Edict of Nantes, promising freedom of worship to Protestants. But Henry later converted to Catholicism, and so the fine motive of religious tolerance was lost. By 1685 the anti-Protestant fervour in France was so pronounced that, under Louis-XIV, the original Edict was repealed. This single act of 'Revocation' – perhaps unanticipated by the French royal establishment – caused a sudden and huge exodus, in which more than 40,000 refugees made haste to Britain with as many more to other Protestant countries.

For those arriving in London, special permission from the Crown was set in place to allow the French-speaking churches and their charities to organise the in-coming families. In the early waves of expulsion and escape, these Huguenots tended to populate areas such as Southwark, St Katherine's-by-the-Tower, Blackfriars and St-Martin-le-Grand between Aldgate and Cheapside, each of these parishes holding 'Peculiars' since the age of monastic rule, and so lay outside the jurisdiction of the controlling authorities of the City council and the livery companies. In this way the incomers were allowed to practise their crafts without hindrance. From the late seventeenth century, thousands more were

1 Scouloudi (1985).

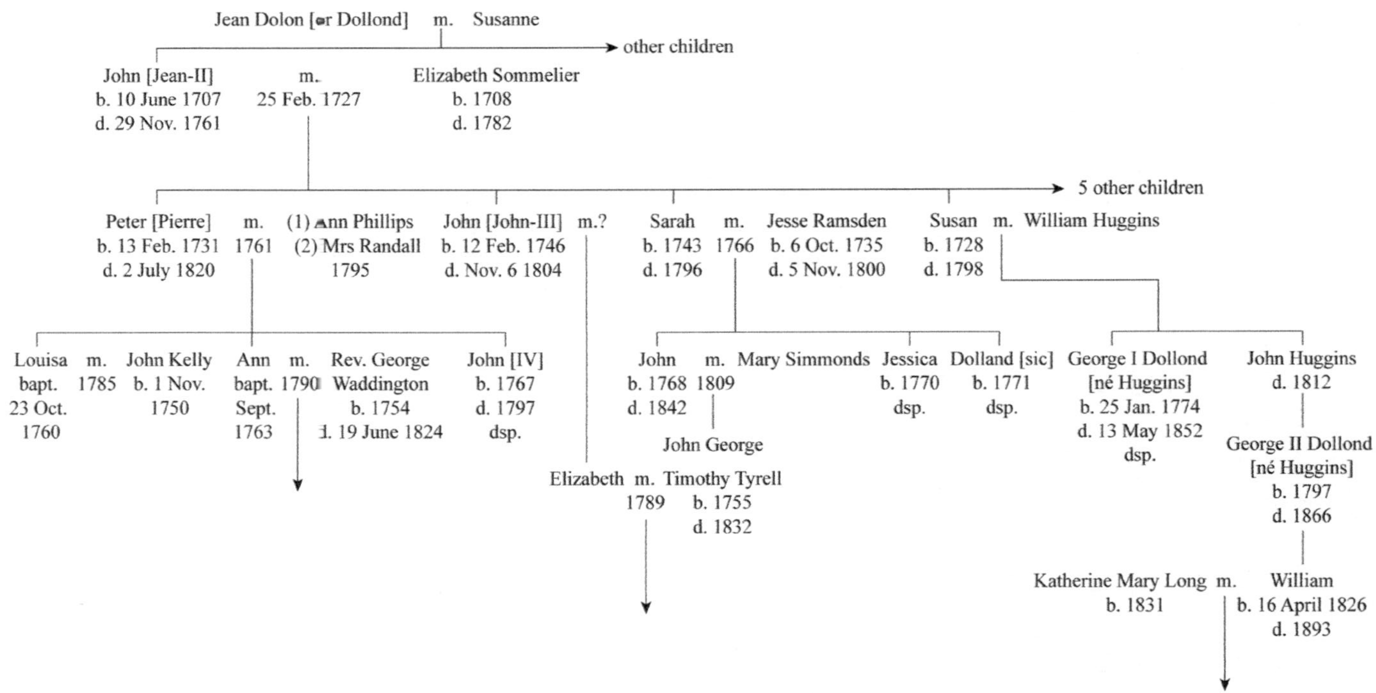

Figure 5.1 Simplified Dollond family tree

domiciled in Spitalfields – a name derived from a hospital (St Mary's Spital) founded in the liberty of Norton Folgate in 1197 – an area which became known as 'Petty France', and it was to this district that John's father (Jean Dolon-I), then still a child, and his parents (Jacques and Anne Dolon) were sent after making their flight from Hautot, near Dieppe, in Normandy.

One hundred years later, when the Marquis de Lafayette and others interceded with Louis-XVI to look more kindly on those who did not hold the Catholic faith, there followed an Edict of Toleration (1787) which allowed Protestants to practise their professions and crafts but not to hold religious services. Undoubtedly, this was a step forward for France but for those families already settled abroad this was hardly a sufficient reason to return to the land of their ancestors. By this time the Dolons – who, long since, had anglicised their family name to 'Dollond' – were into their third and fourth generations, fully integrated into English ways and properly established as respectable business leaders;[2] nonetheless, all alien Huguenots were stigmatised and for that reason alone, the Dollonds were labelled 'strangers' by their host country although this was not a bar to polite English society where advancement was always possible for individuals showing spirit and determination. While it is of some significance to bear in mind that the Dollonds were of Huguenot origin, it is more relevant here to inquire how and why – more than 60 years after their arrival – the family forsook their place of domicile in Spitalfields, in favour of a very different life as makers of scientific instruments in the metropolitan West End.

Settling in Spitalfields

When the Dolons arrived in London at some point during the latter decades of the seventeenth century, they would have passed through one of the established French churches which acted akin to a clearing house. This was a strict rule enforced by the English Government, who required all immigrants to be 'of the church' (whether local parish or permitted French church), on pain of being repatriated for failure to do so. The Dolons were registered at the French church in Threadneedle Street.[3] That they finished up in Spitalfields could be due to their former craft as weavers, since the coastal plains of the Pays de Caux near their Normandy home had, by tradition, been a flax-growing and weaving domain. Stepney, to the east of the City, was already an

---

2 The earliest detected English spelling of the family name as 'Dollond' appears on a baptismal document of Jean Barbet (bapt. 1699), signed by John (*Jean-II*) Dollond's father, *Jean-I Dolon*, who, along with his wife, appeared as witnesses (god parents). *Baptismal Register of La Patente, Spitalfields, 1689–1785*, vol. XI (register MDCCCXCVIII), p. 24 (ENTRY: Barbet).

3 Moens and Fergusson (1896–1916), 23, part IV. Generally, see Gwynn (1971–1976).

Figure 5.2 John Dollond, portrait engraved by J. Poselwhite after Benjamin Wilson, and published by Charles Knight. Museum of the History of Science, University of Oxford, inv. 29135

established district for weaving, and the two adjacent hamlets – Spitalfields and Bethnal Green – were considered appropriate regions for new settlements. By the beginning of the eighteenth century, the working and social life of the weavers revolved around Spital Square where most of the employing masters congregated about their daily business. At first, the Dolons lived in Hunt Street, then Wells Street, where John Dollond was born.[4]

[4] Previous accounts have given the year of John Dollond's birth as 1706, based on Kelly (1804), p. 6. Here, we take into consideration recent research by Gloria Clifton for her *Oxford Dictionary of National Biography* entry: http://www.oxforddnb.com/view/article/49855/7783?docPos=1 accessed 22 December 2011. She cites a page from a family Bible held by a descendant, which clearly reads: 'John Dollond Born June the 10: 1707', and is now held in the Museum of the History of Science, Oxford, which date is reinforced by Colyer-Fergusson (1906), a transcript of TNA: RG 4/4635 fol. 2v. [Editors' comment.]

He was baptised Jean on 24 June 1707 at the Threadneedle Street church.[5] The early years of his life are unrecorded although it is known that in 1727, at 20 years of age, he married Elizabeth Sommelier, daughter of another Huguenot family whose association lay with the Artillery French church, also in Spitalfields. Baptisms of their sons, Peter and John (Pierre and Jean-III) took place on 28 February 1731 and 15 July 1733 respectively, and were recorded at the Threadneedle Street church, although these events may actually have taken place at L'Eglise Neuve, an 'overflow' to the main church, built in Spitalfields in 1688 by royal patent under James-II to accommodate '*si grande foule de gens qui arrivent sans cesse*'.[6] Indeed, such was the rapid expansion of this district that a further eight churches were to be built before 1800.

After their marriage, John and Elizabeth lived in Vine Court, off Vine Street, a short distance to the north of the Spital Market but reasonably close to Spital Square. Living conditions, commercial factors, and perhaps an overwhelming urge that their progeny should lead a more profitable and integrated life than that which could be achieved through the silk weaving trade, are possible reasons for their eventual decision to re-locate to the West End. Socially, there was constant uneasiness among the Spitalfields weavers, and the prospects of a very different life among their fellow Protestants elsewhere in the metropolis, especially in Westminster among the affluent and aristocratic society to be found around the newly built (post-Fire) squares or the district of St James, must have held a certain allure.

Silk Weaving and Beyond

It is not certain precisely where the Dolons fitted in among the silk weavers of Spitalfields. The trade itself comprised several branches, including importers of raw and thrown silk from Italy and later China (although some raw silk was procured locally from the cultivated *Bombyx mori* and imported mulberry trees), 'silkmen' (both buyers and sellers), warehouse keepers, 'throwsters' (fashioners of raw silk by processes akin to cotton spinning), and dyers, together with the weavers themselves who were hand-loom operatives.[7] Masters of the weaving craft had to respond to market demands where whims of fashion could

5 The reader should not pay too much regard to Haag (1846–1859), where the entry in volume 3 for *Jean Dollond* contains a number of errors. For example, the proposal that the Spitalfields Dollonds may have descended from 'Dollon de la Groupillière' who was imprisoned in La Bastille (1685–1688) and then expelled from France should be disregarded; nor is it now thought that there is any connection with Antoine de Dolon who married Rachel de Casamajor at the Hungerford Huguenot church.

6 Scouloudi (1985), p. 526.

7 I am guided here by George (reprint 1984), pp. 178–196.

cause a sudden turn while, at another extreme, they would have to 'hire and fire' labour, according to the variations in demand. The size of a weaving house varied, but even the smaller businesses employed at least ten journeymen and female labour too. In good times, profit was made but rarely did the trade run smoothly. Fluctuations due to interruptions in the supply of materials, over-expansion and subsequent depression, changes in fashion, seasonal variations as well as court mourning, all conspired to cause poverty from the occasional laying-off of the workforce. Nor did the masters always have ready cash with which to pay their suppliers, or pay wages. Sometimes, poverty and desperation led to riotous outbursts, as was the case in the depression of 1719. Moreover, an influx of weavers from the provinces added to the discontent among the local workforce. By the mid-century there were some 500 masters centred on Spital Square owning some 15,000 looms and depending upon a labour force of some 50,000 persons. It is, therefore, not difficult to understand why Spitalfields was sometimes described as a place of toil, poverty and debt. The situation became much worse by the 1760s when weavers marched on St James's Palace to complain about imports of foreign silk. Some richer merchants sought opportunities to open shops in the West End, in order to sell directly to the public without the services of intermediary mercers, and the Dollonds may have sought to take this route.

For those whose parents had money and wished their sons to be educated, private tuition could be found. It seems very likely that John Dollond was encouraged by his father (Jean-I) to maintain the pride and dignity held by his forebears before quitting France.

Role Models at the Spitalfields Mathematical Society

John Canton, a teacher at the school in Spital Square from 1737, was acquainted with John Dollond, possibly through his sons who may have attended this school, but more likely through their common attendance at the local Mathematical Society. This Society had begun informally in 1717 when Joseph Middleton, a retired naval man and local benefactor, formed a regular class at the Monmouth's Head tavern in Monmouth Street and later (post-1725) at the White Horse tavern in Wheeler Street. During the time of Dollond's and Canton's membership (post-1735), meetings were held at the Ben Johnson's Head tavern in Woodseer Street. Other mathematicians, also keenly aware of Newton's *Principia* and *Opticks*, were associated with the Spitalfields religious community. Although the available early history of the Spitalfields Mathematical Society is patchy – sadly a notebook, which survived until the 1920s, is now lost – it is known that membership comprised ordinary workers from the trades but was formed predominantly of local weavers.[8] By and large, the rules of this Society were simple: the maximum membership was based on 'squares', that is, in Dollond's time, total membership did not exceed 'the square of seven' although it had begun as 'the square of eight'. For a few pence, members received instruction in the rudiments of mathematics and, seemingly, in a 'silent hour', they would undertake set problems. Some sense of the standard of the mathematics may be gained from the type of problems set by Thomas Simpson, an inspiring teacher of this Society, who also submitted mathematical problems to the *Ladies' Diary* under a pseudonym. From the sheer numbers of those in attendance with a desire for betterment, the Spitalfields Mathematical Society appears as a forerunner to the popular Mechanics' Institute movement, although that organisation did not make its appearance for another hundred years.[9]

Canton and Simpson were leading lights of the Mathematical Society. In fact, both were incomers from provincial weaving families: Canton from Stroud in Gloucestershire and Simpson from Market Bosworth in Leicestershire.[10] Their influence on John Dollond was recorded by John Kelly, an individual who became known to the Dollond family through his marriage in 1785 to Louisa, Peter Dollond's eldest daughter.[11] Although Kelly never knew John Dollond personally, he wrote a *Life of Dollond* (1804), a work to which we shall return later (§ Chapter 9). In this he painted a picture of an extraordinary man with classical learning, very much capable of applying the knowledge acquired from his mentors to seek out a new life well away from the endless routines of the weaving sheds and garrets. Indeed, we should pause a while to consider the sort of influence that Canton and Simpson imposed upon this talented and bookish individual.

Canton's career had blossomed in 1745, when he took over the proprietorship of the Spital Square school from the master to whom he had been articled since 1737.[12] He not only taught mathematics but was also a keen experimenter, particularly in magnetism and electricity. On one occasion, at a meeting of the Mathematical Society on 1 November 1747, it was resolved that 'an Instrument with a Proper Aparatus [*sic*] for Electrical Experiment' should be acquired and, although not stated explicitly, this decision was likely to have stemmed

---

8 Cawthorne (1929).

9 The Spitalfields Mathematical Society was eventually subsumed by the Royal Astronomical Society. Cassels (1979) and (1980).

10 Canton, John (?1718–1772, www.oxforddnb.com.view/article/4576 accessed 9 July 2011. Other sources give '1712' as baptismal year. Simpson, Thomas (1710–1761), www.oxforddnb.com/view/article/25594 accessed 9 July 2011.

11 Kelly (1804), pp. 47–52.

12 Samuel Watkins, to whom Canton was articled, was of no discernable relation to Francis Watkins.

from Canton.[13] On another occasion, Canton visited Gowin Knight of the Royal Society to purchase a pair of magnets but was so shocked at the asking price that he developed his own method of magnetisation. Enthusiasm, and a connection with the watchmaker John Ellicott (later clockmaker to King George-III), led to his informal attendance at the Royal Society on 2 March 1748; he returned yet again on 15 December 1748, this time by invitation of Henry Baker.[14] Canton was elected Fellow in 1749, and a Copley Medal was awarded in 1751 for his method of magnetisation applied to ships' compasses.[15] Clearly inspirational and generous, Canton became an invaluable contact for Dollond and it was he who arranged Dollond's introduction to this recognised scientific circle. This took place on 27 June 1751 when, before an auspicious gathering, Dollond met Henry Baker (the microscopist), John Smeaton (the civil engineer), Sir John Pringle (later President), Daniel Wray, David Papillon (the apothecary), William Watson (the physician and electrician) and Martin Folkes (the President).[16]

Although Canton had facilitated his exposure to some of the capital's scientific élite, it was Thomas Simpson, the mathematics teacher at Spitalfields, to whom Dollond owed his grounding in the rigours of the subject. Dollond's sons, Peter and John, however, did not have this advantage because Simpson – at the recommendation of Martin Folkes, PRS – had already left Spitalfields to take up a post at the newly formed Royal Military Academy at Woolwich in 1743. A while before his association with Spitalfields, Simpson had published *A New Treatise of Fluxions* (1737), an important contribution to the diffusion of Newtonian infinitesimal calculus, written at a time when few had any inkling of how to treat varying physical quantities. Over time, he became a prolific author, publishing texts on solid geometry and spherical astronomy, articles in the *Gentleman's Magazine* (1736–1738) and *Miscellanea Curiosa Mathematica* (1745–1746). Therefore, it was very much Dollond's good fortune to have come into contact with this exceptional individual and, undoubtedly, it was from his tutelage that he was equipped mathematically to articulate lens theory, as mentioned later in this chapter in connection with Leonhard Euler.

The insight gained and encouragement received was of sufficient effect to cause John Dollond to support his son in a venture to set up his own business in scientific instruments.[17] From the Blundell Collection of Dollond papers at

---

13 Cited by Cassels (1979), p. 242.

14 RS: Ms *Journal*, 21 (1748–1751), minutes for 2 March and 15 December 1748, respectively.

15 Controversially, Canton received a second Copley Medal in 1762 for experiments on the incompressibility of water. Crosland (2007), pp. 54–55.

16 RS: Ms *Journal*, 21 (1748–1751), minutes for 27 June 1751.

17 It is not clear how or from whom Peter Dollond received his initial training, other than by working for his father. The Dollond family appears to have had connections later with the Gilberts whose instrument business at Tower Hill was not too distant from Spitalfields.

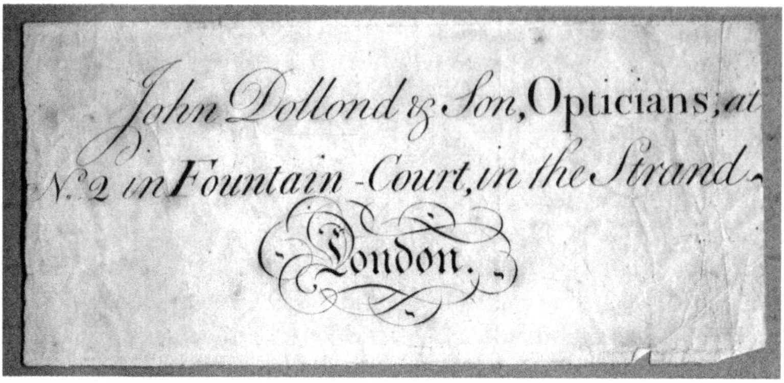

Figure 5.3 Earliest trade card of 'John Dollond & Son, Opticians at Fountain Court'. Museum of the History of Science, University of Oxford, MS Blundell 23

the Museum of the History of Science, Oxford, it is known that Peter ran such a shop in Vine Street, Spitalfields, for some time during 1750–1752.[18] From 1752, he and his father traded as 'John Dollond & Son, Opticians' from a shop in Fountain Court, off the south side of the Strand.[19]

Thus, having gained contacts through his valuable introduction to the Royal Society, it would appear that father and son had evolved a plan to apply the mechanical skills and practical know-how gained from their connection with the weaving business, to other use in the instrument trade. Peter moved to No. 2 Fountain Court ahead of his father who followed in 1753. It transpires that John Dollond did not live at the shop but rather at a rented property in Denmark Court, off the Strand (north side), although not far from Fountain Court.[20]

When John Dollond moved his family to Westminster he began to cultivate new contacts vital to the business. According to Kelly, two further persons of influence were Joseph Harris, assay master at the Mint, and James Short, a

[18] There is a suggestion in *The House of Dollond*, a pamphlet published c.1917, that the origin of the firm began in the reign of Queen Elizabeth when two merchants (Dollond and Carey) were granted licence to manufacture glass. This should be completely disregarded.

[19] Clifton (in *OxDNB*) refers to the Dollonds moving to Denmark Court, Westminster in 1753, although the discovery of an early letterhead in MHSO: Dollond Archives, Blundell item 23, suggests that they traded at 'No. 2 in Fountain-court, in the Strand' (c.1752). Unfortunately, the Westminster rate books (Poor Law) for the Fountain Court and Denmark Court addresses do not reveal the Dollond name, and therefore it may be assumed that the premises were rented.

[20] Peter may have resided alone at the shop until 16 August 1761, when he married Ann Phillips at St Martin-in-the-Fields.

prolific maker of reflecting telescopes; both were incomers to Westminster. Harris, a Welshman from Brecknockshire, had arrived in the capital in 1724 with a commendation to the astronomer Edmond Halley for his mathematical proficiency, and so went on to write a treatise on optics.[21] Short, originally an Edinburgh craftsman, also arrived in the capital with mathematical credentials from Colin Maclaurin, professor of mathematics at the University of Edinburgh. He took the road south to show his work on telescope mirrors (specula) before a body of eminent practitioners, including George Graham, Edward Scarlett Sr and Caleb Smith.[22] Both became Fellows of the Royal Society and, as will be noted in this chapter, it was Short who encouraged Dollond to expand his scientific horizons.

At this particular time, all three – Harris, Short and Dollond – would have been fully aware of the new Jack–Adams refracting telescope, and Richard Jack's claims to have reduced the irksome problem of chromatic aberration; they would also have followed Watkins's objections to Jack's claims through the press, particularly the remark made by Watkins on Jack not having properly understood the significance of Newton's prism experiments (§ Chapter 3). The very way that the deciding competition (*vis-à-vis* the Adams–Jack telescope *v.* Watkins's telescopes) had been organised and judged must surely have accentuated the social differences between those instrument makers (such as Watkins) who had gained their optical skills and knowledge through servitude and those (such as Short and Dollond) who were allowed to practise freely in Westminster without allegiance to any City company. This distinction is raised later (§ Chapter 8). Dollond clearly preferred to associate with the social and intellectual milieu provided by the Royal Society. But although he was tentatively accepted into this polite scientific circle, he was also expected to contribute to their endeavours. An opportunity eventually arose and, as discussed below, from this opening came a seminal paper in which Dollond proclaimed a 'perfect theory' of refraction.

Leonhard Euler (1707–1783)

As a Fellow of the Royal Society, James Short was aware of a paper entitled 'On the perfection of objective lenses of telescopes' by Leonhard Euler that had arrived from the Académie Royale des Sciences et Belles-Lettres de Prusse (better known as the Berlin Academy of Sciences).[23] In it, Euler had proposed a mathematical

---

[21] Harris, Joseph (bapt.1704, d. 1764), www.oxforddnb.com/view/article 12407 accessed 9 July 2011.

[22] King (1955), p. 85; James Short (1710–1768), www.oxforddnb.com.view/article/25459 accessed 9 July 2011.

[23] The Berlin Academy of Sciences, originally titled '*Societas Regia Scientiarum*', originated in 1710 from an initiative of Leibniz who petitioned the Elector of Brandenburg (who later

Figure 5.4 Portrait of James Short, attributed to Benjamin Wilson, c.1758. Museum of the History of Science, University of Oxford, inv. 16498

model by which to solve for achromatic aberration without recourse to Newton's work and so avoiding the implications given in experiment 8 (§ Chapter 4).[24] Although this paper had been presented to the Berlin Academy in 1747, the tardy process of publication and action meant that it was not until early 1752 that James Short invited John Dollond to examine it. Short – being a practical man and doubtless much engaged in his own business, and maybe not having Dollond's facility with the French language in which the paper was written –

became Frederick-III of Prussia), to establish an institution along the lines of the Royal Society of London. Euler joined the *Societas* in 1741 but pressed for an organisational change; hence was formed the '*Societé Littéraire de Berlin*' in 1743, a group having the intention of pursuing science without interference from the Crown. This arrangement was short-lived and after the Silesian war, when Frederick returned to Berlin and began to deliver on his earlier promises, the former *Societas* was reformed (1744) as the '*Académie Royale des Sciences et Belles-Lettres de Prusse*'.

[24] Euler (1749). Euler, as with Newton earlier, became much engaged in understanding the nature of light. Hakfoort (1995).

felt confident in Dollond's ability as 'an excellent analyst and optician.'[25] And so, here at least was an opportunity for him to participate alongside those with whom he wished to align his intellectual goals.

On examining the paper, Dollond found that the author had proposed – rather as David Gregory had done three decades earlier – that the human eye (with its lens and vitreous and aqueous humours) was a naturally occurring achromatic arrangement that potentially could be mimicked by an arrangement comprising an object glass surrounded by water or other liquid, contained centrally within a pair of meniscus lenses.

Such an arrangement, Euler argued, would have several advantages, not the least of which was: (i) that an object would be seen much more sharply due to a reduction in chromatic aberration; and (ii) that the spherical aberration would also be reduced since the overall curvatures of the meniscuses would be much less than that due to a single lens. Other claimed advantages included: (iii) that a reduction in tube length due to increased focal length; and (iv) that larger magnifications, even up to 2,000 times, could be achieved with the proposed objectives by using eyepieces of shorter focal length. For such an arrangement, Euler evolved a condition for the nullification of chromatic aberration based axiomatically on a logarithmic law of refraction.[26] In effect, he solved expressions (in terms of the four unknown radii of curvature) simultaneously for both blue and red images set theoretically at the same distance from the objective; thus, from this, he obtained a condition for achromatism expressed in terms of refractive indices.

Of Euler's model, Dollond complained bitterly that the whole notion was 'destitute of support either from reason or experiment.'[27] Indeed, no lens had been made to suit Euler's specification and so his proposed logarithmic law of refraction had not yet been put to test. However, it was one thing for Dollond to observe that a practical test was wanting but it was very much an affront that he should have asserted that Euler 'lacked reason'. Furthermore, his critique – 'A mistake in M. Euler's theorem for correcting the aberrations in the object-glasses of refracting telescopes', dated 11 March 1752 – was little more than a pompous reminder of Newton's unquestionable authority. Dollond added the relevant passage concerning Newton's experiments with prisms from experiment 8 in *Opticks* (see Chapter 4), in which Newton pronounced the irremovable nature of chromatism. Somewhat brusquely, Dollond therefore informed the continental mathematician:

25 Short (1753–1754), p. 288.

26 Euler (1753–1754a), pp. 287–296.

27 Strictly, Dollond's argument for the rejection of Euler was more complex. That is, he retraced Euler's calculations (respecting the refractive powers of glass and water) and supposed that they did not accord with those from Newton's dispersion law (theorem 1 of experiment 8); hence, he found that there could be no chromatic correction, except in the singular case of a compound lens of infinite focal length.

> It is therefore, Sir, somewhat strange that anybody now-a-days should attempt to do that, which so long ago has been demonstrated impossible.[28]

Indeed, Dollond was so imbued with a sense of the absolute in Newton's writing, that the mere suggestion that 'refraction without dispersion' could be contrived by the mere invention of a logarithmic law to substitute for Newton's dispersion law was treated with utter incredulity and dismissed without due consideration.

Putting aside this overconfident, naïve dismissal of Euler, the quality of Dollond's analysis goes some way towards explaining why Short had allowed him to scrutinise this continental authority. With regard to the Royal Society daring to publish the critique, Short advised Dollond that he would first write to Euler in order to allow the author an opportunity to defend his views. This was a first lesson in scientific etiquette although, as we shall see before the end of the present chapter, Dollond was to be admonished for his persistent lack of courtesy. In explaining this cautionary move, Short wrote to Peter Daval, FRS, the Royal Society's Secretary, on 9 April 1752:

> I have proposed to Mr. Dollond to write to [Euler], shewing him the mistake, and desiring to know his reasons for that hypothesis [a logarithmic dispersion law]; and therefore I desire, that this letter of Mr. Dollond's to me may be kept among the Society's papers, till Mr. Euler has had a sufficient time to answer Mr. Dollond's letter to him.[29]

Short must also have forewarned Euler of Dollond's imminent letter because, on 19 June 1752, Euler replied:

> You have done me a great service, in having disposed Mr. Dollond to suspend the proposal of his objections against my objective lenses until I have had an opportunity to respond, and to you I am infinitely obliged.[30]

Euler was not to be rebuffed so easily, especially by an upstart of whom, at this point, there was no knowledge abroad, nor indeed any credentials in London. Dollond was made to bide his time until Euler's reply, dated 15 June 1752 and addressed 'à Monsieur Monsieur Dollond' – an overly polite form of address in which the double emphasis was usually reserved to mean 'the very most esteemed ...', although perhaps here intended otherwise, with a sense of light-heartedness to the novice starting out in the field. A covering letter to Short dated 10 June 1752 contained the simple message: 'I hope that Mr. Dollond will

28 Letter from Dollond to Short (1753–1754a).

29 Letter from Short to Daval (Short, 1753–1754).

30 Letter from Euler to Short (Euler, 1753–1754a).

Figure 5.5 Leonhard Euler, engraving by Benjamin Holl after a portrait by A.M. Lorgna. © The College of Optometrists/British Optical Association Museum

be satisfied, since I agree with him on the little success that one could achieve with my lenses, by handling them according to the ordinary manner.' It would appear that Euler had assumed Dollond to have carried out practical checks because he wrote:

> Thus since you have taken pains, Sir, to experiment on such lenses, and having done said experiments, I pray that you discern the flaws, which can give rise to the different refrangibility of light rays, that arise from too large an opening; ...[31]

From the tone and content of his response, it is clear that Euler had challenged Dollond to engage further. Meanwhile, he was happy to exchange views, in particular volunteering that the quantities in his own formulæ were far too minute to invalidate the theory; and that he preferred to retain the analogy

[31] Letter from Euler to Dollond (Euler, 1753–1754b).

with the human eye because, he supposed, in time every objection to his theory would be removed. Whether Dollond appreciated that Euler was using the analogue of the human eye as a temporary device in model building is uncertain. He did, however, have cause to re-examine his own critique for mathematical inconsistencies pointed out by Euler.[32] For the moment, Dollond had exhausted entirely what he knew of the subject and, at this stage, he had not experimented (as Euler had presumed) and so had no practical evidence by which to confirm or deny the Newtonian standpoint or, indeed, to refute Euler's mathematical modelling.

Thus embroiled, Dollond became preoccupied with the subject of optical aberrations. Before the actual publication of the held-back critique on 'Euler's mistake', Dollond wrote to the Royal Society on 21 February 1753 announcing that he was now willing to engage in the challenge to improve refracting telescopes. Thus, the content of his first paper, read 1 March 1753, centred upon how – after studying the three- and five-glass telescopes to be found in optical shops – he was able to reduce both spherical and chromatic aberrations by the judicious introduction of additional eye-glasses, an improvement which he claimed to have resulted in successful sales of six-glass telescopes to foreign customers. (Optically, this was accomplished by increasing the field of view, so shifting the coloured edges of an image outward, away from the direct field of vision.) More important than the actual contents, this paper was his statement of intent:

> ... And tho' I am sensible, that you are not unacquainted with the theory contain'd in this letter, yet forasmuch as the subject has never been fully treated by any author, I shall endeavour, as soon as may be, to draw up a more particular explanation of the aberrations of light by refraction; but shall add no more at present, only beg leave to take *this opportunity of subscribing myself*.[33] [Author's italics]

Dollond could not add any more to the subject at this time because he knew little else. He would, however, have been alert to every nuance spoken about the subject. In the meantime, he contributed a further paper describing an improvement to Savery's micrometer, a detail which immediately found an application in Short's reflecting telescopes.[34]

---

[32] Ibid. For an excellent mathematical account of the achromatic object lens, see Brewster (1830), I, article 'Achromatic telescopes', p. 105, col. 1.

[33] Dollond (1753–1754a).

[34] Dollond (1753–1754b).

Samuel Klingenstierna (1698–1765)

When Dollond's critique of Euler finally appeared in print in 1754, it was read far more widely than anticipated. Perhaps surprisingly for Dollond, it was noticed at the ancient University of Uppsala where the eminent mathematician Samuel Klingenstierna had been engaged on the subject of refraction since reading Euler's 1749 paper.[35] Importantly, Klingenstierna had penetrated the problem of lens aberrations and, along the way, had detected a logical inconsistency between Newton's dispersion theorem and the associated experimental results to be found in experiment 8 (§ Chapter 4).[36] News of this discovery reached London early in 1755 via an Uppsala student, Fredrik Mallet, who carried with him a shortened account (in Latin) of a paper already destined for publication by the Swedish Royal Academy.[37] It had been Klingenstierna's earnest intention that this Latin account should be read before the Royal Society. He thus commanded Mallet:

> If it can, in any way, be considered a pretentious criticism of Newton, about whose infallibility you write they are so sensitive, then I wish it should not be done [i.e. not read before the Society].[38]

Klingenstierna had also expressly requested Mallet to acquaint Dollond with the relevant mathematical arguments, particularly those regarding the respective dispersion laws of Euler and Newton, neither of which, in his view, had been subjected to rigorous experimental tests. That Mallet met with Dollond is not in question although it does appear that Dollond retained the Latin paper because it was not read at the Royal Society; moreover, it was still in Dollond's possession several years later when Short needed to examine it.

Although Euler's ingenious dispersion theorem had carried no weight with Dollond, there was no corresponding dismissal of Klingenstierna's account. On the contrary, it is known from the diary of Bengt (Benedict) Ferrner – another travelling Uppsala academic who met Dollond later in

---

35 Heymann (1952); see also article 'Samuel Klingenstierna' in *Dictionary of Scientific Biography* (1970–1980) vol. 7, pp. 403–404.

36 A modern account by Beckman (2002a) describes how the ingenious Klingenstierna transcribed an essentially optical argument into one of euclidean geometry, thereby refuting the generality of Newton's refraction (dispersion) theorem. Klingenstierna then substituted his own theorem for dispersion which, in approximation, corresponds to Newton's law for small-angled prisms.

37 Klingenstierna's paper *Consideratio circa Logem refractionis Newtonianam Radiorum luminis diversi generis in mediis diverisis*, included with his letter to Mallet of 18 January 1755, was based on the formal paper published in Swedish: Klingenstierna (1754).

38 Nordenmark and Nordström (1939), p. 381 (n. 2).

October 1759 – that their conversation had turned to Klingenstierna's work in optics.[39] It is also known – from later correspondence from Dollond to Ferrner – that Klingenstierna's communication of 1755 was the crucial motivation which had led him to investigate Newton's notions of dispersion. Of this, Dollond wrote:

> I acknowledge that Mr Mallet when he was in London shew'd me a Letter which he had receiv'd from Mr Klingenstierna of Upsal, relating to Sr Isaac Newtons Laws of Refraction, in which it was evidently demonstrated that such Laws were inconsistent with the Nature of things. *As soon therefore as I saw Reason to doubt of Sr Isaac, I thought it high time to begin to think for myself.* I therefore immediately enter'd upon the Experiments which I have had the Honour to communicate to the Royal Society of London as may be seen in the Transactions & by which it appears that Sr Isaac Newton was greatly mistaken in his Notion of Refraction. I there shew that I had found two sorts of Glass that diverg'd the Colours very differently in equal Refractions, for the white flint Glass in the Same Refraction would diverge the Colours as 3 : 2 in the Crown ....[40] [Author's italics]

Clearly, then, Dollond was tuned for action. First he queried the practices of local optical instrument makers who already applied flintglass in the reduction of chromatic aberration. James Ayscough was on Dollond's 'visiting list' although it would seem that Ayscough was somewhat taciturn, perhaps not wishing to divulge what he considered to be a trade secret. Others, more forthcoming, were Robert Rew and George Bass. Each of these contacts will now be considered in turn.

James Ayscough, Robert Rew and George Bass

Keeping abreast of developments was an essential task of all in the optical trade and, to this end, Dollond kept his ear close to the ground. He most certainly would have heard of Ayscough's new telescope which had undergone trials at sea during 1754. Therefore, it is assumed that his visit to Ayscough took place sometime during 1755 after Klingenstierna's conversation with Mallet. According to Ramsden (writing from memory), it was about this time that Ayscough informed Dollond of the 'wonderful performance of a telescope in his

[39] Lindberg (1956), quoted in Beckman (2002a), p. 10. *Note.* It was during Ferrner's sojourn in Paris in 1760 that Clairaut and other French mathematicians also came to learn about Klingenstierna's work and, at the same time, of Dollond's work on flint–crown combinations.

[40] Letter of Dollond to Ferrner. Extracted from Nordenmark and Nordström (1938), pp. 48–50. The original letter survives at the Uppsala University Library. UUL: MS A9.m.

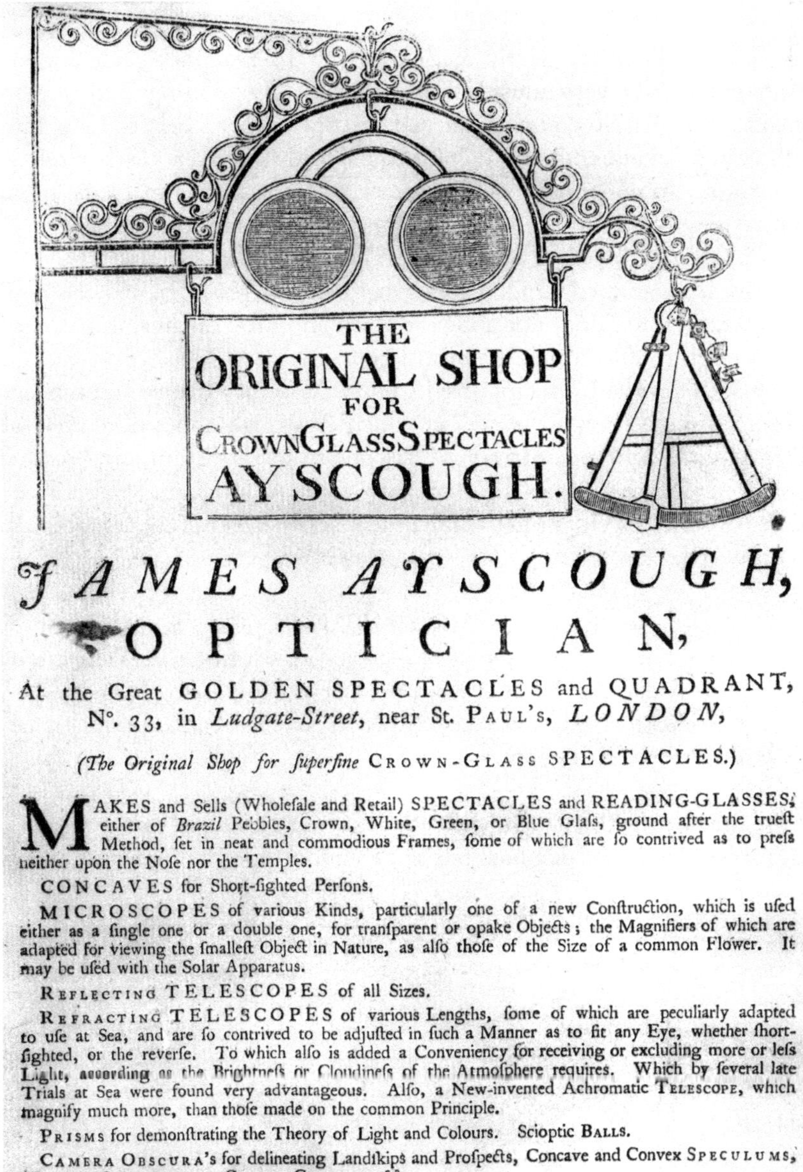

THE ORIGINAL SHOP FOR CROWN GLASS SPECTACLES AYSCOUGH.

JAMES AYSCOUGH,

OPTICIAN,

At the Great GOLDEN SPECTACLES and QUADRANT, N°. 33, in *Ludgate-Street*, near St. PAUL'S, *LONDON*,

(*The Original Shop for ſuperfine* CROWN-GLASS SPECTACLES.)

MAKES and Sells (Wholeſale and Retail) SPECTACLES and READING-GLASSES; either of *Brazil* Pebbles, Crown, White, Green, or Blue Glaſs, ground after the trueſt Method, ſet in neat and commodious Frames, ſome of which are ſo contrived as to preſs neither upon the Noſe nor the Temples.

CONCAVES for Short-ſighted Perſons.

MICROSCOPES of various Kinds, particularly one of a new Conſtruction, which is uſed either as a ſingle one or a double one, for tranſparent or opake Objects; the Magnifiers of which are adapted for viewing the ſmalleſt Object in Nature, as alſo thoſe of the Size of a common Flower. It may be uſed with the Solar Apparatus.

REFLECTING TELESCOPES of all Sizes.

REFRACTING TELESCOPES of various Lengths, ſome of which are peculiarly adapted to uſe at Sea, and are ſo contrived to be adjuſted in ſuch a Manner as to fit any Eye, whether ſhort-ſighted, or the reverſe. To which alſo is added a Conveniency for receiving or excluding more or leſs Light, according as the Brightneſs or Cloudineſs of the Atmoſphere requires. Which by ſeveral late Trials at Sea were found very advantageous. Alſo, a New-invented Achromatic TELESCOPE, which magnify much more, than thoſe made on the common Principle.

PRISMS for demonſtrating the Theory of Light and Colours. Scioptic BALLS.

CAMERA OBSCURA's for delineating Landſkips and Proſpects, Concave and Convex SPECULUMS, MAGICK-LANTHORNS, OPERA-GLASSES, *&c.*

OPTICAL MACHINES for viewing Perſpective Prints.

BAROMETERS, Diagonal, Standard, or Portable.

THERMOMETERS, whoſe Scales are adjuſted to the Bores of their reſpective Tubes; Hygrometers, and Hydrometers; Hydroſtatical Balances; *Hadley*'s Quadrants; Globes; Caſes of Drawing Inſtruments; Scales; Parallel Rulers; Sun-Dials; Rules; Pencils; with all other Sorts of Optical, Mathematical, or Philoſophical Inſtruments.

Figure 5.6 James Ayscough's trade card, undated. Science Museum/Science and Society Picture Library – all rights reserved

possession of a new Invention; [saying] it was about 18 Inches long and would shew Jupiter's satellites clearly & distinctly.'[41] To Dollond, such dimensions must have appeared very unusual, for he replied: '... according to the establish'd principles of optics, it would require a Telescope of 4 or 5 Feet to shew Jupiter's satellites in the manner describ'd.'[42] Of course, only a few years earlier, telescopes of enormous focal lengths were thought to be essential in order to reveal planetary satellites.[43] Thus, the conception of a refracting telescope of incredibly short dimensions together with the sort of capabilities claimed by Ayscough must have been an intriguing, if somewhat baffling, prospect. However, of this encounter, Dollond does not appear to have inquired further: nor, seemingly, did Ayscough volunteer further information.

Ramsden's evidence is supported somewhat by newspaper announcements (running from 12 August 1755) in which Ayscough announced that 'about twelve Months ago he contriv'd a Telescope to use at Sea in any Weather', of which 'it has had the Approbation of many of the Admirals and Captains of His Majesty's Navy, as well as the Commanders of other Vessels'. Apologetically, he also proposed:

> ... it might be thought needless to have Recourse to this Method [of announcement] (which he greatly disapproves of;) but having lately discover'd some low Artifices made Use of in order to imitate it [the telescope]; and having no Patent, relies on the Honour and Generosity of the Publick, and at the same Time calls upon all those Imitators for a publick Trial.[44]

For the time being, then, Ayscough was reluctant to divulge any detail over the optical construction of his new telescope which appears to have incorporated

---

41 According to Ramsden (1789), this conversation between Dollond and Ayscough had taken place 'two or three years previously while I work'd for Mr. Burton', that is, about 1755, but not earlier.

42 Ramsden (1789).

43 It was once thought that the only criterion for discovering new satellites was for telescopes to have large magnifications and since this was governed by the ratio of focal length of objective to that of the eyepiece, it followed that objectives of *great* focal length would produce a *large* magnification. One contemporary telescope was that erected in Wanstead Park in 1717. This was arranged along a 125-foot redundant Maypole (formerly used for festivities in the Strand and reputedly procured by Newton) with the eyepiece mounted at one end. An objective lens (made by the Huygens brothers and donated to the Royal Society in 1686) was affixed to the steeple of the parish church of the Rev. James Pound. Pound (aided by his nephew, James Bradley, later third Astronomer Royal), succeeded in verifying the existence of five of Saturn's satellites. Mills and Jones (1989), p. 176; King (1955), pp. 63–64.

44 *London Evening Post*, 12 August 1755; also *Public Advertiser*, 14 August 1755; and *Whitehall Evening Post and London Intelligencer*, 14 August 1755.

an improved objective after the stamp of that by Chester Moor Hall. Even so, it was only a matter of time before the secret of its construction would become generally known.

The question therefore remains as to how and when John Dollond came by an understanding of the crown–flint combination. Since the answer is not to be found in the formal *Philosophical Transactions* account of 1758, it has to be deduced from elsewhere. It is important to seek out how Dollond came by the notion of a crown–flint combination for an achromatic objective because his claim to originality was soon cast into doubt by those in the trade who knew otherwise. A glimmering of what actually took place may be derived from the Spectaclemakers' Company petition of 1764 in which the king was requested to withdraw the Dollond patent for an achromatic objective because it had been granted on the basis of a false claim (§ Chapter 6). Two testimonies in this document provide clues to what had actually taken place. The first came from Robert Rew, a brass worker at Coldbath Fields, who declared that 'in the year 1755 [he] informed Mr Dollond of the construction of the compound object glass'. The second came from George Bass, the glass dealer at the Fleet Ditch near Bridewell, who informed that he was 'the maker of the [compound object] Glass in the year 1733'.[45] Behind these two statements lies much of the disquiet that was, in due course, to develop into a major unsettling within the optical instrument trade. The course of events leading to this petition are examined in detail later, but it is sufficient here to extract the names of Rew and Bass.

From a quite different source – discrete communications to the Royal Society made in 1789 by both Jesse Ramsden and Peter Dollond – these issues surrounding Rew and Bass appear once more and so it is certain that John Dollond did meet with Rew. Indeed, Peter Dollond related how 'over a tankard of porter at the Fountain public house in Cheapside' his father had conversed with Rew, who had declared to him 'the whole principle of the invention.[46] However, Peter offers no inkling as to what Rew had actually divulged at that meeting.

More revealing are the remarks concerning John Dollond's visit to Bass's workshop provided by Ramsden. He wrote:

> Bass who made reading glasses for me ... has often told me that he made Mr. Hall's object Glasses many Years before Mr. Dollond's patent ... These particulars have often been mention'd to me in conversation with the late Mr. Dollond.[47]

---

45 SMC petition. TNA: PC 1/7/94.

46 Dollond (1789).

47 Ramsden (1789).

Indeed, Ramsden went on to explain that, during Dollond's visit to Bass's workshop – ostensibly to seek out a suitable glass for an order from the Duke of York – Bass had explained something of the properties of flint glass:

> ... He [Bass] shew'd him several [glasses] and Mr. Dollond seem'd to give the preference to one of flint Glass from its clearness and transparency; but Bass told him that the fault of that Glass was, that letters seen through it towards the edges was much more ting'd with Colours than in one made of Crown glass adding that at the same time that he work'd the Concaves for Mr. Hall's object Glasses of that glass, that is the flint glass.[48]

Of this same visit, Peter Dollond had written:

> When he [his father] got home [from visiting Bass] he showed me the glasses, and mentioned the remark that Mr Bass had made; and added that if the difference of the dissipation of the different coloured rays could be easily perceived in looking through the two reading glasses in that way, he had no doubt but that the difference of the dispersion of the colours in the different kinds of glass would be found much greater than he had supposed it could be and on that account he should go on with his Experiments of trying all the different kinds of glass *as fast as possible*.[49] [Author's italics]

Thus, there can be no doubt concerning Dollond's visit to Bass. Moreover, so significant was this occasion that it was easily recalled, even 33 years later.

As for the quintessential experiments being carried out, as Peter Dollond claimed, 'as fast as possible,' that detail would appear at odds with the formal account in the *Philosophical Transactions* in which his father had suggested:

> ... *a considerable time* elapsed before I could ... set about that work; for tho' I was determined to try it at my leisure, for satisfying my own curiosity, yet I did not expect to meet with a difference sufficient to give room for any great improvement of telescopes ...[50] [Author's italics]

The Royal Society paper was written in a style designed to portray an orderly approach while Peter's own recollection, perhaps more realistically, suggested there was a great urge to get on with the job after receiving the vital clues from Bass.

---

[48] Both flint and crown glasses were termed 'white glass' and were subject to duty charges. In 1758, when Dollond and Watkins began their partnership, the production of white glasses in England and Wales ran at 4,500 tons per year. Mitchell with Deane (1962), p. 267.

[49] Dollond (1789).

[50] Dollond (1757), p. 739.

These opposites – the public and private faces of John Dollond – hint at a man bent on a particular quest but acting so as to present a vision of scientific procedure as it ought to have happened rather than the actual course taken. This should be a salutary reminder to the reader that scientific papers rarely represent the true process by which the fruits of research are obtained.

It is worth considering Bass's position just a little further because – as we shall see – it was he who was causal in initiating a defence for Chester Moor Hall as originator of the compound object glasses. On the one hand, John Dollond certainly had the problem of chromatic aberration at the back of his mind at the time that he visited Bass's workshop. So, had he been ignorant of the nature of flint glass, he would have quickly perceived a value in its greater dispersive property, especially when pointed out to him by Bass. On the other hand, if he already knew the value of this particular property of flint glass, from the earlier meeting with Rew, then his visit to Bass may have been expressly to seek out the sort of glass that Bass had used for Hall's work.

In his account, Peter Dollond emphasised that what his father had gleaned from Bass was no more than 'a piece of information'. However, this was very much a *formative* piece of information, which not only motivated him to proceed with experiments but also it enabled him to shorten the research by several months. And yet Peter, dismissively, refused to admit there was any value in what Bass had told his father:

> He [Bass] did not know that such a difference in the quality of two kinds of glass could be applied to any advantage. Mr Bass was a practical optician; not a theorist.[51]

This lack of regard for the practitioner smacks of a superior attitude, perhaps emulating the style of his deceased father, who, it is supposed, believed in the age-old hierarchical view where theory took precedence over practice.

Bass was a very practical man with a lifetime's experience in the optical trade and, although not a theorist, he was surely not unintelligent. Therefore, it was very unreasonable that he should be held incapable of understanding the contribution made by Chester Moor Hall of compounding flint and crown glasses. Even if there had been, at the time, some haziness in understanding the purpose of Hall's original commissions, there had been an interlude of 24 years during which time he would have gained a proper understanding and, more importantly, he would have been the first to respect Hall for his ingenuity. In fact, as Master of the Worshipful Company of Spectaclemakers in 1747–1748 and 1754–1758, Bass was well known among the opticians and well aware of those, like Ayscough, who had followed Hall's procedure for crown–flint

[51] Dollond (1789).

achromatic objectives (§ Chapter 4). So, by the time that John Dollond visited him at his workshop, there had been ample time for him to appreciate the value of using Hall's combination of flint and crown glasses. It was, therefore, iniquitous that Peter Dollond should have cast Bass aside in 1789. However, with Bass deceased by the time that the above remark was made, it was a simple matter to deny him any credit for what he surely must have known (§ Chapter 8).

A Great Culmination: The Refrangibility Experiments

In 1757, John Dollond was in a very strong position to set about a series of very exacting experiments. Three vital pointers were at hand: firstly, a personal communication from Klingenstierna (1755), casting doubts on the dispersion proposition in Newton's experiment 8; secondly, some gleanings from Rew on compound glasses, conveyed in 1755; and thirdly, an impression, from Bass, of the suitability of flint glass, conveyed to John Dollond around 1756. As Dollond himself relates, the pace of research quickened from the beginning of 1757, although it was not until early in 1758 that the project was finished. Viewed retrospectively, three phases were needed before the commercial achromatic telescope was fully realised. To begin with, he needed to demonstrate, convincingly, that Newton had been in error in the dubious experiment 8; he then had to show, with due reasoning, how a compounded arrangement of flint and crown glasses could correct for chromatic aberration; and the final breakthrough came when he devised a *modus operandum* by which spherical aberration could be corrected without upsetting the chosen curvatures for chromatic aberration.

The first stage entailed repeating Newton's water–glass prism experiments and showing that dispersion was not dependent upon mean ray alone (as believed by Newton) but also dependent upon the optical density of the prism. The second step entailed much grinding of prisms in order to determine which of the glasses (Venice, English, common crown, flint and so on) would form a suitable complementary pair. After satisfying himself that flint and crown glasses would answer the purpose, he then proceeded to determine the necessary ratio of their refractions that would permit 'refraction without dispersion', and thence transfer this information from experiments with prisms to lenses of appropriate focal lengths. The final phase came after discovering that, by removing the chromatic problem, he had introduced, detrimentally, considerable spherical aberration The actual breakthrough came when he realised he could make the aberrations of each lens of the combined pair *equal* but *contrary*. Writing of this success, Dollond remarked:

> And thus, at last, I obtained a *perfect theory* for making object glasses, to the apertures of which I could scarce conceive any limits: for if the practice could come up to the theory, they must certainly admit of very extensive ones, and of course bear very great magnifying powers.[52] [Author's italics]

So, by trial and error – working within the limits of what he hoped to achieve – he succeeded.

The outcome of this achievement – his 'perfect theory' – certainly confirmed the fallacy in Newton; it also provided a technical basis for how best to make flint and crown objectives in considerable number. Nowhere in Dollond's account, however, was there any sign of a mathematical treatment by which he had calculated the specific radii of curvature of his lenses. But, of course, his paper dealt with practice rather than theory. Thus, avoiding any theory, he outlined only the difficulties that had been overcome such as: (i) the figuring or shaping of the combined objective by balancing the required focal length with the curvatures of the surfaces; (ii) the choice of pairs of glasses of suitable refracting powers; (iii) the need to ensure that the combination of lenses were axially co-linear so that the rays would travel down the telescope tube paraxially, as desired; and finally, (iv) the need to grind the four lens surfaces perfectly, such that the chromatic correction remained unaffected when spherical aberration was removed. This final detail was the all-important 'simultaneous correction' later mentioned (though left unelaborated) in the patent specification.[53]

Dollond's contribution to scientific advance was *not* in the original invention of the compound object glass – a myth often perpetuated and (as we shall discover in subsequent chapters) curiously permitted by a twist of patent law – but rather in an account of some very skilful experiments, first confirming Newton's error and, thereafter, working towards the production of a perfect achromatic lens. That account was read before the Royal Society on 8 June 1758 and to this day it remains a milestone in the history of the achromatic telescope (reproduced in Appendix 1). These impeccably conducted experiments superficially appear as an exemplary model of scientific writing although, as stated earlier, there were omissions of detail and even concealment. In this respect, Dollond's paper was a fraud, on at

---

52 Dollond (1757), pp. 733–743, quote p. 742 (Appendix 1).

53 From the point of view of scientific advance, Dollond edged towards his goal, intuitively and inductively. By contrast, Euler's mathematical model for dispersion—no matter that this was based on a false analogy with the human eye—was hypothetico-deductive and so offered a Popperean-type 'risky prediction' by way of a condition for the nullification of chromatic effects. However, the test of this condition was not undertaken for some considerable time and when Maupertuis came to make up a combined lens according to Euler's specification, the test failed.

least two counts:[54] There was no mention of how Klingenstierna's ideas inspired and caused him to revise his earlier views; nor was there any indication of how valuable insights into the use of flint glass were gleaned from Bass. The earlier work of Chester Moor Hall was completely ignored in the account although, according to his son, Hall was not known to Dollond *before* the paper was presented. Perhaps most infuriating of all, he failed to describe the particular feature – already extolled in principle in the patent specification but left unelaborated – namely, that of explaining exactly how in practice spherical aberration and chromatic aberration could be corrected simultaneously. But, of course, such 'know-how', was guarded and, in any case, those technicalities were not obligatory in the formal paper.

Towards Letters Patent

At some point early in 1758 – as the refrangibility experiments drew to a close – the Dollonds, father and son, satisfied that they held the secret to a potentially lucrative device, must have considered how to secure and market the invention. Of course, any commercial launch requiring safeguards and exclusivity would need to take a formal route, via a patent. Inevitably then, before proceeding further, they must have borne in mind the following factors: how they alone – but along with their assistant John Berge – could cope with a potentially large influx of trade given that the production of combined glasses would be a time-consuming process; how such a new strand to their business would affect other instrumental developments for which they were just then becoming known;[55] and, most importantly, how – given their financial circumstances – they might afford or obtain the capital outlay for the acquisition of letters patent. If these problems could be overcome and, of course, if successful in their application for a patent, then there would also be an overriding need for a sustained commitment to the project in order to fulfil the obligations bestowed by this royal privilege.

In the end, the Dollonds may not have pondered too long over the issue of risk versus reward. They had made a major scientific breakthrough and knew that it was important to seize this opportunity in order not to lose the commercial edge. From his former experience in the silk-weaving trade, John Dollond probably appreciated that any hesitation to grasp what had come his way – by dint of effort and good

[54] A well-known view of the late Nobel laureate (Sir) Peter Medawar, expressed in his broadcast talk 'Is the scientific paper a fraud?' Medawar (1963).

[55] Barty-King (1986) claims that the business was flourishing with 'run-of-the-mill astronomical and navigational items' by 1756 (p. 30); he refers also to the need to cope with the 'influx of business' (p. 37). If these remarks can be substantiated, as Barty-King suggests, then John Dollond's desire for partnership with Francis Watkins on the grounds that he was in 'indigent circumstances' could well be untrue. [Alternatively, he may have spent more time on his experiments than the Dollond business could properly afford. Editors' comment.]

fortune – could quickly lead to being overtaken by others. As far as the commercial risk was concerned, it was always possible to mitigate such danger by co-opting another, solvent optician who would be willing to co-operate in the venture but under their terms. The question of who this should be is of fundamental interest but moderately easy to answer. At all costs, Dollond would need to avoid the close-tied City opticians who were already engaged in making compound object glasses (§ Chapter 6); nor would he have considered most Westminster opticians since they too had their cliques. Close at hand, however, was a singular optician who, over the years, had shown confidence and leadership when expressing an opinion, whether in objecting to Ayscough's new directions with optical glass for spectacles or in criticising the claims of Adams and Jack for their new refracting telescope. In many ways, Francis Watkins was an ideal person to have on board for what John Dollond and his son had in mind because, most importantly, to all appearances Watkins's finances were secure.

The move to involve Watkins came in April 1758. Drawing on later evidence – supplied by Watkins himself at court in 1765 – it is known that John Dollond approached him with a proposition, the essence of which ran as follows:

> … [he, John Dollond] had Constructed such Object Glasses … and that the same were Intirely New and of his own Invention and not known to any other Person whatsoever but as he … was in *very low and Indigent Circumstances* he proposed … that if [he, Watkins] would Advance the Money Necessary and use his Endeavours to obtain his Majestys Letters Patent for Confirming the Sole Making and Vending of such Object Glasses [then he, Dollond] would in Consideration thereof admit [him, Watkins] into Copartnership ….[56] [Author's italics]

Seemingly, Watkins was impressed with this overture and keen to participate. Here was an artefact of 'public utility' and, therefore, of potential profit from which he could derive a 'private emolument', so ran John Dollond's enticement. Of course, we do not know the degree to which Watkins was enthused although clearly he agreed to the terms proposed. From Dollond's viewpoint, sharing the venture by partnership would reduce the risk and demand on time, thus allowing the Dollond enterprise opportunity to develop and diversify their business in other ways while still enabling the business to gain prestige for an invention singularly claimed as his own.[57] Dollond's claimed 'indigence' may have been due to his having spent much time on his experiments.

---

[56] 'The Several Answers of Francis Watkins one of the Defendants to the Bill of Complaint of Peter Dollond, Complainant', 30 November 1765. Court of Chancery, Clerks Office Pleadings. TNA: C 12/1956, roll 19.

[57] Francis Watkins was financially secure through his sideline interests and so his ability to help Dollond's patent was not an issue as supposed by Sorrenson (2001), p. 52.

A Tiresome Process at Chancery

Letters patent (*literae patentes*) are 'open writings of the sovereign' and as such, they are grants directly from the Crown issued under the Great Seal of Great Britain. In practice, they were privileges of the sovereign, affording some protection to the first inventor of any new contrivance in manufactures and entitling that inventor to make articles according to his invention during a certain period only. As such, they had to follow the law as laid down in the Statute of Monopolies (1624), that is, the relevant law in force at the time of Dollond's application.[58] Under this law, monopolies generally were prohibited except in the case of articles of new invention.[59] Accordingly, an invention had to be material and capable of manufacture (that is, an idea or scientific law could not be patented); it had to be new within the realm of the British Isles and not known to the public at the date of the grant; it had to be of some utility to the public; and the person applying for the patent had to be the true inventor of the article. It is presumed that John Dollond was made aware of these provisos before starting on his venture but Francis Watkins may have held only a layman's understanding of patents although, as we have noted, he was sufficiently aware of their import to have criticised George Adams and Richard Jack in 1751 over their dubious claim to a patent for their new telescope (§ Chapter 3). Watkins did not, however, have any personal experience in the complicated procedure of Chancery although, acting for Dollond, the would-be patentee – he very soon participated in this ancient and convoluted process.

A detail in the court deposition by Watkins of 1765 (quoted above) might easily be passed by without a second thought although it should be observed that an important part of Dollond's proposal was that Watkins should 'use *his endeavours* to obtain his Majestys Letters Patent' [author's italics]. In other words, the intending partner was obliged to take on the role of agent since he was to pay the relevant fees, Dollond otherwise being in 'very low and indigent circumstances'. Of course, the drafting of the application to the Crown – outlining the nature of the invention, its benefit in terms of public utility, and particular claims to preferment – would have rested entirely with Dollond himself. According to the standard process at the time, this would have been written simply, in a plain style, praying that a patent be granted; it would

Sadly, as we now know, Watkins did not pay sufficient attention to the nature of the man with whom he was dealing.

[58] James-I (21 Jac. I c.3): *An Act Concerning Monopolies and Dispensations with Penal Law*, with the relevant subsection (s.6) titled 'Saving for letters patent'. This statute remained in force until 1852 although in 1713 an edict had been issued under Queen Anne ordering the Royal Society to scrutinise all new applications. *RS: Journal Book* C (13 January 1713). It is not apparent that this procedure was carried out with regard to Dollond.

[59] Hewish (2000), pp. 2–3.

also have contained an affidavit sworn by Dollond himself before a master in Chancery for a fee of one shilling and a stamp duty of six pence. As financial backer and agent, Watkins would then have commenced his task immediately after the initial application (the petition) had been accepted.

Throughout the early part of April 1758, Watkins became engaged in meeting with the various examiners at Chancery, possibly through the venerated 'Six Clerks of the Crown' who formerly were examiners of writs but latterly superior clerical officers attached to the Court of Chancery. The procedure was strict, if somewhat tedious, and had evolved over the previous 200 years as a system of checks and counterchecks. Some notion of the process may be derived from Appendix 2 where it will be seen that the petition made its progress through the offices of the Secretary of State, the Attorney or Solicitor General, the Signet Office, the Lord Privy Seal, and finally the Lord Chancellor with intermediary visits to relevant law officers whose task was to legalise each stage until, ultimately, the patent was fully 'engrossed', that is, formally constructed in the peculiar language of law. Awaiting the document at each stage, paying the relevant fees and stamp duties, dealing with the pursebearer, and transferring the copies onwards to the next office – sometimes taking several days – was the burden undertaken by the petitioner or his agent. Thus, by the end of this process Watkins would have incurred an outlay approaching £50. Eventually, the letters patent – all neatly boxed-up and scrolled – would become available for collection. The letters patent in the final form available to John Dollond bore the date 19 April 1758.[60] It then only remained for the patent to be enrolled within the allotted period of three months before trading could commence (Appendix 3).[61]

The Dollond–Watkins Co-partnership

Satisfied with Watkins's completion of this initial and costly business, Dollond then arranged for articles of co-partnership to be prepared. Regardless of whatever handshaking had taken place during their early discussions, it was this document that would spell out the strict rules of practice.[62] To accept, Watkins was to advance £200 and confirm his agreement by signature. This he did on

[60] Patent application. TNA: SO 7/233. The substance of the patent is given in *Abridgements* (1875), No. 721.

[61] *Note:* Enrolling a patent was a fairly new demand and only 15 enrollments had taken place between 1720 and 1733, that is, less than a fifth of all patents then in existence. After 1734, the validity of a patent was made dependent upon lodging a full specification although such specifications tended to remain incomplete.

[62] 'Articles of co-partnership between Francis Watkins and John Dollond for making object glasses for refracting telescopes according to the new method of John Dollond,

29 May 1758. Indeed, since obtaining the letters patent, Dollond moved with great rapidity over these legal affairs. All was now set to commence trading in the new achromatic telescopes because, only six days earlier, he had personally visited Chancery to finalise the business by testifying to a specification for the manufacture of the compound object glass. In the formal language of procedure, the Petty Bag clerk added the following words to the specification:

> *And Be it Remembered* that on the said twenty third day of May in the Year above [1758] written the aforesaid John Dollond came before Our Lord the King his Chancery acknowledged the above specification and all and every thing therein contained in force above written ....[63]

With this bureaucratic chain completed, Dollond's specification was 'so forth *Inrolled*', in other words, extracted from the writ of the Privy Seal onto the Patent Rolls and dated 25 May 1758 – that is, two days after the Petty Bag clerk's endorsement (above). Trading could then begin in earnest and what was hitherto held secret between Dollond and Watkins was now available to public scrutiny.

Only *after* the contracts of co-partnership were signed and sealed did Dollond submit the account of his experiments to the Royal Society. Thus, Francis Watkins had no way of knowing the details of Dollond's experimental work, other than what may have been described to him informally, although he must surely have been shown some of the prepared achromatic telescopes which he had agreed to assemble with compound object glasses according to the business plan. Ten days later – after securing Watkins's signature – Dollond dispatched his formal account to the Royal Society whereupon, on 8 June 1758, it was read along with an introductory letter from James Short.[64] With this formality completed and publication agreed, the entire project was finished and no further contribution was made to the subject from Dollond's pen.

Meanwhile, Watkins and Dollond commenced their trade in the 'New-invented telescope' according to the indentured articles of co-partnership (Appendix 4). These stipulated precisely how the business between the two instrument houses was to be conducted: namely, that telescopes were to be sold at a profit of 'sixty pounds per centum', that prices were to be fixed at 12, 16, 21 and 32 shillings for tubes of 2, 3, 4 and 5 feet, respectively, and that in carrying out each sale, an allowance was to be given for the cost of tubing-up and any other expense; thereafter a quarter of the net profit was to be retained by the

as granted to him by letters patent, 29 [*sic*] May 1758', GL: CLC/444/MS14805/001 (with CLC/444/MS14805/002 the patent).

63 Specifications and Surrender Rolls (1709–1848) of the Petty Bag Office, TNA: C 210/5.

64 Dollond (1757).

seller. 'Books' and 'Records' of accounts were to be available at all times and monetary reconciliation was to take place on the usual Quarter Days, at which point the remaining three-quarters profit per telescope was to be shared according to the reciprocal agreement that whoever sold a telescope should recompense the other. It should be emphasised here that there was no intention, legal or otherwise, that Watkins was to share the patent ownership, despite his laborious trails made through Chancery on Dollond's behalf.[65]

Watkins's Afterthoughts

In the cold light of day a different perspective on Dollond's enterprise percolated through. As mentioned already, nowhere in the Royal Society paper, nor in the application relating to the letters patent, nor the articles of agreement with Watkins, was there any indication of how the compound glasses were actually made. Completely absent in the enrolled specification were any remarks relating to size, preparation and finish of the surfaces, and only the vaguest outline regarding the process of manufacture of the achromatically corrected glasses. It would, therefore, have been very difficult for anyone to reproduce a compound object glass based on the description supplied. Such essential (but absent) information remained a secret to the Dollonds and, naturally, this became a bone of contention among members of the trade who later would argue that this 'insufficiency' was reason enough to invalidate the patent (§ Chapters 6 and 7).[66] Nor, it would appear, was Watkins allowed any insight into the actual optical process by which the paired lens combinations were selected after grinding and polishing. In fact, all the object glasses were to be made and supplied by the Dollonds, not Watkins, and so the 'mystery of the craft' was preserved.[67]

Regardless of the agreements that were in force at the start of the partnership, Dollond had not been entirely forthright. Whatever the plausibility of his argument or affability of his personality it transpires that – during the haste with

[65] Published papers in the literature sometimes erroneously refer to a *shared* patent between Dollond and Watkins. The actual arrangement of partnership was only to share profits; Watkins had no rights in the patent.

[66] Peter Dollond was keen to keep the method of manufacture a trade secret in the early years. By 1769, when Jean Bernoulli Jr visited his workshop, it was known that the method very much depended on a 'good eye', for which Peter Dollond was renowned. The process itself entailed several lenses of crown and flint glass made simultaneously and then, by trial and error, combinations in pairs were chosen which gave the best results. King (1955), p. 160.

[67] The detail concerning the lack of adequate specification was raised as an objection by the defendants to Peter Dollond's 'plea of trespass'. In Watkins's answer was the remark: 'the Glasses now in his [Watkins's] possession which were made either by John Dollond Deceased or the said Peter Dollond during the continuance of the Partnership'. Note 56 (above).

which the agreements and documents were prepared – Watkins overlooked one important detail: trustingly, but foolishly, he did not investigate the authenticity of Dollond's claim to be the first inventor. Perhaps he had been duped; but, if this was so, then so had the Crown although, it must be borne in mind that the patent was issued for an invention described as a 'new invented *method*' rather than an actual *device*.[68] It was upon this particular distinction that some individuals were confused but on which later litigation was hinged. Although the patent documents had passed before the scrutiny of the Crown's law officers at Chancery on more than one occasion, it was not normally considered their duty to investigate the viability or utility or truthfulness of a proposed patent of invention; their only concern was that the proposal itself was lawful under the Monopolies Act and 'not mischievous to the State', meaning that it would not raise prices or hurt the trade. As far as 'novelty' was concerned, the law officers usually trusted an applicant's declaration and rarely checked beyond this. Thus, since the patent had not to be broadcast until after its enrolment, any objection or petitions of revocation had to come afterwards through courts of law rather than be initiated by the Chancery law officers.

An advertising campaign for the Dollond 'New-invented Refracting Telescope' was begun in the *Public Advertiser* on 26 April 1758, that is, only a week after obtaining the letters patent.[69] Shortly afterwards, a visitor – most probably George Bass – called upon Watkins to inform him that his partner was not the true or original inventor since that privilege could be due only to one Chester Moor Hall, a name then quite unfamiliar to Watkins. Thus, at the very outset, Watkins was distressed to learn that he had been deceived by his partner. But, what was to be done? The agreement of partnership was binding and penalties would surely follow should he withdraw on unsure grounds. What followed was an unsavoury business with consequences that extended well beyond Watkins's personal embarrassment (§ Chapter 6). For the time being, however, the partnership was allowed to proceed, tentatively and cautiously, although the business relationship was soured from the very start.

An 'Artful' Advertisement

It would not be correct to suppose that Dollond's new telescope was taken up immediately in any large measure. In fact, the relatively small balancing or purchase payments made by Watkins to Dollond during the first two years do not suggest any remarkable uptake and, in all likelihood, the improved instrument – whatever its merits at the time – still had a long way to go. Nevertheless, the

68 Hulme (1896).

69 *Public Advertiser*, 26 April 1758.

possibility of owning a telescope with the new compound objective, based on a principle denied by Newton, must have had some appeal.

The advertising campaign for the 'New-invented refracting telescope' ran into the following year. By this time, however, criticism was building up which demanded some form of public explanation. This came on 27 February 1759 in a lengthy addendum to the regular advertisement in the *Gazetteer and London Daily Advertiser* in which doubting competitors were reminded of the 'unquestionable Experiments [which] entirely satisfied the learned World of the Reality, as well as the Importance of the abovementioned Improvement'. Homing in on those who had done him some injustice, Dollond remarked:

> ... it will be needless to expatiate on the Advantages that may arise from it [the patent], nor will any Kind of Opposition whatever, that may be made by Ignorant and envious Persons, be in the least regarded. But there having lately appeared in some of the Daily Papers *a very artful Advertisement*, offering an Assortment of Telescopes to public View, in which are several expressions evidently calculated to insinuate some considerable Improvements: It is therefore necessary, in order to prevent the Public from being imposed upon, to assure them, that those Insinuations are entirely false; and that those Telescopes are constructed in the same Manner as the common four Glass Telescope was when first Invented, which was toward the Middle of the last Century. And as for the Combination of Mediums which this ingenious Artist has made use of, has been well know[n] to all optical Workmen in London for many Years past.[70] [Author's italics]

From this statement, it would appear that Dollond was no longer claiming originality for the crown–flint object glasses but, rather, he switched the emphasis of his claim to the acceptance of his refrangibility experiments by the Royal Society. As for the 'artful Advertisement', this may have been made by Benjamin Martin who was strongly averse to crown–flint objectives. Certainly, Martin did not hold a license to sell Dollond's patented telescopes and was at that time using a ploy of advertising his own telescopes as '*Dalone*'s.[71]

From the viewpoint of others in the trade with vested interests in making and selling telescopes, three distinct paths were destined to follow in the wake of the patent: instrument makers could ignore it and so force Dollond to prove the validity thereof through an action in court; they could formally accept it, and so submit to buy rights by licence; or they could revolt by dismissing the new invention as inferior to what existed already. As will be described in Chapters 6 and 7, some instrument makers took the first or second path,

---

70 *Gazetteer and London Daily Advertiser*, 27 February 1759.

71 The erroneous spelling 'Dalone' appears elsewhere: (§ Chapter 8, note 8); also note the spelling 'Dalond' (§ Chapter 8, note 96).

although, in practice, there was no recourse to the second path until the dubious patent was properly put to the test through litigation. Only Martin took the third path by explicitly listing his objections to the crown–flint compound object glass in a special 'Memento' to his latest text.[72]

Even ignoring vociferous individuals such as Benjamin Martin, there were others in the trade who objected strongly to John Dollond's patent, although not on the personal grounds of his Huguenot origin. Meanwhile, Dollond basked in the recognition afforded him by the Royal Society. His meticulous experiments had been a fine achievement that would inevitably lead to vastly improved refracting telescopes. It is not surprising, therefore, that within months of the reading of his formal account, he was proposed for the Copley Medal. The President, George Parker, second Earl of Macclesfield, had been ill and away from London, and so had little time to study precisely what Dollond's experiments had actually shown although, acting on advice supplied, he endorsed the opinion that Dollond was 'justly intitled ... to that honourable token of the Society's esteem which the Council had bestowed upon him.'[73] So on 30 November 1758 the medal was awarded. Granting this seal of approval, however, committed the Society to a certain stance that later caused them a degree of embarrassment in communications with foreign members.

The reward of the Copley Medal did not automatically guarantee Dollond the status of becoming a Fellow although it was a step *en route* to that privilege. In fact, it took two more years before his name was proposed for Fellowship. That came on 19 February 1761, with Gowin Knight as first signatory:

> Mr John Dollond of the Strand, Optician to his Majesty, well known by several papers he has given into the R. Society, and particularly one, of a new Instrument for measuring small angles and another of an Object Glass consisting of two Spherical Lenses of different densities, so contrived, as to correct the Errors arising from the different refrangibilitiy of the Rays of Light, being desirous of the benefit of becoming a Fellow of this Society. We, where names are hereunto subscribed, recommend him, from our personal knowledge of him and his Abilities, as a person well deserving of that honour, and very likely to prove an useful Member of this Society.[74]

Two months on, with only six signatures proffered in support, James Short took action to speed up the balloting process. Short's concern regarding the sluggish process of Dollond's election deserves closer attention because it reveals

---

[72] Martin (1759), pp. 53–83 (Appendix 6).

[73] Lord Macclesfield's speech on Dollond's award of the Copley Medal. RS: *Journal Book of the Royal Society*, 30 November 1758, pp. 204–210 (Dollond citation, pp. 206–209). RS: MM.4.20.

[74] Dollond's election certificate. RS: EC/1761/04.

something further of the persona of the man with whom Francis Watkins had to cope during their partnership. After Klingenstierna's paper to the Royal Swedish Academy in 1754, of which Dollond had held the Latin abridgement since the beginning of 1755, there had been a second paper in 1760, dealing more generally and fully with spherical and chromatic aberrations. The astronomer Bengt (Benedict) Ferrner, who had met Dollond during his European tour of 1759–1760, undoubtedly discussed Klingenstierna's interests with him because, upon his return to Uppsala, he handed Klingenstierna a pair of part-prepared lenses of crown and flint-glass which Dollond had supplied for experiment.[75]

Ferrner's position in this course of events is of relevance here. First it should be noted that Ferrner had been an FRS since December 1760, and since his European tour had been closely in touch with colleagues in London. In particular, he knew directly of Dollond's dependence on Klingenstierna's previous communication (quoted earlier in this chapter) which had caused him to examine afresh what Newton had proposed in *Opticks* concerning dispersion. Secondly, as intermediary, it was Ferrner who recommended to Klingenstierna that he should prepare a Latin version of the 1760 paper (already published in Swedish in the *Transactions* of the Royal Swedish Academy) for the *Philosophical Transactions* of the Royal Society. Hence, a copy was prepared and entrusted to one Jennings (a traveller between Uppsala and London) who delivered it to the Royal Society sometime during August of 1760. There was nothing unusual about this process, although it should be observed the paper was not actually read until the meeting of 2 April 1761.[76]

To all intents and purposes, Dollond had contributed to the very same area as Klingenstierna and had already achieved recognition for his success with the achromatic telescope objective through the award of a Copley Medal. It could be viewed that Klingenstierna's paper had been held back in order to allow Dollond's election to take place unhindered. However, the reading of the paper on 2 April during the actual period of balloting for Dollond's election (possibly at Ferrner's insistence) caused consternation among members who urged for greater clarity of what, respectively, had actually been achieved by Dollond and Klingenstierna. It was against this background that James Short gathered vital evidence from Dollond, namely, Klingenstierna's (1755) paper, which was never actually presented before the Royal Society, and a copy of Dollond's letter to Ferrner (1759) in which he had admitted to Klingenstierna's influence. After having digested the import of these documents, Short reported on 16 April 1761:

---

[75] Ferrner's purchases from Dollond are presently housed in the history of science collection of the Royal Swedish Academy. Rudd et al. (2000), p. 7.

[76] Klingenstierna (1759–1760).

> As soon, therefore, as he [Dollond] saw reason to doubt of Sir Isaac Newton, *he thought it high time to begin to think for himself*, and endeavour to find out the truth by experiments....[77] [Author's italics]

The italicised phrase had appeared in Dollond's original letter to Ferrner. After passing over Klingenstierna's papers, Short then made the following appeal:

> It becomes me, Gentlemen, to observe that whereas I was not appointed to inspect or render any account of these papers to the Society, I hope my presumption may be in some measure excusable, if I therefore declare that my sole motive for acting as I have done, is a regard for truth, and for the reputation of an ingeneous [*sic*] and honest Man at present a Candidate for election into this honourable Society.[78]

With the position somewhat clarified in Dollond's favour – Dollond having achieved *by experiment* what Klingenstierna had indicated *from theory* – Dollond's validity for candidature was somewhat restored. Four extra signatures were added to the ballot register and, by the final ballot of 21 May, the requisite ten signatures were to hand, and so, one week later, Dollond was welcomed into the fold.

A Very Distasteful Business

James Short may well have added emphasis to Dollond's achievement in order to push the election process forwards and yet, perhaps ungratefully, it appears that Dollond had feared that any acceptance of Klingenstierna's theoretical contributions would lessen the value of his own achievement. Upset at the likelihood of having to share the priority and complaining bitterly to Benjamin Wilson, FRS, Dollond let it be known that he did not consider the Swede's contribution to offer any solution that he himself had not already solved. Correctly, in procedural terms, Wilson relayed the sense of Dollond's complaint directly to the Swedish Royal Academy of Sciences.

To understand the context of this sad affair, we must backtrack to the summer of 1760, just after Ferrner's return to Uppsala with the pair of part-prepared crown and flint glasses bought from Dollond. Despite much grinding and polishing of these glasses, Klingenstierna failed to achieve any success at preparing an achromatic doublet. The reason for the failure was attributed to the poor quality of the samples. Not wishing to offend Dollond, Klingenstierna requested – by a letter of 7 February 1761 – two or three further pairs (crown

77 Short (1761).

78 Ibid.

and flint) along with a pair of prisms (crown and flint) of refracting angles 60 degrees and 40 degrees, respectively.[79] Not having received any response after three months, Klingenstierna suspected that something was amiss and so wrote once more, on 30 May, repeating his request and adding (in postscript): 'I hope you will find upon second thoughts that I have not given you any occasion to be angry with me, and to refuse to serve in your Profession.'[80] A question therefore arises: Had there been anything for which Klingenstierna might have had cause to reproach himself?

Most of Klingenstierna's letter – written in French for Dollond's convenience – dealt with the lack of success with the lenses. In translation, this reads:

> You sense already, Sir, where all this is leading. If I am not abusing your kindness, I beg you once again to be so good as to supply me with two or three pairs of glasses for the double objective. But above all, Sir, do not send me any that you have not examined carefully enough to be very certain of their purity, and that they are free from these dreadful veins. I do everything with my own hands, it is my pastime [*orig.* usual pastime], which I have taken on with enthusiasm, since I do not always feel myself capable of the application necessary for other studies; and you will easily appreciate, Sir, how disagreeable it must be to find myself disappointed at the end of some long period of labour. I should wish these glasses to be of 3 inches diameter and 1/4 inch thickness, because the telescope will be 6 feet in length. In procuring them for me you will render me a significant service for which I shall be much obliged to you; also Mr. Jennings will pay you whatever you may ask for them. I hope that you will not refuse this friendship.
>
> Your very humble and very obedient servant
>
> Klingenstierna

Having made firm his expectations as to the quality of glasses required for his experiments, he further reminded Dollond (in postscript) that the corners (refracting angles) of the prisms must be '*très purs, très plans & très-bien polis.*'

However, the actual clue to Dollond's reluctance to respond to the order appears in the very opening lines of that letter of 7 February:

> I certainly did not imagine that my *little remark* about Mr. Newton's rule concerning the dispersion of colours would lead to a discovery so beautiful and so useful as that of your new telescope: but in the hands of such an able man

---

79 Letter from Klingenstierna to Dollond, in French (7 February 1761), in Nordenmark and Nordström (1939), p. 372.

80 Letter from Klingenstierna to Dollond, in English (30 May 1761), in Nordenmark and Nordström (1939), pp. 331–334, p. 334.

> it takes very little to set genius in motion. I congratulate you on your happy successes, and can assure in advance of the gratitude and respect of all physicists and astronomers. [Author's italics]

Klingenstierna well understood that his 'little remark' concerning the error in Newton's 'rule' was the essential prompt that had caused Dollond to revise his ideas and set him on the path to experiment. Indeed, Dollond had every reason to be grateful for this hint and had said so in his subsequent letter to Ferrner. It therefore appears, that on receiving Klingenstierna's letter and order, he became acutely aware that he would lose credit for that part of his claim in the 1758 paper – his 'perfect theory' – in which he had implied some mathematical background although, in fact, having omitted to elaborate upon it.

In defence of his position, Dollond wrote to Benjamin Wilson, FRS, expressing the opinion that Klingenstierna had contributed nothing that had not been solved already by him in 1758. With such a contention, Wilson had no option but to notify his corresponding colleague, Torbern Bergman at the Royal Swedish Academy (by letter dated 23 March 1761) of Dollond's position and open dismissal of the significance of Klingenstierna's contribution. Bergman then alerted Klingenstierna by sending him the relevant remarks that Dollond was broadcasting among members at the Royal Society.[81] After due consideration, Klingenstierna wrote to Thomas Birch, Secretary of the Royal Society, on 30 May 1761, explaining his position and enclosing a separate letter, addressed to Dollond, which he requested should be read before the Society before it was given to Dollond.

Taking Dollond to task, Klingenstierna began:

> Sir,
>
> M. Bergman, M.A. at Upsal has communicated to me a part of a letter from Mr B. Wilson to himself (23 March 1761) which I shall transcribe here because it concerns you and myself.
>
> As this part of the Letter has been communicated to me at your desire, Sir, and contains nothing but your pretensions, I cannot but regard it as your own. And therefore I chuse [*sic*] to address this my Answer to yourself, as the person concerned. I wish you had been pleased to do the same, before you had acquainted the Royal Society with your pretensions; that might have saved some trouble to me, and perhaps to yourself.

Then followed a summary of what Dollond had openly claimed in London:

---

81 Letter of Wilson to Ferrner in Nordenmark and Nordström (1939), p. 326.

> If I guess right at the meaning of the foregoing Letter [from Bergman], you pretend, Sir, and have acquainted the Royal Society with that pretension, that the Discovery of the Things, contained in a Paper of mine, presented [sent] last Summer [1760] to the Royal Society, (concerning the Aberrations of Rays refracted in the spherical Surfaces of Lenses, and the manner of correcting them) belongs not to me, but to yourself: and you offer two reasons to make good that Claim. The *First* is, because the notion of correcting the aberration of a Lens by an equal and contrary aberration of another, belongs to you, and is explained in a Paper communicated by you [in 1758] to the Royal Society ... so that nothing was left to be done by others, but to solve a Mathematical Problem, pointed out in that Paper. (I suppose the Problem of finding the Figures of your double Object-Glasses is meant here.) Your *Second* Reason is, because you have solved that very Problem long ago, as Mr Maskelyn [*sic*] shall at any Time be ready to witness.[82]

Klingenstierna then proceeded with a 1,000-word reprimand admonishing Dollond for his peevishness and pretences (Appendix 5).

With regard to having already solved 'the Mathematical Problem' (mentioned in the above letter), it should acknowledged that Nevil Maskelyne, Astronomer Royal, had been much intrigued with Dollond's success and had indeed prepared his own account of how perfect cancellation of the aberrations could be achieved. In a letter to the Royal Society, dated 22 January 1761, he wrote:

> About two years ago, becoming acquainted with Mr. Dollond's curious discovery in optics ... and learning from him, in conversation, that he had invented a theorem, shewing the quantity of the aberration of the rays refracted through a lens, on account of the imperfection of the spherical figure; by the application of which, he was able to make the aberrations of the combined concave and convex object lenses perfectly equal to, and consequently to correct one another; I was desirous of being more minutely acquainted with this farther great improvement in optics; and Mr. Dollond accordingly readily offered to gratify my curiosity.

Thus, Dollond had been encouraged to release the secret of how the simultaneous correction had worked. Continuing, Maskelyne wrote:

> ... But in the mean while that he [Dollond] was looking over his papers, in order to lay them before me, having leisure, I set about the investigation of a similar theorem myself; which having completed, I interchanged with Mr. Dollond for his theorem, he taking mine, and I taking his. Our theorems, though similar, were not exactly the same; but, by reduction to the same form, I inferred his theorem from mine; which gave me a farther confidence of the exactness of both.[83]

82 Ibid (note 80).

83 Maskelyne (1761), pp. 17–18.

Of course, Klingenstierna was unaware of this communication and, in any case, he was more appalled by the content of Wilson's letter to Bergman, confiding to his friend Ferrner that it was pointless to expect any courtesy from Dollond.[84] Indeed, Dollond's comportment – perhaps not thought befitting a Fellow of the Royal Society – needed correction, even reprimand but whether he was actually admonished – before the fellowship as Klingenstierna wished – remains unknown.

When Klingenstierna's second paper, '*De aberratione luminis …*' (dated Stockholm 20 April 1760) was finally read in London one year later on 2 April 1761, it appeared with an obscure footnote telling that the delay of presentation had been due to an 'unfortunate accident'.[85] That delay, however, had allowed Maskelyne time to prepare an account giving insight to the mathematical theorem employed by Dollond in order to solve *simultaneously* for the spherical and chromatic aberrations. Maskelyne's account, however, was more restricted than that in Klingenstierna's 1760 paper which had dealt generally with the theory of refractions; it did, nevertheless, suggest how Dollond had been able to determine the required curvatures of the crown and flint glasses. In principle, Maskelyne's theorem derived an expression for the angular aberration of a meniscus lens *only*, but his paper concluded as follows:

> With respect to the application of this theorem to Mr. Dollond's combined object glasses, it is evident, that if the aberrations of the convex and concave lenses added together (paying due regard to the signs [i.e., a defined convention] of the theorem) are made equal to nothing, the two lenses will perfectly correct one another: but as there are two unknown quantities unlimited in the equation, namely, the radius of one surface of each glass … there is room for an arbitrary assumption of one of them, at the discretion of the theorist, or artist; which being done, there will remain a quadratic equation, whence there will result two values of the radius, which remains unknown, either of which will produce an aberration equal to that of the other lens.[86]

Hence, with Maskelyne's contribution ready for publication by January 1761, there was no further need for Dollond to publish beyond what he had declared in 1758. Viewed otherwise, Maskelyne had covered up for Dollond's incomplete account. But no matter how much Dollond may have railed against those who attempted to diminish his achievement, there was no way that he could be excused for his lack of courtesy and recognition of Klingenstierna.

---

[84] Klingenstierna's correspondence with the Royal Society and with Dollond is reprinted in Nordenmark and Nordström (1939).

[85] Klingenstierna (1759–1760).

[86] Maskelyne (1761), pp. 20–21.

Conclusions

Undoubtedly, John Dollond, urged by his son Peter, succeeded in the realisation of a successful commercial achromatic telescope although, until recently, much of the historical background to this narrative has been hidden from view. It has therefore been an aim of this chapter to examine afresh the life and actual contribution of that individual to whom credit is usually admitted, although bringing to the fore the names of individuals who steered him towards his goal. Beyond those who challenged him intellectually (Simpson, Canton, Short, Harris, Euler), and those from whom he extracted practical clues (Ayscough, Rew, Bass), there were others who should not be neglected (Wilson, Ferrner, Klingenstierna). Each of these persons were formative to a life whose contribution to science has not been without controversy.

The inclusion of Francis Watkins has also been important in this chapter, as without his financial backing, John Dollond would never have obtained the patent. Indeed, without the patent the Dollond business might never have met with the flourishing success for which it is remembered today. So, by 1761, father and son must have felt justly proud of their achievements. They had pooled their intellect and business acumen – as much out of social desperation as intellectual zeal – to create a high-class enterprise which by the end of the decade had begun to change the family fortune, thus alleviating the uncertainty and unease which had been the burden of their ancestors since their arrival from France at the end of the seventeenth century. But whatever feelings of success and achievement the 54-year-old John Dollond may have prized on his acceptance into the Royal Society, they were short-lived. Quite unexpectedly, at home, on the Sunday 29 November 1761 at 8 o'clock in the evening, John Dollond suffered a stroke and, within the hour, he was dead. Accolades followed in the press, where 'his great Genius and Abilities being suited to much higher Objects' was recalled, adding that 'he hath shone as an eminent Mathematician and Optician, which brought him into the Acquaintance of the learned Part of the World, who held him in great Esteem'.[87]

Whatever distrust or sensitivities existed by those who doubted Dollond's integrity or disapproval of the jealous way in which he guarded his achievements, these were put aside until after his funeral at St Martin-in-the-Fields on 6 December 1761. As we shall note in the following chapter, Dollond's unexpected death brought about a new twist, ultimately freeing Watkins from the burden of partnership and so providing an opportunity to rectify the 'mistake' that the patent should ever have been issued. Whether from guilt or embarrassment, but certainly in seeking justice, Francis Watkins found himself a candidate for Master

[87] *London Evening Post*, 28 November 1761; repeated in the *St James's Chronicle or the British Evening Post*, 28 November 1761 and the *Public Advertiser*, 1 December 1761.

Figure 5.7 Peter Dollond, engraving by John Thomson after a portrait by John Hoppner, R.A., c.1820. © The Trustees of the British Museum, 1865,0708.713

of the Worshipful of Spectaclemakers, the means by which opticians could seek redress for the monopoly imposed by the Dollond patent. Meanwhile, as successor to his father as senior male of the family, Peter took a harder line in developing the business. It soon became clear that he was unwilling to accept his father's former arrangements with Watkins and, as will be discussed further, litigation followed.

From 1762, the situation was threefold. From the point of view of the practitioners in the trade, there were those such as Bass, Watkins and Smith who maintained that John Dollond had been deceptive in gaining what, more properly, was due to Chester Moor Hall; there were others (to be discussed) who objected to the Dollond patent on the grounds of its *insufficiency* or lack of specification, while from the point of view of the theoreticians, mathematicians and experimental philosophers including Euler, Klingenstierna, Ferrner and Wilson, there was dissatisfaction that Dollond should have accepted encouragement and direction from his peers without due acknowledgement.

In London and Stockholm, it became clear that John Dollond was unlikely to show any of the usual courtesies, and was prone to denigrate or insult those whose personal input had been formative to his success. What had simmered in discontent before Dollond's death, spilled over afterwards, hence causing a major upheaval in the trade during the early years of Peter Dollond's proprietorship of the family business.

So, the present chapter has finished on a note of discord, first with the view, expressed by Benjamin Martin, that nothing in the current state of optics was entirely new, and, by Samuel Klingenstierna, that John Dollond did not conduct himself in the manner befitting a member of the scientific fraternity. Nevertheless, the ground covered here, from Newton's error to Dollond's detection and confirmation of it, maps the route taken which led to the commercial realisation of the achromatic telescope. John Dollond's success depended upon far more than the concept of the compound (achromatic) object glass because he was also able to solve the technical problem of how to correct simultaneously spherical and chromatic aberrations, although precisely how this was achieved was kept secret. Overall, Dollond's achievement was creditworthy enough, and the commercial application of the principle of the compound object glass assured a return of interest to the refracting telescope, beginning with the Dollond–Watkins profit-sharing scheme to market the 'New-invented refracting telescope'. Despite the undercurrents of dissatisfaction within the instrument trade, the commendations from certain quarters of the Royal Society, followed by recognition, helped to fix the notion of Dollond's success and certainly boosted the image of the Dollond enterprise as an establishment of very high class. On his father's death, Peter took on family responsibilities and was resolute in his determination to preserve everything for which his father had striven since their years of poverty in Spitalfields. Nor were Peter's efforts relaxed later when others abroad began to misrepresent his father's place in the chronology of scientific achievement, as we shall see in Chapter 8.

Chapter 6

Peter Dollond and his Conflict with Watkins & Smith

Few accounts in the literature reveal much of significance regarding the early development of the achromatic lens. The Dollond family sold out in 1876, although the business name survived in Britain's high streets, from the 1880s as Dollond & Aitchison, finally disappearing in the early twenty-first century when it merged with Boots Opticians in 2009, and its stores were rebranded in 2010.[1] There are some references to Chester Moor Hall but little was known of him. No regard was paid to the optical instrument makers such as James Ayscough, William Eastland or James Champneys, who each employed crown–flint combinations well before John Dollond patented his method of making crown–flint combinations. There was no account of the part played by Francis Watkins, who procured the patent on Dollond's behalf and invested money in Dollond's scheme to market the achromatic telescope. Indeed, none of the issues surrounding John Dollond's patent were straightforward and even more aggravation came to the fore when Peter Dollond took sole responsibility of the family business after his father's death. Where disputations are mentioned in the literature, they tend to emphasise the successful Dollonds for, despite all else, they did make valuable strides in the development of the achromatic telescope. But the blind acceptance of the legal ruling against Watkins and Smith in 1763, or even ignorance of this ruling by later authors, has deprived those curious about the development of the achromatic telescope of any accurate account of what took place.

Much more has been uncovered in recent years and this chapter illuminates Watkins's role in the important battles that took place in the law courts, and within the Royal Society and the Spectaclemakers' Company. His principal action was, as Master of the Worshipful Company of Spectaclemakers, to have the Dollond patent withdrawn through a petition to the Crown in 1764.

Prosecutions regarding new inventions were relatively infrequent before 1770, and judicial precedents to do with patents are barely detectable. Moreover, although it was argued in the Spectaclemakers' Company's petition that Dollond's patent would create a monopoly, scant regard was paid to the

---

[1] Information from Neil Handley, Curator, British Optical Association Museum, November 2011.

Monopolies Act of '21 James I' (1624) under which Dollond's patent fell. Thus, the singular way in which certain legal issues were dealt with provides a valuable insight, not only into the optical trade and its main personalities but also into the contemporary legal system. In parallel with the famed petition of 1764, a legal battle ensued which showed that Watkins did *not* hold with Lord Mansfield's judgement at high court in 1763. At that time, by a moot point only, Peter Dollond, as plaintiff, defeated Watkins and his partner Smith in a critical case to establish the validity of his father's patent. Watkins and Smith had argued that the original patent was not valid, principally due to it having been acquired on unsafe grounds, but after their failure, no other interloping optician was safe. In a test case against James Champneys in 1766, in which Champneys failed in his defence of the action brought by Peter Dollond, Mansfield strongly argued a new principle and so a precedent was created in patent law. But even this did little to ease the minds of those steeped in the honest custom of only attributing true invention to originality rather than bringing a product to commercial fruition. Perceived injustices, voiced within and without the courts, were never fully reconciled in the eyes of Jesse Ramsden, whose intimate knowledge of the Dollonds caused him to believe otherwise. The historical and legal issues raged again at the end of the century and then spilled over in a confused state into the nineteenth century. Attempts to clarify these issues have been portrayed empathetically according to the position of Ramsden or the Dollonds but with little understanding of what actually happened in the courts.

Before we can embark on these issues in detail, we must first examine the series of legal actions as they unfolded over 1763–1766. These cases are complex and, therefore, the account proffered here has been broken into two chapters, with some degree of overlap for completeness and sense. Chapter 6 focuses on those disputes which led to the break-up of the Dollond–Watkins partnership; Chapter 7 then continues by tracing the consequences of the failure by Watkins as Master of the Worshipful Company of Spectaclemakers to put an end to the Dollond patent, and hence leading to the subsequent litigation against James Champneys. The point and counterpoint of argument – not always evident but sufficiently implicit from proceedings and depositions – fought through the highest courts of the land (Kings Bench, Chancery and Common Pleas), provide an insight into the lively manner and style by which both the optical instrument trade and the courts functioned.[2] The reader should note that, unless written otherwise, the surname 'Dollond' refers generally to John Dollond while his son, Peter, is named either in full or as Dollond Jr, although sometimes simply as Dollond where the context is clear.

---

[2] The discussion in Chapters 6 and 7 relates only to the law as it was understood and practised at the time.

George Bass and Some Early Conversations

Until John Dollond had the letters patent in his hands, he and his partner Francis Watkins must have agreed to remain silent about the actual application through Chancery. Such secrecy was normal and not unreasonable; however, under the sovereign's grant of patent, Dollond was obliged to put his privilege into practice from the moment the patent was granted on 19 April 1758. Thus, as a first step Dollond met with the City opticians James Ayscough, William Eastland, James Champneys and Christopher Stedman at Ayscough's shop in Ludgate Street to inform these men – who he knew to be already making and selling telescopes containing the compound object glass – of the new situation with regard to his privilege from the Crown. Indeed, it was an old principal of English law that 'the king must not be deceived in his grant' and, in any case, a clause of 'inconveniency' had been introduced during the reign of King James-I whose purpose was to revoke a patent, not for any lack of novelty or priority but rather for the failure to introduce an invention within a reasonable period.[3] We must therefore assume that Dollond was keen to make his intentions known in order to fulfil his obligations.

Fortunately, there are extant documents which shed light on the history of events. These affidavits were provided to the Court of Chancery in reply to a 'bill of complaint' lodged by Peter Dollond in 1765, that is, a full seven years after the rendezvous at Ayscough's shop, and therefore they provide a retrospective view. As sworn affidavits, they should be trusted today, just as they were trusted when used as working documents in court. Of course, Ayscough may have known more than others about the reasons underlying John Dollond's visit but, in the absence of any testimony from him in 1765 – since he died in 1759 – we must rely on those depositions provided by the other three who had been at Ayscough's shop to hear what John Dollond had to say.[4] An examination and comparison of these affidavits by the present author suggests they are fully in accord over matters of common fact. Thus, Champneys's recollection that Dollond had announced to those present that he had obtained the patent for Chester Moor Hall's invention, and Eastland's recollection of Ayscough's response, which was to remind Dollond of Hall's priority in this matter, are accepted generally as truthful statements. However, implicit in these testimonies, beyond the indisputable truths, are undercurrents suggesting that Dollond's declaration of the privilege of patent was not welcomed and that he had overstepped the mark by claiming the compound object glass as his own invention.

---

3 Seaborne Davies (1934).

4 'The Several Answer of James Champneys, William Eastland and Christopher Stedman, Defendants to the Bill of Complaint of Peter Dollond, Complainant. 29th October 1765'. TNA: C 12/1956/roll 19.

Given that Dollond knew these optical instrument makers were already familiar with the compound object glass – a detail apparently unknown to Watkins at this stage – it seems very odd that Dollond kept this knowledge to himself, and that it was not mentioned to Watkins during the preparatory stage leading to the signing of the partnership agreement. Possibly Dollond feared that Watkins would not co-operate if he doubted the veracity of the claims made in the patent proposal. Therefore, it came as a great surprise to Watkins to discover – very soon after beginning his trade in the new telescopes – that one Chester Moor Hall, a gentleman and barrister, had made such object glasses long ago. This revelation came to Watkins via an unnamed informant and is cited in his deposition of 1765 in response to the same 'bill of complaint' as the City opticians. We may now conjecture that this informant was George Bass, the Bridewell glass dealer and lens grinder who had provided Hall's glasses in the early 1730s.

Bass seems to be the most likely informant because we now know that, upon hearing of Dollond's newly acquired patent, it was he who also made a point of alerting Chester Moor Hall about the same time. Bass was then in his mid-60s and had recently served as Master of the Worshipful Company of Spectaclemakers; thus, as a busy operative within the instrument trade and well known to most opticians, he would be in a good position to hear and spread gossip. The suggestion of Bass as intermediary is reinforced by a detail in a communication sent by Peter Dollond to the Secretary of the Royal Society later in 1789, explicitly mentioning Bass's role in informing Hall.[5] This letter asserted that neither he nor his father had ever heard of Chester Moor Hall or his work on compound object glasses until *after* Bass had alerted Hall. It also revealed that Hall called at their house, *soon after* his receiving a communication from Bass, that is, soon after 19 April (the issue date of the patent) but presumably prior to Dollond's visit to Ayscough's shop. From the conversation that took place, John Dollond learned that Hall had indeed done something along the same lines but very long ago, and owing to his preoccupation with legal studies he had let the investigation lapse. Thus, as reported by Peter Dollond, Hall was 'happy to find that somebody had succeeded and that the invention was likely to become useful'.[6] If this was so, then John Dollond may well have breathed a sigh of relief for having claimed a *fait accompli*, namely, that the invention was his own; at this late stage, however, there was little that Hall could have done to protect his priority, except perhaps to object vehemently had he been so minded. As it turned out, the opposite occurred: the generous Hall offered encouragement.

These conversations – between Watkins and (most likely) Bass (at Watkins's shop); Dollond and Hall (at Dollond's residence); and Dollond and the City

5 Peter Dollond's letter to the Royal Society (1789). RS: L&P, decade IX, item 131.

6 Ibid. See also Bennett (1998), p. 5.

opticians (at Ayscough's shop) – each took place within a matter of days. It should be remembered that the 'truths' declared in the depositions were made under oath and therefore admitted as evidence by the Chief Justice, with all the caveats thereby implied. But, if we accept the additional evidence of Dollond Jr's communication of 1789 respecting the conversation between Hall and his father – which Peter did not personally witness – then we only have his word that Hall had blessed his father's venture. Thus, taking into account the intimacy with which John Dollond understood the subject area for at least two years prior to taking out the patent, it would be very strange that he had no knowledge of Hall's earlier, albeit incomplete, attempts. Something further of this mystery is explained in Chapter 8. For the moment, it is assumed that Dollond had no need to inquire of Ayscough an account of the origin of the achromatic compound object glass because – as we shall learn later from the testimony presented in the Spectaclemakers' petition – he was twice informed of relevant developments: first, via Robert Rew in 1755, and then via George Bass in late 1756 or early 1757.

Additional small details appear on first sight to be of little import, but when taken together tend to show differences between the traditional opticians, on the one hand, and the formidable Dollonds, on the other hand, that were becoming intolerable in the trade. We can now return to Francis Watkins who, until this point, was still ignorant of Hall's visit to the Dollond household and of the meeting Dollond held with the City opticians. How Watkins dealt with his new partner upon first hearing that he had not been entirely forthright over the essential detail of priority is of enormous importance to comprehending the chronicle of legal actions that broke out *after* John Dollond's death.

Nowhere in any of the early dealings with third parties did John Dollond actually admit that the compound object glass was anything but his own invention. This is apparent from his approach to Watkins when he raised the notion of a partnership, from the claim that he made in the formal description to the Crown for the letters patent, and also from letters written to James Short who acted as his intermediary with the Royal Society. Indeed, Dollond appears to have convinced all three that the invention was his, and his alone. Over the days following the meeting at Ayscough's, two further significant conversations with Dollond are noted: first, with Francis Watkins, and second, with William Eastland.

John Dollond's Conversation with his Partner, Francis Watkins

As already described in Chapter 5, Francis Watkins acted as Dollond's agent to obtain the letters patent on his behalf. We shall assume that he acted in good faith; certainly, he accepted the terms of co-partnership to sell telescopes fitted with the new compound object glass. He even paid Dollond £200 to join the profit-

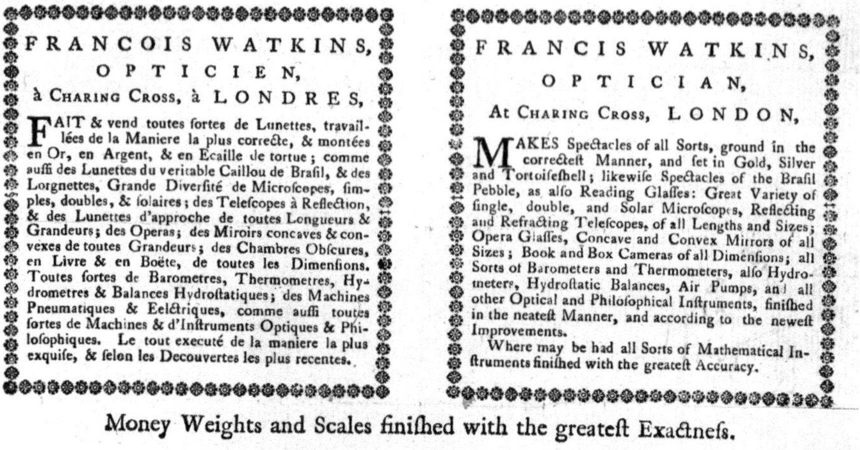

FRANCOIS WATKINS,
OPTICIEN,
à CHARING CROSS, à LONDRES,

FAIT & vend toutes ſortes de Lunettes, travaillées de la Maniere la plus correcte, & montées en Or, en Argent, & en Ecaille de tortue; comme auſſi des Lunettes du veritable Caillou de Braſil, & des Lorgnettes, Grande Diverſité de Microſcopes, ſimples, doubles, & ſolaires; des Teleſcopes à Reflection, & des Lunettes d'approche de toutes Longueurs & Grandeurs; des Operas; des Miroirs concaves & convexes de toutes Grandeurs; des Chambres Obſcures, en Livre & en Boëte, de toutes les Dimenſions. Toutes ſortes de Barometres, Thermometres, Hydrometres & Balances Hydroſtatiques; des Machines Pneumatiques & Eelctriques, comme auſſi toutes ſortes de Machines & d'Inſtruments Optiques & Philoſophiques. Le tout executé de la maniere la plus exquiſe, & ſelon les Decouvertes les plus recentes.

FRANCIS WATKINS,
OPTICIAN,
At CHARING CROSS, LONDON,

MAKES Spectacles of all Sorts, ground in the correcteſt Manner, and ſet in Gold, Silver and Tortoiſeſhell; likewiſe Spectacles of the Braſil Pebble, as alſo Reading Glaſſes: Great Variety of ſingle, double, and Solar Microſcopes, Reflecting and Refracting Teleſcopes, of all Lengths and Sizes; Opera Glaſſes, Concave and Convex Mirrors of all Sizes; Book and Box Cameras of all Dimenſions; all Sorts of Barometers and Thermometers, alſo Hydrometers, Hydroſtatic Balances, Air Pumps, and all other Optical and Philoſophical Inſtruments, finiſhed in the neateſt Manner, and according to the neweſt Improvements.

Where may be had all Sorts of Mathematical Inſtruments finiſhed with the greateſt Accuracy.

Money Weights and Scales finiſhed with the greateſt Exactneſs.

Figure 6.1 Francis Watkins's trade card, 1775. © The Trustees of the British Museum, Heal, 105.108

sharing venture, and so we must accept that he was attracted by the prospects of gaining an edge over his fellow opticians. It must, therefore, have come as a great shock to discover that his partner had been more than economical with the truth when a visitor (probably Bass, as already described) called at his shop to inform him about one Chester Moor Hall who had combined flint and crown glasses to produce an achromatic telescope objective long before Dollond's claim to originality.

Now, if Watkins had truly been ignorant of Hall's priority – a detail also corroborated by Addison Smith in his affidavit of 1765 – then the effect of this revelation must have been very profound indeed. How Watkins actually approached his partner can only be imagined although we do have Dollond's admission from Watkins's testimony of 1765:

> [Dollond] then admitted that such Object Glasses were not his Invention and that the same was the Invention of the said Chester Moor Hall and the said Dollond then added that he did not know that such Invention had ever been made Publick or even Communicated to any of the Trade ....[7]

No matter what Dollond's pretence to innocence may have been, the plain fact remained that a gross blunder of trade etiquette had been made. Not only

[7] 'The Several Answers of Francis Watkins one of the Defendants to the Bill of Complaint of Peter Dollond, Complainant, 30th November 1765'. TNA: C 12/1956/ roll 19.

indignant and very possibly embarrassed at not having checked out Dollond's credentials beforehand, Watkins was convinced that his new partner had absolutely no right to call himself 'inventor', let alone be allowed to advertise this claim, as was the original joint intention. Watkins's understanding was probably based on the common interpretation of the Statute of Monopolies of 1624 in which an 'inventor' was assumed to be the first to bring an invention forward. In describing his position with regard to Dollond, Watkins did not mince words: his testimony to Chancery records that he was 'greatly deceived' by the man in whom he had placed 'Faith and Confidence that he was as he affirmed'.[8] By a separate attestation, Addison Smith reported that his former master had been 'Deceived and Imposed upon' by the man whose claims to invention were an 'untrue assertion'.[9] Here then was the underlying difficulty: Watkins had been affronted and so his relationship with Dollond was at odds from the very outset of their partnership.

Evidence that their working relationship was marred by a strong element of distrust is borne out by the subsequent advertising of the 'New-improved refracting telescope'. According to the reciprocity agreed in the partnership, Dollond and Watkins were to advertise, not jointly, but in a like manner. On 18 January 1759, however, an advertisement in the *Daily Advertiser* (clearly placed by Watkins) ran: 'Mr. Watkins begs Leave to observe, that the Telescopes are found to be of great Use to the Gentelmen [*sic*] of the Army.' Unlike the standard wording used previously, this one failed to acknowledge Dollond as inventor and so it appears as an indicator that Watkins wished to avoid any further commitment in print to an assertion which he could not honestly support.[10]

John Dollond's Conversation with his Enemy, William Eastland

Of the group of City opticians who met at Ayscough's, it is William Eastland on whom we shall concentrate here. The reasons for this will become clear. Eastland must have reflected upon what had taken place at the meeting with other interested parties and, subsequently, became alarmed at the prospect of losing that part of his trade (refracting telescopes) that had become increasingly profitable since learning the formula for making such compound glasses from Ayscough in 1752. Resolved in what he must do, he met with John Dollond for a second time and declared that he would ignore the patent and its associated

8 Ibid. (Watkins).

9 'The Several Answers of Addison Smith one of the Defendants to the Bill of Complaint of Peter Dollond Complainant, 29th October 1765'. TNA: C 12/1956/roll 19.

10 This conclusion was reached after discussions with John Millburn (1925–2005), who first noted what otherwise appeared as an anomalous advertisement.

rights. In explaining his decision – again according to his sworn deposition at Chancery – Eastland testified:

> ... he [Eastland] this Defendant had made and Sold many such Telescopes and should continue to make and Sell the same as usual or to that or the like purport or effect ... [in answer to which] ... the said John Dollond said it was very well for that he did not care who made them or used words to that or the like Effect....[11]

From Dollond's response, it might be supposed that he was willing to turn a 'blind eye' to what Eastland proposed by allowing him to continue making telescopes with compound glasses without application for licence or other agreement. However, we should contrast this with the corresponding remark from Watkins's testimony:

> ... [he] the said Dollond proposed to this Defendant [Watkins] to bring an Action against the said Mr Eastland for making and vending such Tellescopes the same being as he alledged an Infringement upon the aforesaid Letters Patent Granted to the said John Dollond...[12]

Could Dollond really have been so obscure with Eastland as to lead him to commit further infringements and so, ultimately, to bring an action against him? Certainly, his intentions were plain: he wished to sue Eastland in order to establish the rights of monopoly protected by the royal grant of patent; what is more, he expected Watkins to back him in such action.

The apparent conflict between these two statements – of 'not caring' on the one hand and 'wishing to bring an action' on the other hand – is resolved immediately when it is understood that Watkins had already intervened. In fact, Watkins's attitude towards Dollond became very severe following Dollond's confession that Chester Moor Hall was the original inventor. In no uncertain terms, Watkins impressed upon Dollond that since he had already admitted that Hall had precedence then he must not commence any action against Eastland. It was Watkins's sincere belief that because Dollond held no priority *in truth*, then so also '[he, Watkins] was fearful that such Action [as proposed against Eastland] could not be Supported', that is, it could not be supported *in law*.[13] This particular subtlety, perceptively seen by Watkins,

11 'The Several Answers of James Champneys, William Eastland and Christopher Stedman, Defendants to the Bill of Complaint of Peter Dollond, Complainant. 29th October 1765'. TNA: C 12/1956/roll 19.

12 'The Several Answers of Francis Watkins one of the Defendants to the Bill of Complaint of Peter Dollond, Complainant, 30th November 1765'. TNA: C 12/1956/roll 19.

13 Ibid.

was later to turn into an issue of colossal proportion. For the moment, as bound partners, it was important to appreciate that any attempt to take action against Eastland might end in failure, thus besmirching both their names. Watkins prevailed and Dollond desisted: so, with no action taken – neither against Eastland, nor anyone else in John Dollond's lifetime – the partnership survived.

To all outward appearances, the respective businesses – Dollond's in the Strand and Watkins's at Charing Cross – proceeded in their arrangements according to the terms stipulated in the co-partnership. Watkins bought in compound object glasses made by the Dollonds, fitted them into tubes and sold the refracting telescopes. Details from extant ledgers of the bankers Drummonds show that Watkins paid Dollond £30 13s in 1758 and £47 in 1759.[14] It is assumed that these payments were either to do with balancing the books after reconciliation of profits or, perhaps, as direct payments for the compound glasses bought in from the Dollonds. Either way, the figures are not large and, possibly for this reason, Watkins and Dollond alike may have allowed their disagreement to pass by without further fuss.

Beneath this veneer of good practice, however, Watkins may have been seething even though Dollond complied by holding off from exercising his rights of patent. The prospect of having to endure this position for the 14 years' duration of the patent must have vexed Watkins and maybe even more so when the feelings of other traders in the field became apparent to him. Up to this point, Watkins had built a highly commendable business although he was now embroiled in a partnership from which he could not extricate himself without penalty. These were dark years indeed, and worse were yet to follow.

Dissolution of the Co-partnery

Peter Dollond's exceptional skills and business acumen, together with the head start given by his father, helped to establish the shop at The Golden Spectacles & Sea Quadrant in those early years of the 1750s. Following his father's death on 29 November 1761, Peter became successor to his father's estate (including the rights of patent) by 'letters of administration' since his father had died intestate.[15] It followed, therefore, that Francis Watkins became his partner, legally bound by default. However, there were signs that Peter did not wish to maintain such a connection: indeed, the two instrument makers were soon set on a path leading

---

14 Drummond's ledgers. RBS: MS (microfilms): Dr/427/37 (1758), fol. 402 and Dr/427/39 (1759), fol. 504.

15 Cited in TNA: C 33/427.

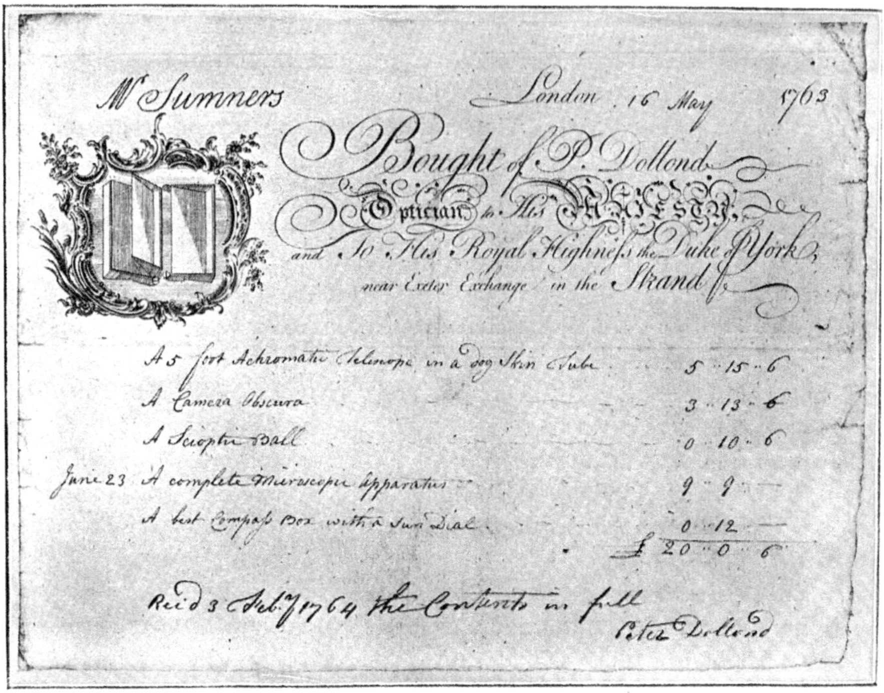
Mr Summers London 16 May 1763
Bought of P. Dollond
Optician to His MAJESTY,
and To His Royal Highness the Duke of York,
near Exeter Exchange in the Strand

	A 5 foot Achromatic Telescope in a dog Skin Tube	5 · 15 · 6
	A Camera Obscura	3 · 13 · 6
	A Scioptic Ball	0 · 10 · 6
June 23	A complete Microscope Apparatus	9 · 9
	A best Compass Box with a Sun Dial	0 · 12
		£ 20 · 0 · 6

Rec'd 3 Feb. 1764 the Contents in full
Peter Dollond

Figure 6.2 Peter Dollond's billhead showing the achromatic triplet as his sign, 1763. © The Trustees of the British Museum, Heal, 105.37

to their separation.[16] Difficulties had first arisen when Peter Dollond proposed to augment the price of telescopes in order to achieve a clear 200 per cent profit. Such a huge increase – originally stipulated at a margin of 60 per cent in the partnership agreement – was sheer profiteering and Watkins refused to comply.

As is often the case in partnerships, disputes, conflicts and changes of vision arise which make it impossible to continue under the terms of the original agreement, however lucrative and perfect the business plan may have seemed at the start. Such was the case when Peter Dollond – freed from his father's influence – wished to promote the business without hindrance. His reasons for this are not difficult to find. In the early 1760s he had improved the achromatic telescope by the simple expediency of reversing the glasses of the telescope

[16] One sign, pertinent to the direction in which Dollond was moving, appears in Mortimer's *Universal Director* for 1763 p. 53, in which he is listed as 'Optician to his Majesty and the Duke of York, and sole Maker of the Refracting Telescopes, invented by the late Mr. John Dollond who obtained his Majesty's Royal Letters Patent for the same invention'. This entry was probably submitted well in advance of its appearance in print and therefore well before of the King's Bench hearing in May 1763.

objective, that is, by placing the crown glass facing the incoming beam with the flint glass on the inner side rather than vice versa, crown glass weathering better than flint glass;[17] other testing of achromatic combinations continued until about 1765, by which time he had announced an achromat with three compounded glasses (achromatic triplet).[18] It is for reasons such as this that Peter Dollond eventually earned the title 'father of practical optics'.[19] Watkins took no part in any of this research although, under the terms of the original partnership, he would stand to reap any profit resulting from any increased sales. From Dollond's standpoint, therefore, it became necessary to rid himself of the burdensome partnership, a hangover from his father's time.

It is the purpose of any agreement of partnership, then as today, that the interests of each party should be safeguarded. Eventualities, such as irreparable disagreements, had been catered for in the original articles of co-partnership although, given Peter Dollond's new horizons, he sought to close the partnership down rather than to vary the agreement. To achieve this, either he or Watkins had to instigate proceedings towards cessation according to the formula for 'Parties in Difference' written into the original binding. Not wishing to appear the blameworthy partner, therefore, Dollond seized upon the notion that Watkins had lapsed in balancing the accounts since his father's death. Indeed, the Drummond's Bank ledgers do indicate that Watkins made no further payments to the Dollond account after the 1759 payment. However, the allegation that he had defaulted in payment thereafter was flatly denied by Watkins who responded that it was he who 'did not seek for one Moiety of the said Tellescopes made and sold by him since the last accounting with the said Complainants said father'.[20]

Unfortunately, the records of 'Reckonings Dealings and Doings', stipulated as one of the 'proper books' in the co-partnership document, do not survive. Thus, it is not possible to determine from that source whether Watkins had lapsed in account reconciliation. From the later document of legal settlement in the actual dissolution process, however, it appears that the balance of payments should have been the other way round for it was actually Dollond who owed money to Watkins. In his affidavit, Watkins stated:

---

17 More precisely, Sorrenson (2001) pp. 38–39, explains that the flint glass (originally) was plano-concave with the flat side outwards to meet the incoming light, whereas the improvement employed an aspherical concave flint glass placed on the inner side.

18 Dollond (1765), p. 54, see Jaecks (2010).

19 King (1955), p. 161.

20 'The Several Answers of Francis Watkins one of the Defendants to the Bill of Complaint of Peter Dollond, Complainant, 30th November 1765'. TNA: C 12/1956/roll 19.

> ... [that] after the said Intestates [John Dollond's] Death Neglecting and refusing to settle with this Defendant [Watkins] in the same manner his said late Father had done and *the said Complainant* [Peter Dollond] *being Indebted to this Defendant* and this Defendant thinking his Interest in the said Letters Patent of but small value as it appears to be Obtained upon an Untrue Allegation of the said John Dollond and this Defendant not Choosing to have any Law Suit with the said Complainant touching the matters Depending and Unsettled between them did at or about the time in the said Bill Ment[d] agree to accept of the said Complainant a Bond bearing Date the Sixth day of May one thousand and seven hundred and sixty three....[21] [Author's italics]

If Peter Dollond was 'indebted to [Watkins]' as asserted here, then technically the original premise for the dissolution was false. Not wishing any confrontation, Watkins refrained from any counter action but rather took a higher moral path by allowing the partnership to be abrogated as Peter wished.

On the same day that the partnership ended (6 May 1763), a new 'article of agreement' was set in place. This stipulated that in recognition for relinquishing his part in the partnership and 'all rights and interest in the patent' Watkins would receive a £200 settlement – a sum equal to that given by him to John Dollond on joining profit-sharing scheme – plus a further £30 '... being the Defendants Share of the Profits of the said Copartnership to the day of the date of the said Articles'. The latter detail is of some importance because it vindicated Watkins's position, namely, that he was correct and scrupulously accurate in his earlier assertion about Peter Dollond's 'indebtedness' in the matter of balancing the accounts. Thus, it would appear that he had been presumptuous in commencing the dissolution on a basis that Watkins was at fault. With regard to the £200 returned 'in full Satisfaction of all this Defendants Right Tithe and Interest in the said Letters Patent...', Watkins's only remark was that had the patent truly been valid – which he firmly held was not the case – then Peter Dollond's new profit margin of not less than £200 per centum would have yielded an estimated profit of not less than £800 for that year (1763) alone.[22] Indeed, Peter had been excessively greedy while Watkins had to content himself that he had taken the moral high ground by accepting absolutely no more than that part which he had invested into the partnership at the outset.

News of this dissolution must have reverberated around the various instrument shops throughout the City and Westminster. Those who made up the new telescopes may not have been surprised; others, alarmed that Peter Dollond would now threaten any of those who infringed the patent, could only wait to see what next would transpire. In the meantime and quite understandably, Watkins

---

21 Ibid.

22 Ibid.

made it plain to fellow opticians (and customers too) why he had agreed to withdraw from the partnership. Of course, he would wish to quash any rumours that might have arisen over inappropriate accounting, but by returning to the issue which had irked him from the very start, namely, John Dollond's claims to invention, he had, effectively, cast aspersions on the whole Dollond family and their business:

> … he [Watkins] hath *given out* that the said John Dollond was not the Sole or first Inventor of the said Object Glasses and that the same were several years before the said Patent was Granted Publickly Sold in England and this Defendant [Watkins] doth now Insist that the said Object Glasses were not the Invention of the said John Dollond but of the said Chester Moor Hall Esquire who many years before the said Letters Patent were Granted *Communicated the same to several persons of the Trade* and the same were many years before the Date of the said Letters Patent Publickly Sold in London as aforesaid and in Particular by the said William Eastland.[23] [Author's italics]

Having thus 'given out' what he knew about the deceased's false claims, the articulated 'wrongs' by which the patent had been obtained quickly became common knowledge among the more alert members of the trade. It was on this basis of understanding that Eastland, Stedman and Champneys (the City three) and Watkins and Smith (the Westminster pair), for different emphasis of reasons, chose to ignore Peter Dollond's subsequent threats.

First Action at King's Bench: A Case of Trespass

Furious with Watkins over his public declaration that the patent was invalid due to Chester Moor Hall's apparent priority, Peter Dollond was forced to act quickly in order to defend the privilege bestowed by his patent. Positions were set, and only by asserting his ascendancy in a public way could he establish his position. In short, a battle had to be fought through the court. By choosing to ignore the terms agreed by the bond of 6 May 1763, set up at the closure of the Dollond–Watkins partnership, which forbade Watkins to sell any more telescopes containing the compound object glass, Watkins and his partner Addison Smith were deemed to have defaulted. According to Peter Dollond, Watkins and Smith held a large stock of these glasses which must have been bought before John Dollond's death. It was, therefore, an easy matter for them to continue making up the telescopes and quite in order to do so from the standpoint of Watkins who declared the patent to be invalid. But what was

---

23 Ibid.

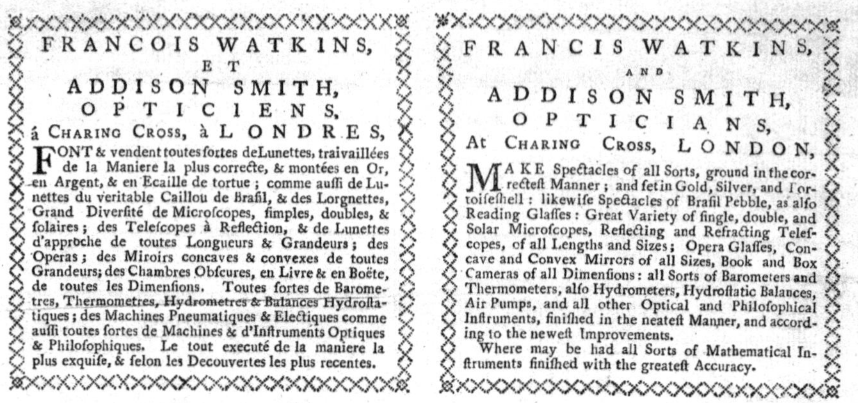

FRANCOIS WATKINS,
ET
ADDISON SMITH,
OPTICIENS,
á CHARING CROSS, à LONDRES,

FONT & vendent toutes ſortes de Lunettes, traivaillées de la Maniere la plus correcte, & montées en Or, en Argent, & en Ecaille de tortue ; comme auſſi de Lunettes du veritable Caillou de Braſil, & des Lorgnettes, Grand Diverſité de Microſcopes, ſimples, doubles, & ſolaires ; des Teleſcopes à Reflection, & de Lunettes d'approche de toutes Longueurs & Grandeurs ; des Operas ; des Miroirs concaves & convexes de toutes Grandeurs; des Chambres Obſcures, en Livre & en Boëte, de toutes les Dimenſions. Toutes ſortes de Barometres, Thermometres, Hydrometres & Balances Hydroſtatiques ; des Machines Pneumatiques & Electiques comme auſſi toutes ſortes de Machines & d'Inſtruments Optiques & Philoſophiques. Le tout executé de la maniere la plus exquiſe, & ſelon les Decouvertes les plus recentes.

FRANCIS WATKINS,
AND
ADDISON SMITH,
OPTICIANS,
At CHARING CROSS, LONDON,

MAKE Spectacles of all Sorts, ground in the correcteſt Manner ; and ſet in Gold, Silver, and Tortoiſeſhell : likewiſe Spectacles of Braſil Pebble, as alſo Reading Glaſſes : Great Variety of ſingle, double, and Solar Microſcopes, Reflecting and Refracting Teleſcopes, of all Lengths and Sizes ; Opera Glaſſes, Concave and Convex Mirrors of all Sizes, Book and Box Cameras of all Dimenſions : all Sorts of Barometers and Thermometers, alſo Hydrometers, Hydroſtatic Balances, Air Pumps, and all other Optical and Philoſophical Inſtruments, finiſhed in the neateſt Manner, and according to the neweſt Improvements.

Where may be had all Sorts of Mathematical Inſtruments finiſhed with the greateſt Accuracy.

Figure 6.3 Watkins & Smith trade card, 1768. © The Trustees of the British Museum, Heal, 105.107

arguably in order for Watkins and Smith was treated as sheer defiance by Peter Dollond. Thus, on the grounds of trespass, simultaneous actions were brought against both these opticians.

In law, there was nothing unusual about this course of action: 'trespass' covered any transgression less than treason or felony and, in the case of Watkins and Smith, their trespass had been against property, that is, the patent. The relevant choice of court through which to file this action was King's Bench since the alleged infringement had taken place against a patent of the Crown. This was the most superior court in the land, taking precedence over other courts of common law, such as the court of Common Pleas. At this time, King's Bench had as its Lord Chief Justice the Right Honourable William Murray, an eminent and distinguished individual, better known in political spheres as Lord Mansfield.[24] Mansfield, reputedly, had the best and fairest legal brain in the land and, furthermore, he took an active role in the on-going task of refining the law, especially in that applied to patenting which had begun to take on a growing importance. There is, therefore, a sense in what follows of an attempt to be fair in judgement while, at the same time, adding something very new to the meaning of the term 'invention'.

In English law, *trespass* implies 'violence' which may be *actual* or *implied* and the law always implies violence although none is actually used. This was the situation with regard to Watkins and Smith whose trespass had been committed – it was claimed – with the intention to *injure* the plaintiff.

[24] William Murray (1705–1793), www.oxforddnb.com/view/article/19655 accessed 1 July 2011.

When the day of trial arrived during Trinity term of '4 Geo.III' (the fourth year of the reign of George-III) only four weeks after the dissolution of the partnership – that is, on 12 July 1763 – there was, in Peter Dollond's place, John Walton his attorney, and in the place of Francis Watkins and Addison Smith, John Ramsden, their attorney.[25] Smith and Watkins were dealt with separately, and in that order. Both cases opened with a charge of 'counterfeit' which, in Watkins's case, ran:

> ... that between 7 May 1763 and the day of exhibiting the bill of the said Peter ... [that] unjustly and unlawfully without Licence or agreement ... the said Francis well knowing the promises but devising and *maliciously intending to injure* the said Peter and further to deprive him of the benefitt profitt and Advantage of the said Invention and the said Letters Patent after the death of John Dollond ... [did make] 500 object glasses ....[26] [Author's italics]

Bringing an 'action on the case' meant that *damages* could be recovered for 'loss or injury' directly resulting from the act complained of (that is, the trespass). But the damage had to be proved and so it was up to Peter Dollond to provide evidence for his counsel and, in turn, up to the jury to consider the weight of argument with, of course, Mansfield's guidance.

So that there was no doubt in the minds of the jurors, it was important to set the scene by explaining the privilege of patent that Peter Dollond had gained from his father. The preamble thus ran:

> ... [John Dollond hath] by application and study and at considerable expense at last invented and brought to perfection a new method of making the Object Glass of Refracting Telescopes by compounding Mediums of different Refrangibility of Light as well as those which were produced by the Spherical surfaces of the Glass were perfectly Corrected that that Improvement which had been in vain sought after by several others above a Century then past would be of great Benefit to the public in General but especially to the Navy where good refracting Telescopes were of the utmost importance ...

---

[25] The 'Law Directory' in *The Universal British Directory 1793–1798* (1799, fac. edition 1993), vol. 1 (part 2), gives John Walton and Joseph Walton at Girdlers Hall, 39 Basinghall Street and John Ramsden (unrelated to Jesse Ramsden) at 64 Margaret Street, Cavendish Square.

[26] Plea of Trespass. *Dollond* v. *Smith* and *Dollond* v. *Watkins*, Michaelmas Term (1763). TNA: KB 122/320, fols 109–110 (Smith) and fols 111–112 (Watkins), resp.

and, continuing:

> ... [that] his Execs Admins and Assigns might lawfully have and enjoy the whole profit benefitt comodity and Advantage from time to time coming growing accruing and arising by reason of the said Invention for and during the said fourteen years according to the form and effect of the said Letters Patent .... [27]

Watkins and Smith had denied Peter Dollond his lawful rights and so this was the basis upon which Peter's attorney wished the plea of trespass to proceed.

With this charge in hand, Watkins's and Smith's attorney (John Ramsden, acting for each similarly though individually) prepared a defence upon the sure belief that Chester Moor Hall's priority would negate entirely the premise set out in the plaintiff's plea. In summary, Ramsden's argument was threefold: that his clients had a right to continue making the compound object glasses because they were the invention of Chester Moor Hall, not John Dollond; that should it be necessary, Hall himself could attest in court to what he had made in and since 1733; and that there existed a certain (but unspecified) publication describing Hall's invention, thus corroborating the priority. On this basis it was argued to be a matter of straightforward deduction that the Dollond patent must be 'void and of no Effect as the same were obtained upon Untrue Suggestions'.[28] Nothing further seemed necessary and so, Ramsden pleaded 'Not guilty' to the charge on behalf of his clients, each in his turn. [29]

Prior to the trial – in the course of seeking evidence for their defence – Watkins and Smith, together with the Strand mathematical instrument maker John Bird who knew Hall as a former customer, visited Hall at his home in Rochford, Essex, in 1763.[30] From subsequent actions and correspondence, it appears that Hall had allowed his visitors to examine several of his crown–flint object glasses, even donating one to his visitors, possibly intended for use as exhibit at the trial;[31] that he had spoken of a publication referring to his invention;[32] and that he had

---

27 Ibid.

28 'The Several Answers of Francis Watkins one of the Defendants to the Bill of Complaint of Peter Dollond, Complainant, 30th November 1765'. TNA: C 12/1956/roll 19.

29 Expressed in the formal language of the court: 'Francis Watkins Attorney comes and says that Francis Watkins is Not Guilty of the premises above laid to his charge in manner and form as the said Peter above Complains ... and he [F.W.] puts himself upon the Country [i.e., the Jury] and the said Peter doth the like.' Note 26, fol. 112.

30 Extract of letter from Francis Watkins to Jesse Ramsden. RS: L&P decade IX, letter 138 (n.d.). RS: MS L&P decade IX, letter 139 (4 June 1789). Addison Smith was at pains to state the year '1763' no less than three times in his letter to Jesse Ramsden.

31 Ibid.

32 Possibly, Smith (1740); see Chapter 4.

actually attended court (probably by subpoena) to act as witness on behalf of Watkins and Smith. As it happened, Lord Mansfield did not consider Hall's priority as original inventor to be in any doubt but, rather, he held there to be a distinction between the commercial act of 'vending' and Hall's private act of 'invention'. In principle, it was ruled that profit by patent was to do with public benefit and Hall had not succeeded with his invention as such or, at least, he had not been successful in developing it to the point of commerce. Thus, although Hall was present at court, he was not called upon to explain any of his attempts to involve instrument makers in its commercial application.

With this direction from the Chief Justice, Mansfield's jury inevitably returned that Addison Smith was 'Guilty' of the charge in hand and then, the process was repeated for Watkins who was also pronounced 'Guilty'. In passing, we can see how Mansfield had improved the process of law by involving ordinary merchants as jurors.[33]

With regard to the carefully prepared evidence, with Hall held in readiness to bear witness – possibly with an account of the publication alluded to by Watkins but not known today, and maybe with John Bird in attendance to add authenticity, little of this preparation was needed after all. Of course, the outcome must have been an immense disappointment to John Ramsden (the defending counsel) who had prepared his brief solely on the basis that the concept of a compounded object lens was not Dollond's privilege but, rather, that of the original inventor as commonly understood and imagined to be enshrined in the Monopolies Act of 1624. Sadly, perhaps, a much more sophisticated argument had been wanting: for instance, in the absence of Ayscough who surely had taken to the grave the sort of clinching evidence needed to prove that Hall-type objectives had already been adopted commercially, it might have helped the defendant's case had (say) William Eastland been subpoenaed to testify along these lines although, of course, he might well have baulked at being placed in such a perilous position. But even Ayscough had only hoped for customer loyalty because he could neither afford the time nor cost to develop and pay for a patent. When, eventually, the fines and costs were levied – on 'the Monday next 15 days after the morrow of St Martin' (29 November 1763) – Lord Mansfield considered the defendants' positions very leniently, setting damages at only one shilling each, with 40 shillings added for costs to the court and Peter Dollond's costs set at £22 19s and £24 9s against Smith and Watkins, respectively.[34]

To an outsider it might have been supposed that these rulings had settled the dispute for all time – after all, the judgement had been passed in the highest court. So, on the one hand, Dollond had won the right to reap the benefits

---

[33] Official list of jurors. GL: Dollond papers, CLC/444/MS14805/001.

[34] Plea of Trespass. *Dollond* v. *Smith* and *Dollond* v. *Watkins*, Michaelmas Term (1763). TNA: KB 122/320, fols 109–110 (Smith) and fols 111–112 (Watkins), respectively.

of his father's patent while, on the other hand, Francis Watkins – though far from happy with the outcome – probably took solace from the fact that the fine was minimal. Lord Mansfield undoubtedly realised what effect his radical judgement – based as it was on an entirely new principle favouring the world of commerce rather than that of the lone inventor – would have on the sensitivities of the trade. To judge in favour of the patent was to confirm the validity of a monopoly and so, inevitably, restrict the trade of the opticians. Although the case against individuals (determined by the hearings against Watkins and Smith) had been lost, there still remained one ancient route by which the trade could assert its discontent; that is, it was perfectly valid for a City company to lodge an appeal directly to the Crown – by whose Great Seal of the British Isles the patent had been issued in the first place – that the said patent was invalid and so should be withdrawn.

Historically, control of the trades within the City limits was governed through the various City companies. While some of the larger companies, such as the Grocers' and Ironmongers', had begun to relax their ancient laws by the end of the seventeenth century, there were others – mostly the smaller companies – who were still committed to enforcing control. In the earlier eighteenth century, as in previous times, a master could be reported for selling inferior goods or an inspecting Warden could confiscate goods deemed to be of poor quality.[35] The Spectaclemakers' was among those Companies still feeling that masters and freemen should be reminded of the more important parts of their duties, and its Quarterly Court selected those opticians who were due for inspection. But nowhere in this reaffirmation of the laws, by-laws and procedures was there any specific rule which could allow the company to exercise its control over an individual holding a king's patent. Somehow, Dollond's threat to company members had to be blocked and therefore an attempt had to be made to have the patent withdrawn. To put such a plan into action would entail a petition to the sovereign, and since Francis Watkins was fully aware of all the wrongs of the Dollond patent, and sufficiently motivated to lead such an action, it was inevitable that he should lead the Spectaclemakers' Company on a campaign for justice.

Serving at the Quarterly Court

In the earlier part of his career, Francis Watkins had very little to do with the Spectaclemakers' Company protocols, save that of dutifully paying his 'quarterage' and attending the 'Quarterly Courts' when he was obliged to do so,

[35] Clifton (1993a), p. 347.

for example, just before taking on an apprentice.[36] Nevertheless, during 1757 – while George Bass was Master – he did gain some experience as an 'Assistant' to the court and, later that same year, at a meeting on 5 October, he was elected 'Underwarden'. Unfortunately and inexplicably, he chose to absent himself from that meeting and thus news of his election was relayed to him via one of the company's beadles who summoned him to attend the very next day. For whatever reason, whether through disinterest, reticence, illness or even modesty, he failed to attend. If this was a blatant refusal on his part then it was most unusual and without precedent in the company's annals and so this particular episode may well reflect upon the nature of the man himself.

At the meeting held the following day (6 October) Bass issued the following instruction to be relayed to Watkins at Charing Cross:

> ... at a Quarterly Court this day holden for the Worshipful Company of Spectacle Makers at the Sun Tavern in Ludgate street London you are hereby required to give your answer in writing by the Beadle of the Company whether you will accept of the Office of Underwarden for the Year ensuing and enter into Bond according to the By Laws of this Company and of notice given you thereof the 5th Instant and on your Refusal that this Court may proceed to the Election of another in Your place 6 Oct. 1757....[37]

Whatever the initial reason for his reluctance to attend, Watkins did accept Bass's ultimatum. Had he taken up the position of underwarden in October as requested he would have served alongside Peter Eglington, the second underwarden.[38] As was the custom, however, he took up the position at the year-ending meeting (29 December 1757) and thence served for three years alongside James Tomlinson.[39] Immediately thereafter, from January 1761, Edward Nairne was substituted for Watkins.

A couple of years on, with threats of lawsuits hanging over certain members dealing in compound object glasses, someone was needed to protect the Company and its members against Peter Dollond, an outsider. As already stated, Francis Watkins was the ideal candidate since he had recently confronted him in court and, very probably, he was motivated to conclude the unfinished business

---

36 Watkins appeared at the Quarterly Courts in June 1750 (before taking on Addison Smith) and June 1753 (prior to taking on Henry Pyefinch). He also attended the meeting of June 1751. GL: SMC Court Minutes, CLC/L/SH/B/001/2.

37 Minute 5 October 1757. GL: SMC Court Minutes, CLC/L/SH/B/001/2.

38 The erroneous mention of Watkins's name alongside Eglington in the SMC Court Minutes of 6 October 1757 appears due to the court secretary having prepared the page of attendees before the meeting had actually begun.

39 For some unknown reason, Watkins was absent at the Quarterly Court of June 1759. GL: SMC Court Minutes, CLC/L/SH/B/001/2.

with his enemy. Something along these lines must have been the situation during the summer of 1763 because, by October, while the first King's Bench trial was underway (that is, with Mansfield's judgement already given but costs still pending), Watkins was elected to serve as Master of the Worshipful Company of Spectaclemakers for the following year. Samuel Johnson then replaced Nairne as underwarden; and Watkins's assistants, for his year of office (1764), were Peter Eglington, George Bass, John-II Howe, David Deane, William Eastland, Thomas Ribright, Edward Nairne and Edward Scarlett Jr. Taken together, this court represented as strong and reliable a force as any that had been recorded in the minutes over the previous decades. Eastland's place in the court of assistants at this particular time is relevant because it was he whom John Dollond had in mind for legal action at the outset of the life of the patent in 1758. Five years on and still making the compound glasses, he was (once again) in a dangerous position and so it is more than likely that he proved a useful ally to Watkins in his determination to settle an old score with the common enemy. Bass and Scarlett Jr, too, were likely to have been greatly supportive in this quest.

On 29 December 1763 the company's court convened at its customary venue, the Sun Tavern in Ludgate Street, with Francis Watkins presiding for the first time. The Dollond issue may well have received an informal airing, though nothing was recorded in the minutes. It was the Master's duty to conduct the company's business in a proper manner, although precisely how it should proceed in this case was still unclear. Nevertheless, Peter Dollond's disturbance to the trade was plain for all to see. He had established a monopoly by exercising his patent rights and was threatening to bring action against selected members of the trade. It was clear that he intended to pursue these rights, regardless of the basis (false or otherwise) by which his father had acquired the letters patent. Although his threat of action was a bone of contention among aggrieved individuals, this detail alone could not stand as an argument; his claimed monopoly offered a *necessary* cause for action; and his declared intention to pursue his rights offered a *necessary* and *sufficient* cause in law for the patent to be made void. Here, at least, was the germ of an idea that could be articulated and presented to the Crown's law officers in due course. Although unrecorded in the Company's minutes, these matters must surely have dominated discussion during the early part of 1764.

Headway was made at the next Quarterly Court (29 March 1764) when it was formally moved that a petition should be drawn up and sent to the king, requesting his majesty to withdraw the Dollond patent.[40] As the ultimate source of jurisdiction, the king, through his Privy Council, was accustomed to receive complaints from persons who alleged that they were unable to obtain redress or amendment in the usual courts. However, not since the time of John Garill 100

---

40 Minute 29 March 1764. GL: SMC Court Minutes, CLC/L/SH/B/001/2.

years earlier, in 1664, was there a case when the whole body of a City company rose up by a petition. In that case, it was the company of Gold and Silver Wyre Drawers who objected to a patent that had been proposed for a process of casting silver ingots to make silver lace. Fortunately for the wire drawers, the Lord Chancellor had 'stayed' the progress of the granting process at the Great Seal in order to hear all objectors from the trade. Whether the Spectaclemakers' or their attorney were aware of the Garill case was of little relevance because Dollond's patent had already been granted under the Great Seal and, therefore, compiling a case against it would have to be argued with extra rigour. As Master – with overriding control of the agenda – there seems little doubt that Watkins orchestrated the decision to petition the king, particularly since it was he who authorised the spending of £20 from the company's coffers for an attorney to prepare the petition.[41]

It should be noted, however, that the Spectaclemakers' Company could not petition the king directly without first obtaining approval from the Lord Mayor, the aldermen and the common council.[42] Evidence for this consent is wanting but since the petition went ahead it is assumed that Watkins was successful in this municipal part of the process.

The Petition and its Signatories

Drawing up a petition in a formal manner in order that it might receive proper attention from the Crown's lawyers at Privy Council – the body through which all applications of a non-political nature were passed – was a task requiring professional attention. To this end, the Spectaclemakers' Company engaged Edward Grubb, a City attorney whose chambers were located in the Old Jewry.[43] Grubb executed his commission with diligence and produced a fine document of some 1,500 words. Most eloquently – but in the strict language of the law – he articulated the feelings of the Spectaclemakers' Company and reasons why Peter Dollond should be stopped. With deference to the monarch – to whom it was addressed – the petition requested that King George-III might exercise his privilege to vacate (*annul*) the letters patent originally awarded by his late grandfather, George-II. The document began in the following vein:

41 Minutes (1731–1785). GL: SMC Court Minutes, CLC/L/SH/B/001/2.

42 Burn (1785), vol. III, p. 300 (entry 'petition'). Under other circumstances, the king could be petitioned by seeking the consent of at least three judges or the major part of a grand jury at the assizes or sessions whereas in London permission was sought through the Common Council headed by the Lord Mayor.

43 The Minutes of the Spectaclemakers' Company refer to the attorney by surname only. His full name derived here from Mortimer's *Universal Director* for 1763, has been independently verified by Lesley Whitelaw (Law Society and Temple archivist). Grubb's surname is inscribed at end of the SMC petition document (note 44).

> Peter Dollond son and administrator (under colour of the said patent) hath threatened to bring actions against your petitioners and any others of the trade, who shall make and sell the said glasses ... and the said Peter Dollond is now attempting to establish a monopoly in the said glasses for his own benefit by untrue a colour of the said exclusive grant.

This led on to the specific detail concerning Peter Dollond's desire for an unreasonable profit, which had so irked Watkins prior to the dissolution of the partnership:

> Peter Dollond under Colour of the said Grant is endeavouring in Effect and Consequence to Ingross to himself the Exclusive Right to almost the whole Branch of making Telescopes in General – And if the said Grant were to be Permitted to Continue in force it will not only be greatly Detrimental to your Petitioners in Particular by Depriving them of a *Legal* and much the most beneficial Branch of their Trade but likewise very Prejudicial and inconvenient to Your Majesty's Subjects in general for this Obvious reason that the said Peter Dollond under Pretence of his having the Sole and Exclusive Right of making and Vending the said Glasses Sets his own Price upon the Refracting Telescopes made by him, and Actually Sells them at a much higher Rate than what they have been always Sold at by others in the Trade and in fact at a Profit of not less than Two Hundred Pounds per Cent.[44]

Other details – such as John Dollond's refraining from taking legal actions due to a fear of loss of credit should the patent have been revoked owing to its deceptive claim on originality and the inadequacy of the specification – were not forgotten. It was, in short, a well-constructed fair and honest document and one upon which Watkins set great store.

The petition was delivered from Grubb's chambers into Watkins's hands by June 1764. The next step – before forwarding it to the office of the Privy Council – was for petitioners to add their names. The task of collecting signatures, if not undertaken by Watkins himself, may have fallen to Addison Smith, for it was his name that headed the list. The names collected appear in order of increasing distance from Charing Cross, and so suggest that the petition was taken to individuals rather than relying on their coming to a single venue. Watkins's signature did not appear on the petition, probably because of his neutral position as Master but, in any case, it would not have been wise to sign this document given the other concurrent litigation in which he was embroiled.

In total, 35 signatures were collected. Seventeen of these were well-known members of the Spectaclemakers' Company, including much-respected opticians like John Cuff and James Burton, while the remainder included 'foreigners' to

---

44 Petition to His Majesty. TNA: PC 1/7/94.

the company such as the mathematical instrument makers John Troughton Senior and John Bird, both of whom would give strength to the cause due to their reputation generally and connections with the Royal Observatory. It was not expected that the young Jesse Ramsden would be on the list since he was on familiar terms with the Dollond family and, in any case, he was destined to marry Dollond's sister, Sarah; however, it might have been expected that James Tomlinson would have signed, although, as co-Warden to the Company, he chose, like Watkins himself, to remain neutral.[45] Other prominent instrument makers that would have been welcome signatories – Adams, Ayscough, Cole, Gilbert, Gregory, Johnson, Lincoln, Nairne, Pyefinch, Scatliff and Sterrop – did not appear because they were already bound to Dollond for the sale of telescopes employing the patent object glass.[46] Two vital names, however, were Robert Rew and George Bass, each of whom added a personal note beneath the signature.[47]

With the signatures added, the petition was dispatched to the Privy Council, a standard procedure for anyone holding a grievance over a patent.[48] The Council had its own judicial committee that dealt with the king's business and it was among these law officers that the petition was read on 22 June 1764 and, four days later, referred onwards to the Attorney General. It should be noted that, as principal counsel to the Crown and leader of the Supreme Court, the Attorney General was *ex officio* leader of the Bar and therefore his position was crucial to the whole procedure of the judiciary, as it still is today. Here, indeed, was an important petition – not merely from a complaining individual but from the body of a City company, therefore deserving attention of the Attorney General himself.

It was very unfortunate for Watkins – as Master of the Worshipful Company of Spectaclemakers – to find that action by the Attorney General was delayed due to concurrent political issues and extraordinary governmental manoeuvrings.[49] From careful reading of Watkins's deposition to Chancery

---

45 The remainder of the signatures included: Tycho Wing, Samuel Scatliff, Joseph Linell, William Eastland, George Ribright, John Eglington, David Deane, Benjamin Martin, John Bennett, Nathaniel Hill, Joshua Bostock, Samuel Wright, James Jameson, Joseph Hitch, John Cox, Peter Eglington, William Cole, Francis Morgan, John Cleare, David Drakefield, John Cooke, John Burton, Robert Featley, William Ford, John Davies, John Clack.

46 *Gazetteer and New Daily Advertiser*, 18 July 1764.

47 Petition to His Majesty. TNA: PC 1/7/94.

48 Hulme (1917), pp. 191–192.

49 Several authors have touched upon this important petition, but since its discovery over 100 years ago none have progressed beyond the mere mention of the dates and actions (which are clearly marked on the document). Thus, until now, there has been no interpretation of the lack of progress in terms of the removal of the Attorney General from his office. Therefore, to avoid confusion, all earlier attempts to understand the petition are completely ignored here. See note 50.

regarding Dollond's bill of complaint in 1765, comes an aside relating to the state of the Privy Council's affairs. Specifically, Watkins himself pointed out:

> ... the Necessary Proofs were laid to support the Allegations in the said Petition but the said late Attorney General *being removed from that office before the said Proofs were duly Inquired into* | no report has yet been made ....[50] [Author's italics]

Indeed, Charles Yorke, the incumbent Attorney General, was pressed to resign from his office at the height of the famous Wilkes affair on which he was advisor to the government during the initial stages.[51] Thus, when he departed his post in November 1763 – under great strain and agitation – there appears to have been a significant hiatus before Sir Fletcher Norton took up that office; moreover, William Murray – the very same Lord Mansfield who was Chief Justice at King's Bench – also became embroiled in aspects of the Wilkes affair. In consequence, the petition – clearly deemed to be of far lesser importance than sorting out the internal affairs of government – was shelved for an even longer time than the mere gap while the Attorney General's post remained unfilled. For the Spectaclemakers' Company these complications arose at a very awkward time and, sadly for Watkins, the petition remained in abeyance.

Seeing the rights of the ordinary optical instrument traders whom he represented – and, not least, his personal desire for justice – slip away due to this lethargy of government must have perturbed Watkins greatly. According to the account by deposition of 1765, he went on to suggest:

> ... that if the Proceedings of such reference [i.e., the scrutiny of the Attorney General] were revived the said Letters Patent would be declared void in Law and ordered to be suppressed.[52]

No such revival took place. Thus, whatever glimmer of hope there may have been at first when the petition was conceived and submitted, it vanished entirely over 1764–1765 and with it was lost forever the possibility of the patent being vacated.

---

[50] 'The Several Answers of Francis Watkins one of the Defendants to the Bill of Complaint of Peter Dollond, Complainant, 30th November 1765'. TNA: C 12/1956/ roll 19. This short remark has considerable import and has not been noticed previously.

[51] Charles Yorke was appointed Chief Justice of the court of Common Pleas in 1762 and appears later in the all-important Champneys case (§ Chapter 7). Yorke, Charles, www.oxforddnb.com/view/article/30237 accessed 1 July 2011; John Wilkes (1725–1795) was the radical politician who was tried for seditious libel in 1764: see Wilkes, John, http://www.oxforddnb.com/view/article/29410?docPos=1 accessed 15 January 2014.

[52] 'The Several Answers of Francis Watkins one of the Defendants to the Bill of Complaint of Peter Dollond, Complainant, 30th November 1765'. TNA: C 12/1956/ roll 19.

Without an Attorney General in office, the king's business was suspended. It now appears that, during that gap, a new ruling had been forced by the Privy Council's law officers who may have advised Peter Dollond (as patentee on behalf of his late father) that, as a first step, he should justify his own position by bringing an action for damages in a court of common law. The intention of such an action meant that any failure to succeed could then be construed as suitable grounds for withdrawal of the patent; conversely, if he succeeded then the need to consider the petition would vanish entirely. If this was the case then it would have marked a radical departure from the custom. It was certainly the tactic employed in 1775 when the Privy Council wished to block petitioners from the Board of Ordnance who sought for a patent for the boring of brass to be revoked.[53] Thus, the less transparent case of the Spectaclemakers' petition may well have been the first instance of the application of this new principle; indeed, it does seem that such a shift in the legal process took place at this time because it fits exactly with the course of action taken.

Whatever additional intricacies lay behind the shelving of the petition in the summer of 1764, members of the Spectaclemakers' Company could no longer rely on this means as a direct route by which to invoke the attention of the Crown. Matters to do with the Dollond patent were not resolved by the Spectaclemakers' Company during Watkins's year as Master; nor would they have been resolved had he been elected for a second year of office. That had been a possibility because some members of the court of assistants were willing to extend his Mastership by proposing him for a second term along with two other nominees (James Tomlinson and Peter Eglington). However, the company's democratic court – at their ballot on 3 October 1764 – decided that Tomlinson should follow on.[54] Thereafter, Watkins took no active part in the Company's affairs.[55] On the one hand, therefore, Watkins's year of office had come to nought; on the other hand, Peter Dollond had benefited from the vacant office of the Attorney General.

Whether it was in a fit of pique upon hearing that a petition – instigated by Watkins under the cover of the Spectaclemakers' Company – was to be paraded among the instrument makers, or, whether it was upon learning that it was he who must first assert his position in line with the new requirement of the Privy Council (as suggested above) against any one individual who infringed the patent, or maybe a measure of both reasons, Peter Dollond reacted by taking out

---

[53] Case of Wilkinson's patent (1775). The Board of Ordnance (by letter of 11 February 1775) Application to revoke Patent 1063 (boring of brass). TNA: 1/11, bdle 55. Order of 16 June 1779. Discussed in Hulme (1917), particularly pp. 191–192.

[54] Minute 3 October 1764. GL: SMC Court Minutes, CLC/L/SH/B/001/2.

[55] Except, that is, for SMC minute of 2 October 1765 when Watkins (along with Bass, Cuff, Howe and Scarlett) was proposed in the choice of persons to examine and audit Edward Nairne's accounts.

two actions through King's Bench. The first of these, in June 1764, was against Martha Ayscough (James Ayscough's widow) and the second, in July 1764, was against Francis Watkins. Both had sold telescopes containing the patent object glass and so both were considered culpable. By putting an end to Martha retailing telescopes containing the compound object glass, it was presumably thought that her supplier, Eastland, would cease production too; Watkins's position, by contrast, was far more serious because he had ignored the earlier ruling at King's Bench. By attempting to quash their respective trades in telescopes containing the compound object glass, Dollond was exercising the first ruling at King's Bench and, therefore, fulfilling his obligation to the Crown by showing that he would not tolerate defiance of his rights of patent.

Martha Ayscough and Francis Watkins's Trespasses

The action brought against the widow Ayscough was based on trespass, with the specific accusation that she did '[cause] so to be done a great Number of Glasses of refracting telescopes after the same Manner and Invention mentioned in the said Letters Patent ....'[56] To have acted in such a manner against this widow – who probably had struggled financially since her husband's demise in 1759 – was brutal, but this was commercial war. However, it does appear that Dollond showed some mellowing because of Martha's plight, because he stopped the proceedings 'at the earnest entreaty | request | and desiring' of Martha who sought an alternative solution to facing court and likely ruin. So, without shifting his position, an order was imposed, binding her '... not to use | put in practice | Counterfeit | Imitate | Resemble | Vend | or make any Glasses of Refracting Telescopes upon the same Method ...' otherwise, disregarding the bond, a penalty of £500 would levied.[57] Thus, on 13 June 1764, Martha Ayscough – in a somewhat uncertain hand – signed the binding, so placing herself at Peter Dollond's mercy. One of the two witnesses to Martha's signature was George Bass, a man who surely felt great sympathy and compassion for her position, and who may have been chosen as supporting witness by Martha herself. The second witness, probably chosen by Dollond, was Thomas Whitford, an instrument maker who, like Dollond himself, had been admitted to membership of the Spectaclemakers' Company as a 'foreign brother'.

Shortly after the action taken against Martha Ayscough, Watkins was summoned to King's Bench to answer a charge of trespass. Unlike Martha's case, however, there would be no leeway; nor, as adversary and instigator of

---

56 Dollond's copy of Martha Ayscough's bond. GL: Dollond papers, CLC/444/MS14805/001 (paper 3).

57 Ibid.

the Spectaclemakers' petition would Watkins have expected any latitude. Once more, therefore, Watkins had to place his trust in the fair-minded judgement of Lord Mansfield.

There was no doubt of Watkins's guilt; in Peter Dollond's mind this was an 'open and shut' affair. Watkins had transgressed the previous judgement at King's Bench and so had denied the patentee of his privileged rights and profit. However, at an early stage in the proceedings of 12 July 1764, Watkins announced:

> ... *the validity of the aforesaid Letters Patent would shortly be Inquired into and determined* this Defendant [Watkins] not doubting but that the same would thereupon be repealed and declared void in Law and being obtained on an Unfair Suggestion and allegation and also as no full or perfect Specification of such Invention has yet been Inrolled as aforesaid....[58] [Author's italics]

At this formal disclosure Mansfield was forced to consider all likely consequences stemming from the two principal objections: namely, that the Spectaclemakers' Company had only just then submitted a petition to the Crown requesting that the patent be vacated, secondly, that the patent specification – a mandatory feature of all patents at Chancery since 1734 – was inadequate.[59] On the first point, there was no reason to suppose that attention to the petition was not already underway at Privy Council and so Mansfield had to consider the likelihood of an outcome in the Spectaclemakers' favour; indeed, any judgement made in his court could, very quickly, be overruled by the Attorney General. On the second point, the argument was somewhat weaker because few patent specifications of this period had the rigour that we might expect today although the reasoning offered does indicate that instrument makers could set a standard by insisting that a specification should be 'full, accurate and useful'; moreover, the specification had, in fact, been enrolled (although this may not have been the case when Watkins was preparing his plea to Mansfield). These were the grounds on which Watkins believed he had the liberty to continue making-up and selling the new telescopes. Thus, not wishing to pre-empt a judgement that might emanate from a higher level of the judiciary system, an impasse was reached. Rather than dismiss the case entirely, Mansfield suspended the proceedings against Watkins.

---

[58] 'The Several Answers of Francis Watkins one of the Defendants to the Bill of Complaint of Peter Dollond, Complainant, 30th November 1765'. TNA: C 12/1956/roll 19.

[59] Between 1711 and 1734, less than a fifth of the patent applications were followed up with a specification; after 1734, however, the validity of a patent was made dependent upon lodging a full specification within a stated time (usually four months but later six months). Correspondingly, an appropriate clause was inserted in the grant of letters patent as soon as the petitioner had complied.

Although unsatisfactory from Peter Dollond's standpoint, Mansfield did assist in resolving his difficulties with Watkins by substituting a 'rule of court' by which Watkins's trade in compound object glasses must be terminated for all time. This was a legal device by which control could be exercised over the plaintiff and defendant alike. Under the proviso of simple but very strict conditions, the 'rule of court' stipulated that Peter Dollond must purchase back all remaining stock of compound glasses in Watkins's possession, and that Watkins should 'immediately and absolutely' (that is, from 12 July 1764) suspend all sales of the patent object glasses. To ensure that all ran smoothly, James Short was brought in as mediator 'to make such satisfaction to Mr Watkins as Mr James Short of Surry Street in the Strand shall think proper'.[60] How successful Short may have been as 'go-between' in reconciling the practical aspects arising between the antagonistic pair is unrecorded. Nonetheless, this stop-gap measure – according to evidence presented in the following chapter – must have worked well enough because Watkins later testified under oath that he had, indeed, abided by Lord Mansfield's rule of court.

With Mansfield's 'rule of court', the present chapter closes by noting that Watkins escaped conviction and fine, and Peter Dollond achieved his goal. To draw attention to his position, Dollond then prepared an announcement so that there remained no uncertainty in the public's mind about who exactly was licensed to sell his 'New-invented refracting telescopes'. On 18 July 1764, only six days after the second King's Bench trial, the *Gazetteer and New Daily Advertiser* published the following:

> By His Majesty's Royal Letters Patent, Mr DOLLOND thinks proper to acquaint the Public, that his new invented Refracting Telescopes (which have been found so useful to the Public in general, and to the Navy in particular) are now sold at the following Places only, viz. G. Adams, B. Cole, in Fleet-street; M. Ayscough, S. Johnson, Ludgate-Street; M. Sterrop, St Paul's Church-Yard; C. Lincoln, E. Nairne, H. Pyefinch, in Cornhill; H. Gregory, near the India-House, Leadenhall-Street; J. Gilbert, Tower Hill; D. Scatliff, near Bell-Dock, Wapping; and by the Patentee, P. Dollond, Optician to His Majesty, near Exeter Exchange, in the Strand.[61]

None of the 11 names on the above list appeared as signatories to the Spectaclemakers' petition. The appearance of Martha Ayscough's name indicates that she subsequently held a licence after the *faux-pas* of infringement.

---

60 'The Several Answers of Francis Watkins one of the Defendants to the Bill of Complaint of Peter Dollond, Complainant, 30th November 1765'. TNA: C 12/1956/ roll 19.

61 *Gazetteer and New Daily Advertiser*,18 July 1764 and repeated 19, 21 and 23 July.

The absence of Benjamin Martin from the list confirms that the telescopes he sold in 1762 – termed 'Dalond's refractors' and priced the same as the bonded opticians (one-guinea a foot) – was no more than an artful ploy and certainly not a misprint. The absence of Watkins, Smith, Champneys, Eastland and all other opticians on the Spectaclemakers' Company petition, tells much about the stranglehold that Peter Dollond exercised on the optical trade. At this stage, there had been no formal outcome from the Attorney General's office regarding the Spectaclemakers' petition; nor had Mansfield's judgement at the first King's Bench trial against Watkins and Smith been put to the test. Thus, to assert his authority absolutely, Peter Dollond still had one more hurdle to clear; he had to show, convincingly, that no one could flout the patent or defy the validity that had been sanctioned in 1763 at Mansfield's court. As we shall now read in the following chapter, this was the purpose of the test case against James Champneys.

Chapter 7

Peter Dollond and his Further Disputes with Opticians

Unlike the first King's Bench trial against Watkins and Smith, which brought Peter Dollond the satisfaction of knowing that his father's patent was secure in principle, the second trial had resulted in the imposition of a 'rule of court' rather than a judgement on their trespass. This decision had been taken by the Chief Justice, Lord Mansfield, while awaiting an outcome at Privy Council of the Spectaclemakers' Company's petition to the king. As described in Chapter 6, the petition had been shelved during the removal of the Attorney General from office although the continuing law officers were not entirely inactive during this period. From the subsequent course of litigation – as described in the present chapter – it would appear that Peter Dollond was instructed to test the validity of the patent in a court of common law by suing for damages any one of those who had infringed the patent. By following this procedure he would, in effect, have subjected the patent – already sanctioned by Lord Mansfield – to its first full legal test, since neither the proceedings against Martha Ayscough, which concluded with a bond, nor the second case against Watkins, which terminated with the rule of court, constituted a reaffirmation or confirmation of Mansfield's new principle. Only if such a test case should fail would it be necessary for the Privy Council to reconsider the Spectaclemakers' Company's petition. Conversely, if Dollond should succeed in a test action, then the patent would be established and the petition could be dismissed forever. The burden of proof, therefore, lay squarely in Peter Dollond's hands.

The thinking behind the above procedure – conjectured here to account for the halted action over the Spectaclemakers' petition – did, in fact, become standard practice in the Privy Council later in the century and appears to have taken place for the very first time with regard to the Dollond situation. Before a test action could be placed through the court, it was important to gather as much evidence as possible. The target for the test action had already been declared as James Champneys and so, in preparation for that event, Dollond required sworn affidavits from each of the five opticians whom he knew or suspected to have defied the rights of patent. Having thus secured as much evidence as possible, he was then in a secure position to proceed against an individual in his quest to have the patent re-affirmed. Thereafter, he could exercise his rights as he thought fit, a prime example of which being the action he brought against Henry Pyefinch, discussed below. But in order to understand the process we shall commence at the same point where Peter Dollond began, namely, with the bill of complaint that he issued in 1765 against each of those instrument makers he believed to have infringed the rights of the letters patent.

A Bill of Complaint

To begin an ordinary suit in law, Peter Dollond had first to file a bill – in this case a 'bill of complaint' – addressed directly to the Lord Chancellor.[1] Details of original bills were rarely preserved by the courts or attorneys although, from normal procedure, it is known that this one would have contained the title, giving the description of the court and the names of the plaintiffs and defendants; the address to the Lord Chancellor (who, as Keeper of the King's Seal, was also Master of the Rolls); the statement of the case; and a conclusion, being a prayer asking for the relief for which the bill had been filed in the first place. Bills were even filed where the plaintiff did not know whether the real facts would support his underlying case: that is, they were filed with the intention of finding out facts (sometimes called 'interrogatories'). Such a 'fishing bill' was the ploy here, where Dollond was keen to extract information from those culprits or interlopers whom he believed to have defied the patent. This process was addressed through the Court of Chancery whose function – as the relevant court for all matters concerning letters patent – was 'to hear and examine the petition and complaints of suitors and to give them a remedy by the king's writ fitted to their case'.

Thus began, on 20 September 1765, Peter Dollond's suit against James Champneys, Christopher Stedman and William Eastland, whom we might describe as the 'City three', each of whom had been associated with the Ayscough establishment. The same bill was also filed against Francis Watkins and Addison Smith, the 'Westminster pair'.[2] From the entry books of decrees and orders issued at Chancery, it is known that:

> ... [on this day] it was ordered that the Plt [plaintiff Dollond] have process of Subpoena retble [returnable] immediately to Compell these Defts [defendants] to appear and Answer the Plt[s] Bill.[3]

---

[1] The word 'bill' was sometimes used synonymously with the word 'suit' but strictly, a bill is only the first step in a suit. The legal procedures described in the present chapter were altered by the Judicature Acts (1873–1875).

[2] Complications arise in locating particular cases due to the filing system originally adopted. Records were usually filed by type: thus, 'bills' and 'answers' were filed separately, even though they refer to the same case; likewise, 'judgements' exist separately but, unfortunately, these are wanting in the Champneys case.

[3] Bill for the year '5 Geo-III'. TNA: C 33/423(2), fol. 460.

This bill, although in principle a statement of claim with interrogatory questions, also had the power of a writ of summons. Therefore, on peril of a penalty from the judge, it would have been a brave person who would dare to ignore this command. The process of litigation to prevent these five opticians from engaging in the manufacture or sale of compound glasses during the life of the patent had commenced and, once begun, it would follow its own peculiar course, first leading to a formal injunction, but ultimately towards a test case.

The five opticians – in legal parlance, the 'deponents' – delivered their respective 'answers' to Peter Dollond's plea, in indirect speech (as required), via the public office of the Court of Chancery. First came depositions from the 'City three', each signed and countersigned by John Madocks, their attorney, and submitted 29 October 1765; then came the depositions of Watkins and Smith, duly signed and countersigned by Edward Salisbury Jones, their attorney, and submitted 28 November 1765.[4] The actual bill of complaint containing details of Peter Dollond's original plea – a document undoubtedly prepared in order that it might be argued before a judge – is no longer extant; fortunately, the depositions of the five named opticians – of which those by Watkins and Eastland are extensively employed here – survive and in sufficient detail to deduce the nature and composition of the original complaints. Therefore, we are able to understand the whole issue as it had unfolded from the very start.

Champneys, Eastland and Stedman

From the closely identical responses of the City three, the basic facts and events over the history of the dispute appear in agreement, although each deponent offered some further personal detail. For example, Champneys described that Eastland had sold several compound object glasses to a Captain Richard Prideaux – a detail corroborated later by Jesse Ramsden (see Chapter 8); Champneys also had believed (mistakenly) that the patent was valid only to the point of John Dollond's death. The principal claim, by Eastland and the others, was that the 'pretended' specification of the patent was lacking because it did not offer a 'full and intelligible discovery of the nature of the invention', moreover

[4] Names of attorneys and relevant dates are taken from the five (imperial size) parchment sheets that comprise the 'answers' of the City three and Westminster pair. Francis Watkins's deposition was read on 28 November 1765 and his attorney dealt with it two days later. The official receipt of the deposition at the 'public office' was also inscribed; in the case of Watkins, the document was sworn before J. Lamey at the Court of Chancery, Clerks Office. This is archived under 'Pleadings 1758–1800'. TNA: C 12/1956/roll 19.

it failed to provide an account of 'the manner how it is performed'.[5] The latter detail was of particular importance to the trade because, as Champneys put it, 'there is no substance by which to teach opticians the manner in which they are to obtain the due degree of Refrangibility.'[6] For a patent to pass into the public domain after its 14 years' duration the specification was supposed to be sufficiently explicit to enable an outsider to understand and make up or copy the invention. However, until this time, specifications were very inadequate, even intentionally imprecise, although they were expressed in a concise though superficial way. This assisted a patentee who wished to reserve a modicum of secrecy about an invention. Such a blurring was rarely questioned by the law officers who signed the enrolment document but it was precisely this lack of detail which became an issue among this group of opticians.

Thus, quite apart from the questionable way in which John Dollond had achieved what he did without acknowledging Chester Moor Hall or other insights gained directly from the trade, the 'City three' made their case in response to Dollond's bill in terms of an 'insufficiency of specification' and a lack of 'recipe for reproduction'. These were the 'full and intelligible' criteria demanded by Eastland in his deposition. For example, the Dollond patent had included the following description:

> ... the Convex Glass is made of a medium or substance in which the Difference of refrangibility is not so great as in the substance which the concave is made of....[7]

This, of course, is correct in principle but, as Eastland pointed out, the specification does not go on to indicate by 'what spherical surfaces the Glasses are to be formed in order to teach opticians the manner in which they are to obtain the due degree of refrangibility which is proper for each Glass'. In his estimation, therefore, the patent had failed. This argument – not unlike that given by Champneys – was expanded and used in full force as follows:

> ... this Defendant [Eastland] saith that if the said John Dollond was possessed of a New Invention in making refracting Telescopes | which this Defendant is so far from admitting that he believes he was not | such a New Invention consisted not in making the Object glass of two or more Glasses put Close together | wherefore one is concave and the other convex (that having been practiced long before the said letters patent was granted) | but such new invention must consist (if such

[5] 'The Several Answer of James Champneys, William Eastland and Christopher Stedman, Defendants to the Bill of Complaint of Peter Dollond, Complainant, 29th October 1765'. TNA: C 12/1956/roll 19.

[6] Ibid.

[7] Ibid. From John Dollond's patent, quoted by Eastland.

> there was) in the particular formation of the convex and concave Glasses and the mediums or substances of which they are made in which particular the said pretended specification is *totally silent* and therefore it hath not set forth how the said pretended Invention is to be performed ....[8] [Author's italics]

Minutiae mattered and so being 'totally silent', as Eastland pointed out, could only mean that the patent must be dismissed as 'null and void'. At least, this presumption was the basis upon which he and the others felt they could legitimately ignore the patent. As they were soon to discover, however, this had little impact with the law as it was about to be interpreted.

A further detail of considerable significance, described by both Stedman and Champneys, concerned how the late Ayscough had requested Eastland to prepare object glasses. Taking Champneys's version here:

> ... In the year one thousand seven hundred and fifty two ... Mr James Ayscough then of Ludgate Street did apply to William Eastland ... to work or make for him ... some object Glasses for Refracting Telescopes the Use or Effect intended by which object Glasses was ... for compounding of Mediums of diff[t] Refractions for which purpose one of such Glasses to be white Flint glass Concave & the other Crown glass Ground Convex, And the said Mr Ayscough then wrote the said William Eastland on a piece of paper the principle on which such glasses Acted in words and figures or [to] the Effect following: '*Quere* Suppose a Crown Glass Flatt on One side and Convex on the other sufficient to make it 2 foot focus then with a flint Glass flat on one side and Concave on the other sufficient by adding together make it 3 foot'.[9]

The *Quere* was a crude recipe for an achromatic flint–crown glass combination; it did not specify how Eastland was to correct for spherical aberration although this would have been left to Eastland's own ability and judgement. Importantly, these focal lengths offered suitable measures by which a glass grinder could achieve reasonable results.

The year 1752, when Ayscough first engaged Eastland to carry out this commission, is not stated elsewhere although it was mentioned later by Jesse Ramsden (see Chapter 8). It is a date of some significance because it confirms that Ayscough's own compound object glass – presumably based on the same formulation as that above and, presumably, on Chester Moor Hall's own crown–flint combination – was already known and sold commercially five or six years before Dollond's patent had been registered. Even if John Dollond's prying had been deterred by Ayscough when the two met a year or so later,

8 Ibid. (Eastland).
9 Ibid. (Champneys).

there was certainly ample time for him to piece together crucial know-how. Furthermore, Champneys added the following to his testimony:

> Mr Ayscough sold them publically in his shop & that Wm Eastland, too, continued selling these Telescopes ever since the said year [1752].[10]

Most of the above evidence, and more, is contained in Eastland's deposition to Chancery and since his rank in the trade was better established than Champneys or Stedman, it is given special credence here.

Watkins & Smith

Some of the 'answers' provided by the 'Westminster pair' were different from those of the 'City three' although the complaints to be answered were the same. Watkins was senior to Smith and so it is Watkins's deposition that is mostly considered here, although much the same response may be found in that of Addison Smith. Emphasis was placed on Chester Moor Hall's priority since it was the revelation of his name as 'first inventor' that had been spread by Watkins in an attempt to exonerate himself for the break-up of his partnership with Peter Dollond back in May 1763 (§ Chapter 5). John Dollond was not the 'first inventor' and, for this plain fact, Watkins reasoned that he could not have infringed the patent; at least, he hoped that under the scrutiny of a court of common law, the patent would be declared invalid, thereby vindicating his position. As we already know from Chapter 6, the outcome was not as Watkins had wished. Also, just as with the 'City three', Watkins and Smith drew attention to the inadequacy of the patent by insisting that John Dollond had obtained it on an 'untrue Allegation as also no full and perfect Specification thereof being involved....'[11] But, of this particular detail, Lord Mansfield avoided all scientific and technical matters, thus the question of its *sufficiency* never arose in court.

One complaint posed by Peter Dollond's attorneys centred upon the allegation that many compound glasses had been made without licence. At the second King's Bench trial held on 12 July 1764 it had been ruled that Watkins was 'not to sell any more without licence and Consent', but the question remained as to how many such object glasses had actually been made and sold illegally since that date. Unsurprisingly, Watkins's replied: '[he] positively denies that he hath at any time since the making of the said [Lord Mansfield's] Rule of Court made

10 Ibid. (Champneys).

11 'The Several Answers of Francis Watkins and Addison Smith one of the Defendants to the Bill of Complaint of Peter Dollond, Complainant, 30th November 1765'. TNA: C 12/1956/roll 19.

Figure 7.1 William Murray, 1st Earl of Mansfield, engraving by William Holl, after a portrait by Sir Joshua Reynolds. © The Trustees of the British Museum, 1857,0214.272

or sold any Object Glasses of which the said John Dollond was the sole or first Inventor.'[12] On the same complaint, each of the 'City three' had merely answered that they were not obliged to offer any answer to that question.

Other aspects of the history of events were recorded in the depositions. With regard to the settlement money paid to Watkins by Peter Dollond as part of Lord Mansfield's plan of dealing with the disputants out of court, Addison Smith explained that:

> ... he doth not believe that the said Defendant Watkins would have accepted so small a sum as Two hundred pounds even if paid in ready money if he had not had the fullest conviction that such letters patent were void and could not be supported by Law ....[13]

---

12 Ibid. (Watkins).

13 Ibid. (Smith).

Indeed, Watkins accepted only what he had laid out by way of initial investment into John Dollond's scheme to hold a 50 per cent share in the marketing of the 'New-invented telescope'; to accept anything beyond this would have been tantamount to placing himself in the pocket of the Dollonds. On questions such as this – relating to the 'Westminster pair' alone – the attorney for the 'City three' always cross-referred to the 'answers' provided by Watkins and Smith, thereby demonstrating his ignorance of any dispute that did not relate directly to his clients.

By far the most serious allegation – from Watkins's standpoint – was that Peter Dollond had alluded to his complicity in an 'unlawful combination and confederacy'. Indeed, if any such conspiracy could be divined at all then, in law, this was an indictable offence. Unsurprisingly, Dollond again pointed a finger accusingly at Watkins whom he knew to be the underpinning, motivating and co-ordinating force behind those wishing to discredit the achievement of his father. The very fact that Watkins was Master of the Worshipful Company of Spectaclemakers – a position obtained with an ulterior purpose – was evidence enough for Dollond to assert that Watkins had conspired to cast aspersions on his business. Mindful of this likely objection – or, more likely, on legal advice from Edward Grubb, the attorney at the time of drawing up the petition – Watkins had been careful not to associate his own name with the petition by signature. In any case, as Master of the Company, he chose (or was obliged) to remain neutral.

Neither did any of the 'City three' lend their names to the petition, but since none of them held office within the Spectaclemakers' Company, the accusation of confederacy in Dollond's bill was irrelevant. Consequently it was left to Watkins to answer this particular complaint. This he did, beginning with the issue surrounding the absence of the Attorney General from office:

> ... It is well known among the opticians and mathematical instrument makers generally that such Object Glasses were not the Invention of the said John Dollond deceased that several of them lately applied by Petition to his Majesty to suppress such letters patent which Petition was referred to his Majesty ... before whom necessary proofs were laid to support the allegations in the said Petition but the said late Attorney General being removed from that office before the said proofs were duly Inquired into no report has yet been made upon such reference but *this Defendant* [Watkins] *saith* that he having with the Petition and the Proofs offered to verify the same and also the saith Complainants evidence to support the said Letters Patent *this Defendant doth* therefrom verily believe that if the Proceedings from such reference were revived the said Letters Patent would be declared void and ordered to be suppressed and *this Defendant denies* all Manner of Unlawfull Combination and Confederacy wherewith he is charged without that, that there is any other matter cause or thing in the Complainants said Bill of Complaint contained Material or Effectual in the Law for this Defendant to make

> answer unto and not herein and hereby well and sufficiently answered avoided or traversed or denyed is true to the Knowledge and belief of the Defendant ... which matter and things this Defendant is ready to over maintain and prove as their Honourable Court shall Direct and Humbly prays to be hence Dismissed with reasonable costs and charges in this behalf most wrongfully sustained [signed] Francis Watkins.[14] [Author's italics]

Thus, even in November 1765 (when Watkins submitted this 'answer') the shelved petition was not entirely dismissed because he still hoped that it could be revived. Indeed, it was the only thread onto which he could hold. As for the 'unlawful combination and confederacy', Watkins flatly denied this allegation, perhaps confident in the knowledge that Peter Dollond could not, conceivably, have any material evidence against him personally.

Having secured as much factual evidence as was likely from the five deponents, Dollond was then in a position to apply to the Court of Chancery for an injunction to be served on those who had flouted the patent. However, one very important step had to be taken before proceeding with the plea for an injunction; that is, he had to satisfy the court that the patent was proven by bringing a test action against one of the interloping opticians. In fact, Dollond had already selected his target for a law suit, even before issuing the bill of complaint. Of course, neither Smith nor Watkins would have been suitable candidates for examination due to their legal position already having been determined at King's Bench under Lord Mansfield; neither was Stedman nor Eastland suitable, and therefore only James Champneys remained a likely candidate for interrogation in court.

Test Case against Champneys

Why Peter Dollond should have chosen to take a suit against James Champneys rather than Eastland or Stedman is not immediately obvious. On the one hand, Champneys was not a member of the Spectaclemakers' Company but rather the Stationers' Company; also, he was relatively young, only having begun his trade a few years earlier, and certainly not having the same clout as the well-established Eastland. On the other hand, Eastland, being longer established, might well have reacted more assertively than Champneys had he been selected. Indeed, he would probably have learned – via Watkins – how John Dollond had wished to sue him back in 1758. Stedman's position appears quite separate and, in fact, he was later dropped altogether from any embargo, possibly because he was innocent of any infringement or, quite possibly, working as an employee of

14 Ibid. (Watkins).

Martha Ayscough who had already acquired a permit to sell Dollond's 'New-invented refracting telescopes' by a separate agreement. Superficially, therefore, it appears that Champneys was targeted because of his greater vulnerability. As Champneys himself explained in his deposition to Chancery:

> ... since John Dollond's death the son was attempting to establish a monopoly for his own benefit which this Defendant holds is Contrary to Law as the said Pretended Invention is not a new Invention of the said John Dollond and whereby this Defendant will be hindered and prevented from carrying out his Lawful Trade and this Defendant further saith that the making & selling of Reflecting & Refracting Telescopes is much the most valuable & beneficial branch of this Defendants business ...[15]

Indeed, Champneys did not hesitate to assert that Peter Dollond 'did intimidate him' and act in such a way as '[to prevent him from] carrying on his lawful business'.[16] However, in open defiance and provocatively, Champneys persisted in undercutting Dollond, as will be revealed below.

It may well have been the severity of Peter Dollond's 'intimidation' that caused Champneys to relocate from Red Lion Court off Fleet Street to a shop in Cornhill, well away from the vicinity of the Dollond business. This move took place during 1764, at the first outbreak of acrimony. Dollond's threats, however, followed him and two lawsuits were served: namely, a writ of *latitat* at King's Bench ('returnable on the first day of Michaelmas now next'), and a writ of *capius* at Common Bench ('returnable on the Morrow of All Souls now next').[17] These actions were issued during August 1765, that is, only a few weeks *before* Champneys submitted his deposition to the already described bill of complaint. If it seems strange that he should suffer two writs for the same offence, it should be noted that, in law, their scope was very different. On the one hand, *capias* (meaning 'as you take') was an ancient and customary means of commencing an action at common law by issuing a fictitious writ of trespass. Of the several possible applications of *capias*, it seems that *capias ad respondendum* (meaning 'that you take to answer') was the likely form applied;[18] on the other hand, *latitat* (meaning 'he has hid') was generally reserved for a person likely to leave (or threaten to leave) the district, in which case a defendant having no material wealth and unable to provide security could then be apprehended and imprisoned in order to ensure his appearance on the allotted day for the hearing.

---

15 Note 6 (Champneys).

16 Ibid.

17 Ibid.

18 For *capias ad respondendum* Champneys would not have been arrested but merely required to appear in person at the hearing.

The procedure of a double writ was moderately standard. With the fictitious trespass in place (by *capias*) and Champneys's presence assured (by *latitat*), the case could proceed and the allegation of trespass could be dropped for the real cause of action or, as in this case, retained.[19] With these legal instruments in place, Champneys was destined to follow an unavoidable and inexorable path.

More important than the means by which Champneys was brought to court was his actual response to being served with the lawsuit. From his answer to Dollond's bill of complaint we note that he had already signified his willingness to 'Try the Actions or either of them with the said Complainant', meaning that he was fully prepared to meet Dollond's challenge of Champneys.[20] But, more interesting still was his course of action from the moment that the summons had been issued (August 1765) to just before his trial at Common Pleas (February 1766): namely, he ran an advertising campaign in the *Gazetteer and New Daily Advertiser* stating that: he constructed refracting telescopes using a compound object glass based 'upon the principles of Chester More [*sic*] Hall Esq; whose invention of the compound Object Glass has so much improved this Telescope' and that he sold (achromatic) telescopes much cheaper than those by Dollond, indeed undercutting him by 6 s (2 feet), 10s 6d (3 feet) and 14s (4 feet) for telescopes at these lengths.[21]

On both these counts, Champneys had blatantly acted in defiance of the Dollond patent because of Lord Mansfield's redefinition of an invention in terms of commerce only. Therefore, any claim that Hall may have held (or any claim that Watkins and Smith argued that he held) to the invention was effectively thrown out of court at the hearing of *Peter Dollond* v. *Watkins and Smith* in 1763. With regard to undercutting the price of the 'New-invented refracting telescope' – which sold by licence at 'one guinea a foot' – Champneys was overtly 'depriving' (*injuring* in legal parlance) Peter of his right of patent.

Legal proceedings relating to the Champneys case are now lost although, from the above analysis, it is clear what had to be argued in court.[22] The case

---

19 In her doctoral thesis, Robischon (1983) considered the points of difference between *capias* and *latitat* as follows: *capias* helped to 'speed up' the trial and *latitat* was a preventative measure because Champneys 'threatened to move abroad to Amsterdam'. Although Champneys's second apprentice, John Cuthbertson (*appr*. Stationers' Company 1761) moved to Amsterdam sometime after 1768 and, about the same time, had a partnership with his former master, no evidence is known to suggest that Champneys intended to leave England.

20 Ibid.

21 The actual advertising campaign ran in the *Gazetteer and New Daily Advertiser* from 13 August 1765 to 1 February 1766.

22 Champneys's case was originally archived under the old style: PRO 'H.6 Geo III, 626 Middx' and has been cited as such by several authors in recent times despite the call mark being defunct. Moreover, this document was lost more than 100 years ago and so could not have been studied by those who refer to it as seen! Today, the call mark is: TNA: CP 40/3667 fol. 626 but the relevant folios are still absent from the binding.

REFRACTING TELESCOPES
With a compound OBJECT GLASS
CONSTRUCTED upon the principles of Chester More Hall, Esq; whose invention of the compound Object Glass has so much improved this Telescope, as renders it far superior to any of its kind; as it magnifies much more, and objects are seen through it with that clearness and distinctness, which makes it not only more usefull at sea, but by land also, as all objects appear through it in their natural colours, and not blended with others: a fault not to be remedied in those made with a single Object Glass. These telescopes are made and sold by JAMES CHAMPNEYS, Optician, facing Tom's coffee-House, Cornhill, London, at the following prices, being near 20 per cent cheaper than those sold at other shops, viz.

	l. s. d.
A two foot	1 16 0
A three foot	2 12 6
A four foot	3 10 0

And all other lengths proportionably cheaper.
N.B. Good allowances to merchants, and others who buy to sell again.

Figure 7.2 Example of one of several advertisements by James Champneys showing that he blatantly undercut Dollond's prices for the achromatic telescope: *Gazetteer and New Daily Advertiser* from 13 August 1765 to 1 February 1766

was heard at the Court of Common Pleas – referred to as 'Common Bench' by Champneys in his answer to Dollond's bill of complaint. This court was detached from King's Bench (*Aula Regis*) and, normally, was reserved for civil disputes, often concerning debts or land but not generally for appeals concerning royal patents. In this case, *the patent itself was not under scrutiny*, and so it was not the king's business; in other words, the case was treated as *an action between private persons*. The proceedings took place early in 1766.[23] From a public announcement in the *Gazetteer and New Daily Advertiser*, it is now known that:

[23] *Gentleman's Magazine*, 36 (18 February 1766), p. 102.

> On Tuesday last [i.e., 18 February 1766] came on before the Right Hon. the Lord Camden and a special Jury, a cause wherein Mr Peter Dollond, an eminent optician in the Strand, was plaintiff, and Mr James Champneys, a mathematical instrument maker in Cornhill, was defendant. The action was brought for imitating the plaintiff's patent for making object glasses of refracting telescopes, by compounding mediums of different refractive qualities; when, after a long trial, it appearing to the satisfaction of the Court and Jury, that the plaintiff had been greatly injured, and put to great expence [*sic*] by unlawful combinations to defeat him of the benefit of his patent, a virdect [*sic*] was given in favour of the plaintiff, with £150 damages.[24]

The presiding judge, the Right Honourable Charles Pratt, Lord Camden, had been Lord Chief Justice at Common Pleas since 1761 and although a man well capable of making up his own mind, his only recourse in this action was to apply the previous ruling against Watkins and Smith at King's Bench in 1763 when Lord Mansfield had dismissed the argument in favour of Chester Moor Hall.[25] Thus, despite the evidence presented at court by John Bird and Robert Rew (who were known to have been witnesses at Champneys's case), Camden – in his own peculiar way – reminded Champneys and those at the trial that:

> ... it was not the person who locked up his invention in his scrutoire that ought to profit by a patent for such invention, but he who brought it forth for the benefit of the public.[26]

Thus, for a second time, the argument in favour of Hall's priority was quashed. Despite the missing folios in the judgement rolls for this case, the outcome is known from the above newspaper announcement and, more directly, from the list of actions at Common Pleas during the Easter Term of '7 Geo-III' (seventh year of George-III's reign: 1766). With the fine of £150 imposed, the case closed.[27] By any reckoning, this was a huge sum and probably intended as a warning to others who dared to question this judgement. It is not difficult to imagine what effect this would have on a young man independently operating a small business.

---

24 *Gazetteer and New Daily Advertiser*, 20 February, 1766.

25 Charles Pratt (1714–1794), First Lord Camden, www.oxforddnb.com/view/article/22699 accessed 1 July 2011.

26 *Gentleman's Magazine*, 36 (18 February 1766), p. 102. *Note:* This quaint expression is often (erroneously) attributed to Mansfield.

27 The correct figure of £150 also appeared in the *Public Advertiser*, 20 February 1766. Larger sums (£204 or £250) were reported elsewhere and may have included the added costs and charges for injuries (see *Gentleman's Magazine* [note 23] and Ranyard (1886a), pp. 460–461).

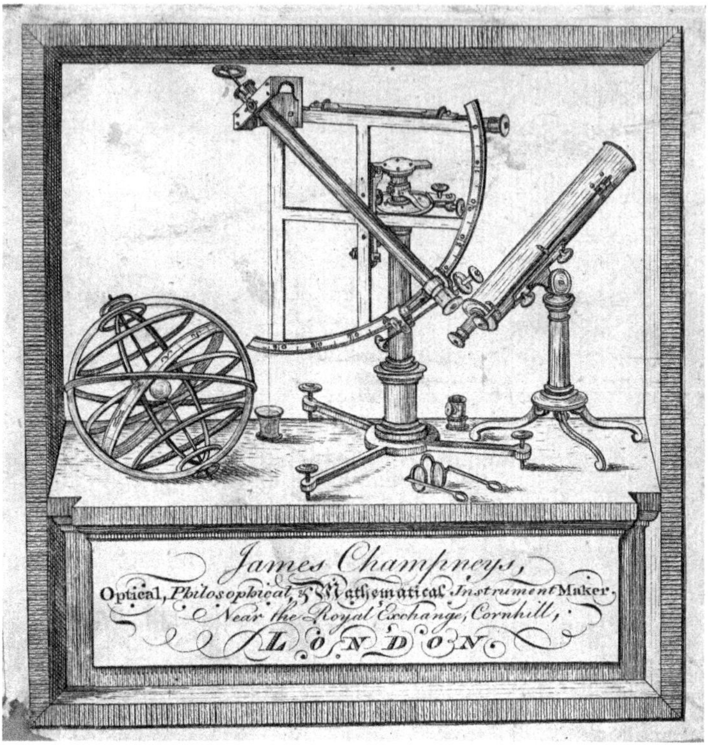

Figure 7.3 James Champneys's trade card, 1763. © The Trustees of the British Museum, Heal, 105.2

Indeed, Champneys was ruined and bankruptcy followed in early 1772.[28] By March 1772, he was forced to sell off the lease to his shop premises and his entire household goods:

> ... his neat and genuine Household Furniture, Linnen [*sic*], China and Books, which were all new within 12 months, and little the Worse for Wear. And Tomorrow will likewise be sold, his entire Stock and Utensils in Trade; consisting of Optical and Mathematical Instruments, amongst which are four large engraved Plates of the Charts of the White Sea; sundry Turning Lathes, and other Utensils.[29]

---

[28] An announcement of Champneys's bankruptcy appeared subsequently in the *Westminster Journal and London Political Miscellany*, 18 January 1772.

[29] *Daily Advertiser*, 20 March 1772.

Champneys's debts were still apparent in 1775 when Peter Dollond – as formal recipient of Champneys's estate and so legal administrator under a Commission of Bankrupt, that is, one of the assignees – advertised for creditors with proven debts to apply at his house for a second dividend.[30]

Camden's elegant judgement was quoted subsequently in the action of *Boulton & Watt* v. *Bull*, and therefore it became a precedent in patent law. Be that as it may, with the *Dollond* v. *Champneys* case settled and found in favour of the plaintiff, Dollond had no qualms about using his established rights. All this became very plain in the year following Champneys's trial when progress was made towards finalising the injunction against his enemies, thereby closing any further potential loopholes.

An Injunction is Placed

With the case against Champneys successfully completed, Peter Dollond appeared before the Court of Chancery with a request for an injunction to be placed against those whom he believed had infringed the patent. Watkins, Smith, Eastland and Champneys – four of the original deponents – were named on the application of 17 July 1767 but not Stedman. Also present at the hearing were the defendants who raised two familiar objections: namely, that John Dollond was neither the first nor sole inventor of the compound object glass, and that he had not enrolled a specification within the permitted period. The first point was no more than the original objection – first raised by Watkins and Smith at their trials in 1763, then duplicated in the petition of 1764, and finally at Champneys's trial in 1766; the second point, however, is interesting because gone was the earlier objection about the inadequacy of the specification and in its place was a new assertion to do with a failure to register the specification within the permitted (three calendar month) period. This switch may have been due to the sufficiency argument being rejected (or being redundant) at Champneys's trial; thus, the alternative complaint here may have been a substituted 'last ditch' attempt in the hope of causing the patent to fail on procedural grounds or, otherwise, simply to maximise Peter Dollond's discomfiture. Whether or not this was the case, John Dollond had, in fact, refrained from enrolling the patent until as late as 12 July 1758, that is, only six days before the expiry of the permitted enrolment period, probably thereby hoping to prevent opticians then from gleaning any detailed knowledge of his invention until the very last minute. Watkins et al. or

30 *Daily Advertiser*, 21 June 1775. Peter Dollond's advertisement referred to Champneys as 'optician, dealer and chapman', the final term meaning 'one who buys and sells', a term applicable to peddlers.

their attorney had not known of this late enrolment and it was therefore a detail which Peter Dollond's attorney was forced to explain, even if to no avail.

From the only extant record of this hearing – based on notes of the proceedings made by the secretary of the court – the reaction to the 'false suggestions' ran as follows:

> … that the Def^{ts} [defendants] pretend that the Pet^{s} [petitioner's] late Father was not the first and sole Inventor of such Object Glasses and that the s^{d} [said] John Dollond did not within the time limited by the s^{d} L^{res} [Letters] Patent inroll or cause to be inrolled in this Court a Specification or Description of his s^{d} Invention and upon those and such like false suggestions the Def^{ts} pretend that the s^{d} L^{res} Patent are void and that they are not bound thereby nor excluded from making and vending such Object Glasses for w^{ch} [which] such L^{res} Patent were granted and actually have and do clandestinely in Defiance of the s^{d} L^{res} Patent make and vend greater Numbers of such Glasses whereby the Pet [Petitioner] is greatly prejudiced and intirely deprived of all Benefit under such L^{res} Patent….[31]

The accusation that these makers did 'actually have' and 'clandestinely make and vend' great numbers of these glasses had, of course, been firmly denied in each of the answers to the earlier bill of complaint.

Peter Dollond's attorneys argued that because they had already recovered a 'verdict at law' against Watkins for making and vending object glasses within the description in the letters patent, they were, therefore, at liberty to call for a ban (*injunction*) to be placed on each and every one of those who had defied the patent and, therefore, prevent any future likelihood of profit by the sale of compound object glasses:

> Therefore in Regard the Def^{ts} [defendants] do still continue to make and vend the s^{d} [said] Object Glasses It was prayed that the Def^{ts} their Servants Workmen and Agents may be restrained by this Inj^{con} [injunction] of the Court from making using exercising or vending the Invention for wch [which] such Lr^{es} Patent were granted to the Pet^{r} s^{d} late Father John Dollond Whereupon and upon hearing of Hm^{js} Att^{y} Genl [His Majesty's Attorney General] of [the Privy] Council with the Def^{ts} Fran Watkins and Addison Smith and Affid^{t} of Notice of the Motion to the Def^{ts} Wm Eastland and James Champneys and the Answers of the Def^{ts} read and what was alledged by the Council for the s^{d} Partys It is ordered that an Injunction be awarded to restrain the Def^{ts} their Servants Agents and Workmen from making using exercising or vending the Invention for sch the s^{d} Lr^{es} Patent were granted to the s^{d} John Dollond dec^{ed} until the Hearing of this Cause
> Friday 17th July [1767] [initials] S M S.

---

31 TNA: C 33/427(2) fol. 296.

With the injunction in place, Peter Dollond lost no time in broadcasting the outcome for others to note. On the following day, the following announcement appeared in the *Gazetteer and New Daily Advertiser*:

> Last term Mr Peter Dollond, optician, of St Paul's church-yard, obtained an order of the High Court of Chancery, for an injunction against Messrs Watkins and Smith, near Charing Cross; Mr James Champneys of Cornhill, London; and other opticians; to restrain them from imitating his new invented refracting telescope, for which he has His Majesty's patent. This, it is hoped, will effectually secure to Mr Dollond the sole exercise of this valuable invention, which he hath not before been able to obtain, notwithstanding diverse verdicts, both in the Court of King's-Bench and Common Pleas, in his favour.[32]

This injunction did, indeed, bring closure to the exasperating pedantry of his enemies. One individual, however, not subject to the injunction was Henry Pyefinch who, up to this point, held an arrangement with Peter Dollond to sell 'New-invented refracting telescopes'. An issue cropped up in 1767 which, once again, caused Peter Dollond to take action through the courts although, on this occasion, it could be that this was an overreaction that should never have gone to court. It was, nevertheless, the duty of the patentee to prosecute anyone who stepped beyond the bounds or conditions set by the working agreement or licence to market the patent.

Further Discord: Pyefinch and Dollond

Peter Dollond's tenure of this patent had depended upon his being either an executor or administrator or assignee to his father's estate. This was the customary and formal provision laid down in the original letters patent in order to ensure its continuity. Since Peter could not be an executor – his father having died intestate – it fell upon him to ensure his proper appointment as administrator. That procedure was carried out through an ecclesiastical court and, therefore, there was no doubt that the ownership of the patent was his and his alone. However, the expression of doubts over the validity of the patent – originally cast by Watkins and Smith at their trial in 1763 – had necessitated these rights being re-affirmed at all subsequent hearings. Such a re-affirmation would also have taken place at the start of the case against Champneys at Common Pleas. Given this background, Lord Camden had no difficulty in

32 *Gazetteer and New Daily Advertiser*, 18 July 1767. Editors' note: both Eastland and Champneys emigrated to the Netherlands after this, where they continued to produce telescopes: Zuidervaart (2011), and Zuidervaart, PC, September 2011.

serving, as requested, an injunction upon four of the five offending instrument makers, that is, excluding Stedman whose position may have been covered by working for Martha Ayscough). Knowing of the lengths to which Peter Dollond was prepared to go to uphold his monopoly, any other trader wishing to vend Dollond's refracting telescopes was well advised to seek permission before embarking on any private enterprise. In practice this entailed entering into a bond which stipulated the conditions under which the maker, Peter Dollond, and the prospective seller could trade. From the account that follows, it may be deduced that such a bond or working agreement existed between Dollond and Henry Pyefinch.

Henry Pyefinch struck a trading agreement with Peter Dollond within a year of leaving Watkins's employment, that is, he was named along with ten other licensees in the *Gazetteer and New Daily Advertiser* for 18 July 1764 (§ Chapter 6).[33] It is even possible that Dollond enticed him from Watkins's employ; certainly, a contract with Dollond would be an attractive proposition for a young instrument maker setting up on his own account; but equally, since this was a difficult time for Watkins and Smith (with the impending action at King's Bench), it could be that Watkins dismissed Pyefinch for divulging information to Dollond or, maybe, urged him to leave in order to allow the newly formed partnership of Watkins and Smith to flourish following the recent entry of Addison Smith into the family on his marriage to Elizabeth Walder (Watkins's wife Clarinda's daughter). It is, nevertheless, curious that Pyefinch dared to hold a contract with Watkins's enemy although, since he worked well away from Charing Cross, he would not be encroaching on the clientele of his former master. We should also remember that Pyefinch did not support the Spectaclemakers' Company petition because, by that time, he held a contract with the Dollond establishment.

The account presented here depends upon both factual and deducible evidence arising from two depositions presented to the Court of Chancery during 1768: firstly, a bill of complaint by Henry Pyefinch against Peter Dollond (during the Easter term) and secondly Dollond's response to Pyefinch's bill (during the Michaelmas term).[34] So, from sometime during 1764, a working arrangement had existed whereby Pyefinch was to be allowed special 'workman's rates' – the term 'wholesale' not yet being in vogue – for the purchase of refracting telescopes for resale.[35] A little more can be said about this working

[33] *Gazetteer and New Daily Advertiser*, 18, 19, 21 and 23 July 1764 (see conclusion to § Chapter 6).

[34] Bill of complaint of Henry Pyefinch (10 May 1768). TNA: C 12/2075/4; Peter Dollond, Answers to the bill of complaint of Henry Pyefinch (17 Nov. 1768). TNA: C 12/2071/11.

[35] The term 'workman's rate' (a special discounted rate) appears in Pyefinch's bill (May 1768), note 34. The loose term 'royalty' (sometimes used by some modern authors) does not appear in any of the documents covered here.

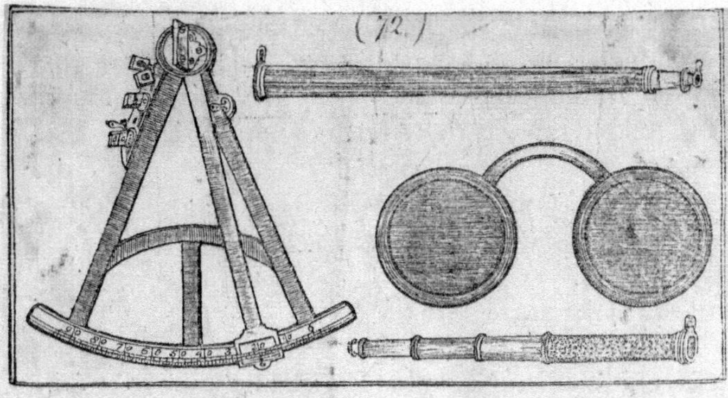

A CATALOGUE

OF

OPTICAL, PHILOSOPHICAL, and MATHEMATICAL INSTRUMENTS,

MADE AND SOLD BY

HENRY PYEFINCH,

At the GOLDEN QUADRANT, SUN, and SPECTACLES, No. 67, between *Biſhopſgate-Street*, and the *Royal-Exchange*, in *Cornhill*, *LONDON*.

	l.	s.	d.
A PAIR beſt ſilver ſpectacles for the temples	0	15	0
Beſt ditto, with double joints	1	4	0
Brazil pebbles, in ſilver frame	1	11	6
Ditto, in a double joint frame	1	18	0
Pebbles in beſt ſteel frame	1	1	0
Ditto in a double jointed frame	1	4	0
Beſt ſteelſpectacles for thetemples	0	6	6
Ditto —— ——	0	5	0
Ditto —— ——	0	2	6
Double jointed beſt ſteel	0	9	0
Beſt ſilver ſpectacles for noſe	0	7	0
Tortoiſeſhell and ſilver —	0	3	6
Spectacles for the noſe —	0	1	0
Spectacle caſes from 1s. to	1	4	0
Concave glaſs for ſhort-ſighted perſons, in horn ——	0	1	6
Same in tortoiſeſhell ——	0	3	0
In tortoiſeſhell and ſilver	0	5	6
Ditto ——	0	10	6
In Pearl and ſilver ——	0	12	0
Reading glaſſes mounted in horn, tortoiſeſhell, tortoiſeſhell and ſilver, &c. from 2s 6d to	1	11	6
Opera glaſſes, from 6s. to	1	1	0
Ditto, nourſa ſkin and ſilver	1	11	6
Larger 2l. 2s. and ——	2	12	6
Refracting teleſcopes for ſea and land, with achromatic moving object glaſſes, 20 inches	1	11	6
Two feet ——	2	2	0
Three feet ——	3	3	0
and ſo in proportion to any length			
Refracting teleſcopes, with ſix glaſſes, two feet long	0	18	0
Three feet, ditto ——	1	4	0
Four feet, ditto ——	1	11	6
Refracting teleſcopes, with 4 glaſſes, 3 feet painted tube	0	5	0
Three feet in mahogany	0	7	6
Ditto ——	0	10	6
Refracting teleſcopes for pocket, of all ſorts and ſizes, from 5s. to ——	1	11	6
Reflecting teleſcopes, for the pocket ——	2	12	6
Fourteen inches long ——	5	5	0
Eighteen inches ——	7	7	0

	l.	s.	d.
Two feet long ——	10	10	0
Ditto, with rack work, finder, &c. ——	16	16	0
Beſt compound microſcope	6	6	0
Ditto, with joint ——	7	7	0
Ditto, with three pillars —	4	4	0
Small ſort, ditto ——	2	12	6
Wilſon's microſcope ——	2	2	0
The ſame adapted for opake objects ——	2	10	0
Solar microſcope compleat	5	5	0
Ditto with magelliſcope ——	8	8	0
Opake microſcope ——	2	2	0
Priſms, 8s. to ——	0	18	0
Magic lanthorn ——	1	1	0
Painted figures for ditto —	0	5	0
Others from 7s. 6d. to ——	0	10	6
Concave and convex mirrors of all dimenſions from 9s. to	31	10	0
Cylindrical concave —	1	1	0
Ditto, larger ——	1	11	6
Cylindrical convex, and its ſix pictures ——	2	12	0
A pocket camera ——	0	10	6
Ditto, larger ——	0	15	0
Camera obſcura, in form of a book, the beſt ——	4	4	0
The ſame adapted to view prints	5	5	0
Diagonal mirror, or optic machine ——	0	15	0
Ditto, for two people —	1	11	6
Ditto, in form of book, —	1	15	0
The beſt coloured views paſted, at per dozen ——	0	18	0
MATHEMATICAL INSTRUMENTS.			
Hadley's octant ——	1	10	0
Ditto, Ivory Ar. and Nonius	[illegible]	[illegible]	[illegible]
Ditto, braſs index ——	2	12	6
Ditto in ebony ——	3	3	0
Small braſs octant ——	2	12	6
Twelve inch ditto ——	5	5	0
Eighteen inch ditto edge bars	10	10	0
Davis's and all other quadrants, from 5s. to ——	0	12	0
Theodolites, from 5l. 5s. to	25	0	0
Circumferenter ——	2	10	0
Plane table, ball, ſocket, and ſtaff	3	10	0
Gunter's chains and meaſuring tapes, of all prices			
Caſes with drawing Inſtruments, from 7s. 6d. to ——	9	9	0

	l.	s.	d.
Protractors from 1s. 6d. to	4	4	0
Proportional compaſſes —	1	5	0
Ditto larger ——	1	10	0
Pantographer in ebony ——	2	12	6
Pantographer, all braſs —	4	14	6
Gunner's callipers ——	1	11	6
Parallel rules, from 3s. 6d. to	2	12	6
Horizontal ſun dials, 6s. to	10	10	0
Ring dial univerſal ——	0	12	0
Ditto, ſix inches diameter —	0	18	0
Nine inches, beſt ——	5	5	0
Ditto on pedeſtal, very curious	20	0	0
Azimuth compaſſes —— and all other ſorts and inferior prices	5	5	0
Artificial magnets, to lift any weight not exceeding three hundred	1	1	0
	2	2	0
	3	3	0
Perambulator, or meaſuring wheel ——	6	6	0
Aſtronomical quadrants, from 15l. 15s. to ——	150	0	0
Globes 17 inches, the pair	6	6	0
Twelve inches ——	3	3	0
Nine inches ——	2	2	0
Six inches ——	1	11	6
Three inches, for pockets	0	9	0
PHILOSOPHICAL INSTRUMENTS.			
Beſt barometer ——	1	11	6
Ditto ——	2	2	0
Barometer with thermometer	2	12	6
The ſame with thermometer and hygrometer ——	3	3	0
Beſt ſort —— ——	4	4	0
Fahrenheit's beſt thermometer	1	11	6
Ditto —— ——	1	1	0
Botannic thermometer —	1	1	0
Ditto for ——	0	16	0
Beſt hydrometer, with all its weights ——	1	1	0
Hydroſtatic balance ——	1	16	0
An electrical machine ——	3	13	6
Ditto, large ——	5	5	0
Electrometers for ditto —	0	10	6
Air pump all braſs double barrel	3	13	6
Ditto ——	5	15	6
Ditto, mahoganyand braſs, large	7	7	0
Ditto, the moſt compleat	9	9	0
Apparatus for ditto, from 1l. 11s. 6d to ——	9	9	0
A pirometer to try the expanſion of metals ——	2	12	6

Planetariums, Orreries, Armillary Spheres, and all other Mathematical, Optical and Philoſophi-

Figure 7.4 Henry Pyefinch's trade card, undated. Science Museum/Science and Society Picture Library – all rights reserved

arrangement because Pyefinch's purchases comprised object glasses already fitted into Dollond's tubes and, quite distinctly, object glasses fitted into tubes supplied by Pyefinch himself, possibly bearing his own mark or finish. This detail is clarified by Dollond's answer to Pyefinch's complaint:

> ... [he, Dollond] always fixt the Object Glasses he sold to [Pyefinch] into Tubes sent to him by [Pyefinch] for that purpose and never sold to [Pyefinch] any such Object Glass by themselves or unfixt in Tubes although many such Tubes were the property of and belonging to [Pyefinch] ....[36]

So, every telescope that left the Dollond workshop was complete. The need for Pyefinch to submit his own tubes for assembly at Dollond's was not written into the formal bond although it may have arisen from a verbal agreement during the conversation alluded to in Pyefinch's deposition. The significance of this is made apparent later.

A Bond is Broken

Modifying any part of the optical arrangement of a Dollond refracting telescope, even if for repair or improvement, constituted an infringement of the bond and selling any telescope containing a compound object glass that had not come from the Dollond workshop was deemed counterfeit. In both these respects, it appears that Pyefinch had broken the bond and, despite his pleading, Dollond would not forgive and forget. Of course, Pyefinch may have grumbled discreetly among those close to him that there was a certain unfairness in Dollond's action, but in open court – at the session during the Easter Term of 1768 – he complained bitterly to Lord Camden that ever since the bond was set up he had 'strictly adhered to the Agreement' and had 'not in any respect broke [the condition] or Counterfeited'.[37] Protesting his innocence, he added that his agreement with Peter Dollond was 'the principal if not the only Inducement or Consideration [for] entering into such Bond'. Justifying himself further, in expressions advised by his attorney, he concluded that the allegation of 'combining and confederating with diverse persons at present unknown' was clearly contrived 'to injure and oppress him'.

At this juncture, one may wonder – as the bewildered Pyefinch also claimed to wonder – why Dollond might have wished to exercise his prerogative by demanding a penalty if, as Pyefinch insisted, he had truly respected the terms of

---

36 Peter Dollond, 'Answers to the bill of complaint of Henry Pyefinch (17 Nov 1768)'. TNA: C 12/2071/11.

37 Bill of complaint of Henry Pyefinch (10 May 1768). TNA: C 12/2075/4. In both actions Messrs Edmonson and Potts of Bury Street were Dollond's attorneys.

the agreement. However, the first sign that all was not well with the arrangement occurred when his supplies were cut off abruptly. Of this, Pyefinch complained:

> ... [he had been] greatly Injured in his Trade and hath lost many of his Customers for wont of being able to supply them with Refracting Telescopes with Object Glasses therein....[38]

There was no escape for Pyefinch from the allegation that he had conducted an underhand act by selling counterfeit refracting telescopes and a lawsuit to this effect was brought at Westminster in the Court of Common Pleas. Had this proceeded successfully to a conclusion in the plaintiff's favour, then Pyefinch would have suffered a penalty of £500, as evidenced by Martha Ayscough's extant binding.[39] It was, no doubt, the prospect of a fine and possible bankruptcy that caused Pyefinch to take counter measures.

Pyefinch Retaliates

Not only had Peter Dollond proceeded against Pyefinch, but also – according to Pyefinch – he had set out to 'harass and perplex' him by acting on another detail regarding an overdue payment, for which a penalty of £150 was imposed.[40] Given these circumstances, there was little that Pyefinch could do except to take a counter action on the grounds that Peter Dollond's proceedings were 'unjust and vexatious'. Thus, in his counter action – via a bill of complaint issued through Lord Camden at the Court of Chancery on 10 May 1768 – Pyefinch flatly denied the two main aspects by which he was charged: the making of object glasses after the form of the late John Dollond or Peter Dollond, and the selling of such refracting telescopes containing the object glasses, except those purchased from Dollond himself for resale according to their agreement.

Despite his denial of Peter Dollond's allegations, there had been a single occasion when an achromatic refracting telescope not conforming to the Dollond specification was sold mistakenly. Rather than avoid this issue – a detail already known to Dollond and undoubtedly the trigger for his action which was bound to be stated in court – Pyefinch openly admitted this unfortunate incident to Lord Camden. Speaking as 'orator' in the third person – as was the custom for all bills at Chancery – Pyefinch, respectfully contrite, announced that:

---

38 Ibid.

39 Martha Ayscough's bond. GL: CLC/444/MS14805/001 (paper 3).

40 Neither judgements nor proceedings of the case at Common Pleas (alluded to in Pyefinch's bill of complaint) have been located and this might suggest that the case did not come to court.

> ... your Orator doth by a means admit your Orator hath never sold more than one such Telescope which was an old one and made before such first action commenced against your Orator ... the same was sold by Mistake and through the Inadvertency of your Orator's servant by such servant without your Orator's knowledge Privily or Consent and without of any View to injure the said Peter Dollond or hurt the sale of his said Telescope Invention and the said Peter Dollond can have sustained only a very trifling if any Damage thereby ....[41]

Whether the damage was trifling or not, it was upon this single incident – possibly resulting from Pyefinch's blundering apprentice, John Meredith, a nephew of his wife – that Dollond had pounced.[42] Of course, Lord Camden may have raised his eyebrows at the man who would impose a huge penalty for what was supposed to be merely an 'inadvertent' sale. But, in law, the evidence of a single instance, however inadvertent, was deemed to be 'the tip of an iceberg'. There were, however, other matters over which Dollond and Pyefinch were at odds.

An issue unclear to Pyefinch – upon which the counterfeit charge stood – was that the bond did not allow him to 'tube-up' his own telescopes. Pyefinch argued that because John Dollond's original patent had not specified anything concerning the construction of the telescope itself, but only the nature of the object glasses, then it was very unjust that the bond should not allow for his own workmen to make up the telescopes by using object glasses bought in from the Dollonds. In this detail, Pyefinch was absolutely correct; the original patent was not to do with the end product, namely the object glass, even though John Dollond had claimed it as his invention, but only the method of making it. That Pyefinch was aware of this subtle difference may have rankled with Peter Dollond. Nevertheless, the essence of what they agreed in a conversation (that is, before the bond was drawn up) was otherwise than what was written down. As Pyefinch put it: 'it [the bond] was contrary to the Express Agreement and true intent and meaning'. Foolishly, perhaps, Pyefinch added his signature to a document that it appears he had not read. In court, he complained that the bond was 'an imposition' because it had been:

> ... Artfully and Fraudulently so drawn by the said Peter Dollond or his Attorney who prepared the same and was an undue Advantage taken by them of your Orator's Attorney to inspect such Bond before your Orator executed the same or to permit your Orator's Attorney [John Holt] to be present at the Execution thereof ....[43]

---

41 Bill of complaint of Henry Pyefinch (10 May 1768). TNA: C 12/2075/4.

42 According to Pyefinch's Will (note 51), his widow Elizabeth's father was Thomas Meredith of [?]Kington (indecipherable), Herefordshire. So, John Meredith may have been a younger brother or a nephew.

43 Bill of complaint of Henry Pyefinch (10 May 1768). TNA: C 12/2075/4.

The implication here is that some fast practice had taken place. Being bamboozled by Dollond's clever attorney, who may have urged the unwitting Pyefinch to sign in the absence of his attorney, has echoes of the time when Francis Watkins was persuaded to accept John Dollond's terms of partnership far too hastily. From Pyefinch's perspective, he simply believed that:

> ... by the express Agreement of the said Peter Dollond your Orator was to have the sale of such Object Glass for Refracting Telescopes and of such Refracting Telescopes with Glasses put therein and fitted thereto by the said Peter Dollond as your Orator should buy of him ....[44]

Pyefinch conveyed – or wished to convey – a degree of confusion in his understanding of the absolute nature of the bond. But Dollond had the upper hand and insisted that he alone should tube-up all the telescopes as written in the contract, and not as Pyefinch claimed through any earlier conversation. Despite these confusions, there were sufficient grounds for Pyefinch to advance a formal complaint through the court.

In summary, Pyefinch, in his plea to Chancery, requested that Lord Camden should make Peter Dollond answerable on two counts: first, concerning counterfeit, he wished to know precisely the facts upon which Dollond based his damages:

> [Specifically,] how many [telescopes] and when and how ... and what was their value or amount thereof and the Damage he hath sustained and whether such amount is not Vastly Exceeded by the said Five hundred pounds penalty ....

Secondly, concerning the principles upon which Dollond conducted his business, he wished to know:

> ... whether such his unjust and vexatious proceedings are not calculated with such unfair view and intention or with what other View or Design ....

To force this clarification, he also requested that two writs be issued: a writ of injunction, thus restraining Dollond from proceeding further with the on-going suit against him through Common Pleas, and a writ of subpoena, thus commanding Dollond to answer for himself and, failing in any of this, that he should surrender the bond with its penalty of £150 cancelled. Camden obliged.

[44] Bill of complaint of Henry Pyefinch (10 May 1768). TNA: C 12/2075/4.

Dollond Answers Pyefinch

Six months later, on 17 November 1768, Dollond – appearing for the first time as defendant – was called to answer the charges brought about by Henry Pyefinch as complainant, under the same judge (Camden) and in the same court (Chancery). When asked to explain his position, Dollond denied that he had any obligation to supply Pyefinch in the same manner and at the same rates and prices that he supplied others of the trade although he said he had no objection to doing so. Indeed, he told the court that he had supplied Pyefinch in a privileged way until discovering that he had 'abused his indulgence'.[45]

Dollond further explained that the condition in the bond – for which a penalty of £500 could be levied – was not meant to deny Pyefinch from selling object glasses bought of him either by themselves or fitted into tubes, but was a condition only to be applied in case of counterfeit. With regard to the patent, Dollond conceded that this referred only to the object glasses, not the tubes. For this reason, he did sell the object glasses without tubes to other persons in the trade although this was not the case with Pyefinch for whom he always fitted up the tubes, whether his own or those supplied by Pyefinch. In terms of the numbers supplied for resale within the trade or for foreign consumption, he reported that he had supplied 107 refracting telescopes and 96 object glasses.

Concerning the element of distrust that had arisen from the inadvertent sale of an 'old telescope' – as already admitted by Pyefinch to the Chief Justice – Dollond had known of this sale via the actual purchaser, one John Phipps of the City. This telescope was no ordinary compound objective, that is, concave flint and convex crown only, but seemingly an instrument containing three combined glasses. He knew this by purchasing the actual instrument off Phipps and hence it was considered fundamental evidence. What may have happened (although pure conjecture here) was that Pyefinch may have opened up one of the ordinary achromatic telescopes and attempted to improve the optics by adding an extra lens, perhaps after the fashion of the new triple achromats. Such examination and experimentation by an instrument maker, done privately, could be supposed praiseworthy because it showed intelligent curiosity. The mistake of selling this modified instrument, however, provided Dollond with an opportunity to stretch his argument to one of 'very considerable damage', along with the usual coda that he had intentionally set out to 'hurt and interfere with the Sale of this Def^{ts} said patent Glasses'. But as for the bond being a 'fraud' or an 'imposition', Dollond simply replied:

[45] Peter Dollond, Answers to the bill of complaint of Henry Pyefinch (17 November 1768). TNA: C 12/2071/11.

> ... [it was] exactly the same as the condition of the Bond in this Deft[ts] former answer ment[d] to be given by Mrs Ayscough and with no other view or design whatsoever and therefore the same was not any fraud or imposition ....[46]

Peter Dollond also denied the connotation – suggested by Pyefinch – that the bonds were drawn up to take unfair advantage and the charge of 'combination and confederacy' was refuted as being 'unjust'.

Adamant in his position and unwilling to accept that Pyefinch had merely sold an old telescope by mistake, Dollond further alleged:

> ... the truth is that the Compl[t] [complainant] hath made and Sold and disposed of Several Telescopes with Object Glasses therein of the same Mak[ing] and Construction of the Invention for which such patent as aforesaid was Granted to this Def[ts] [defendant's] said Father and the Complainants Bill and the Deft[t] hath thereby broke the condition of the Bond in the Complainants Bill for that purpose ment[d] ....[47]

Of this final assertion, Dollond admitted that he could not supply the kind of factual answers that Pyefinch had sought although it had been brought to his attention that Pyefinch had sold one further telescope, likewise counterfeited, to a Thomas Bidwell. But although the recipient was named, this information was hearsay only and still needed to be proved. Of course, Peter Dollond wished to exaggerate by suggesting that the damage was widespread, incurring losses to him 'not vastly exceeded by the five hundred pounds penalty'.

The lack of further report through the judgement rolls would suggest that this case was dismissed. Had the case been carried through to Pyefinch's detriment then, in all likelihood, he would have become bankrupt immediately. This does not appear to have happened but, instead, he went on to develop his own method of constructing and making refracting telescopes with object glasses made on lines that now have a familiar ring. A summary for letters patent taken out by Pyefinch in December 1770 ran as follows:

> A grant unto Henry Pyefinch, of Cornhill, London, optician, of his new invented method of constructing & making refracting telescopes with object glasses therein composed of *two or more different glasses or mediums, separate from each other*, and so placed in the tube that by altering the distance of one from the other the glasses

46 Martha Ayscough's bond was produced as evidence in court. A note penned thereon reads: 'This paper writing was shown to Thomas Whitford [optician] at the time of his Examination taken in Chancery on the part of Peter Dollond (Deft) at the suit of Henry Pyefinch (Complt).' [Signed John Spooner]. GL: CLC/444/MS14805/001 (paper 3).

47 Note 36. Dollond (Answers).

> will have greater or less magnifying powers, without destroying the effect arising from the refractive qualitys of the several mediums, whereby it may be made to suit any eye, weather, &c., ...[48] [Author's italics]

Clearly, Pyefinch was talented enough to obtain this patent and it is probable that Dollond had wished to discontinue the bond because he had detected the direction that Pyefinch was taking.

It is interesting that Pyefinch's patent appeared while the Dollond patent had nearly two years yet to run. From Peter Dollond's standpoint, his father's patent had served him well and, by 1770, his trade and clientele in achromatic instruments were thoroughly established. Sadly, although Pyefinch had begun to make his mark in the optical field, he did not live long enough to develop a long-lasting reputation.[49] Before his death in 1779, in addition to John Meredith, he also employed two more apprentices: Richard Collett in 1768 and Howell Hughes in 1772.[50] After his death, Meredith remained in the business but his widow Elizabeth turned to a different trade, in china pottery, though operating from the same address.[51] Pyefinch may have wished his son, Henry-II, to follow in his footsteps but, with the family circumstances much changed, the younger Pyefinch simply took the opportunity also to turn to another trade.[52]

Conclusions

All Peter Dollond's legal actions were undertaken to defend his father's patent and to confirm the rights and privileges which had been granted to his father in 1758. The motivation for taking the actions described in this chapter, between 1763 and 1767, were designed to secure the patent for the remainder of its term, which would end in 1772: after all, the achromatic telescope objective was a principal dimension upon which the business now thrived, and so the litigation was pursued, partly out of respect for the memory of his late father but mainly because he was obliged to defend the patent on behalf of the Crown. Over the previous three generations, the Dollond family had made a remarkable journey, first by escaping from religious persecution in their homeland, then by extricating themselves from the poverty that beset them as immigrant refugees, and, finally, by achieving recognition among the scientific and social élite. Peter's own directions in life had been formed out of his father's drive and ambition but

48 Pyefinch, patent 976 of 28 December 1770. TNA: C 210/2.

49 For example, a three-draw tube achromatic telescope bound in green ray skin by Pyefinch was illustrated in *Tesseract Catalogue*, 79, Spring 2005.

50 Clifton (1995), p. 225.

51 Will (Pyefinch). TNA: PROB 11/1057/426/*f.* 354.

52 Clifton (1995), p. 225.

his father's sudden demise in 1761 left him with a heavy burden of responsibility. As head of family, and at only 29 years of age, sole breadwinner and only recently married, he had to provide for three younger sisters, Elizabeth, Mary and Sarah, and a brother, John, only 15 years old, although assisted perhaps by his elder sister, Susan. John Dollond had died intestate and, only a few years earlier, his circumstances were described as 'indigent'. If this was so, then the income derived from the business had to be radically increased. One way of doing this was to raise the profit margin for telescope sales to 200 per cent but Watkins, as business partner in the sale of these telescopes, refused to comply. The original agreement between John Dollond and Francis Watkins had stipulated a profit margin of 60 per cent and Watkins considered this adequate. Peter then sought an alternative path to fortune which entailed ridding the business of all connection with Watkins. Thus, the partnership was abrogated and Peter became free to develop his optical work and maximise his profits without further obligation; Watkins, too, became free of a burden which had troubled him ever since learning that John Dollond was not the original inventor of the compound object glass.

Dissolving the partnership had not been of prime importance from Watkins's standpoint. It was, however, an opportunity for him to redress the situation caused by his mistake in supporting John Dollond financially to obtain a patent which – according to custom if not exactly according to the Monopolies Act of 1624 – had been obtained on an untrue basis. Only after beginning to sell the 'New-invented refracting telescope', was it discovered that Chester Moor Hall might have a prior claim to the invention. The optician George Bass knew the situation full well, not only as former Master of the Worshipful Company of Spectaclemakers but as the person who had ground the original glasses for Hall. It was for this reason that Watkins advised John Dollond not to apply his rights of patent against William Eastland who had made similar object glasses. Peter Dollond did not expect, following the dissolution of the partnership, that Watkins would publicise just how John Dollond had deviously presented the idea to him as his own invention. Watkins had been hoodwinked and, foolishly, without checking, he had been led to procuring the patent on Dollond's behalf and, moreover, to investing £200 in the scheme.

The state of affairs in 1763–1764 was unsettled and strongly underpinned by a wish that the trade should return to its former position whereby its organisation and control were governed by the Spectaclemakers' Company. Unfortunately, the damage had been done and, at first, there seemed to be no way that the Company could disallow the patent which, as a monopoly, was bound to affect the trade. Watkins had tested the strength of Peter Dollond's determination by continuing to make up the new telescopes after being forbidden to do so. An action at King's Bench quickly confirmed that Watkins would not be permitted to defy the patent. Of course, stronger measures were needed, particularly in light of Peter's threats to other members of the trade. It was, however, conceived that the situation could be salvaged by requesting King George-III to vacate the patent. Little weight would be attached to such an appeal from Watkins alone although with the full weight of the Spectaclemakers' Company behind him, he could hope for some retribution. To this end, therefore, Watkins was elected Master of the Company for 1764 during which time an attempt was made to have the patent annulled through a petition to the Crown. Concurrent machinations of government spoiled this plan and, by a separate route, Peter Dollond brought an end to each of the known interlopers' activities by indictment, and, more generally, by proving Lord Mansfield's original judgement against Watkins and Smith (1763), in a test case against James Champneys in 1766. Champneys had blatantly defied the patent by advertising that he made and sold Chester Moor Hall's object glasses thus raising the issue of Hall priority afresh. Peter, as plaintiff, won; the case for Hall was lost, and Mansfield's original argument for a patent as a commercial device was established.

With regard to Chester Moor Hall, such was the intensity of feeling among those engaged in the litigation of the mid-1760s, that this uncomfortable part of scientific and trade history has never been forgotten, not even today. It should be remembered that Lord Mansfield did not dispute Hall's position as original inventor but rather he shifted the concept of invention to one where only commerce mattered. Hall had made no overt attempt to register his invention by patent and, therefore, Lord Mansfield would not allow credit for what, effectively, remained hidden from view. At the time, neither Watkins and Smith, nor their attorney, perceived that Chester Moor Hall was anything but the first in the field but nor did they imagine that 'public benefit' would take priority over tradition whereby the first to succeed in some novel advance was considered the inventor. Not even the Monopolies Act of the previous century was clear on this point. Thus, the dismissal of Watkins and Smith's defence in 1763 meant that any secondary argument – such as the inadequacy of the specification – was never given consideration in court. Ordinary practitioners may well have railed against John Dollond and his patent but he would care little because – as Ramsden pointed out later – there was no one else in London who was as passionate as he in his desire to improve optical instruments. Indeed, it was precisely because of this that he received the blessing and support from within the Royal Society.

So, by never openly admitting any credit to Hall at the time of the lawsuits, or at any time afterwards, Peter Dollond managed to conquer his adversaries. The family survived the lean years after their father's death and, by 1766, they had moved to new premises in St Paul's Churchyard at which time the younger brother, John, having reached 21 years of age, joined in the partnership. Together, 'P. & J. Dollond' improved upon the manufacture of the refracting telescope and established new heights in what was technically achievable. By the end of the century, few individuals remained to tell of the legal battles with any accuracy although some remnant of the intense feelings of that epoch did survive. Peter Dollond, ever watchful, continued to defend what he believed to be his father's priority, even long after the patent had expired. An issue arose in 1789 which illustrates just how important his father's contribution had been in turning round the fortunes of the business. In structuring the history of events to suit his purpose, however, Jesse Ramsden objected to Peter's lack of recognition of Chester Moor Hall, and so, once again, the near-forgotten arguments of the 1760s were resurrected though without any further solid evidence. These subsequent and consequential events are now taken up in the following chapter where some signs of a shift in allegiance of members of the Spectaclemakers' Company, from Watkins to Dollond, took place in the aftermath of the great upheaval that unsettled and changed the outlook of the optical instrument trade forever.

Plate 1 Detail of engraving around the mirror of a Culpeper-type compound microscope by Nathaniel Adams, c. 1735. Museum of the History of Science, University of Oxford, inv. 31636

Plate 2 Compound prism demonstrating the principle involved in John and Peter Dollond's more ambitious triplet achromatic lens. © National Museums Scotland, T. 1955.21

Plate 3 John Dollond patented his independently invented achromatic doublet telescope lens in 1758. It incorporated a new lens made of a 'sandwich' of two lenses, one of crown and the other of flint glass. This telescope, made by John Dollond's son Peter in 1765, incorporates an objective made to the new design. © National Museums Scotland, T. 1935.53

Plate 4 Detail of the signature 'PETRUS DOLLOND/Ano. Dom. 1756/ LONDINIUM'. © National Museums Scotland, T. 1935.53

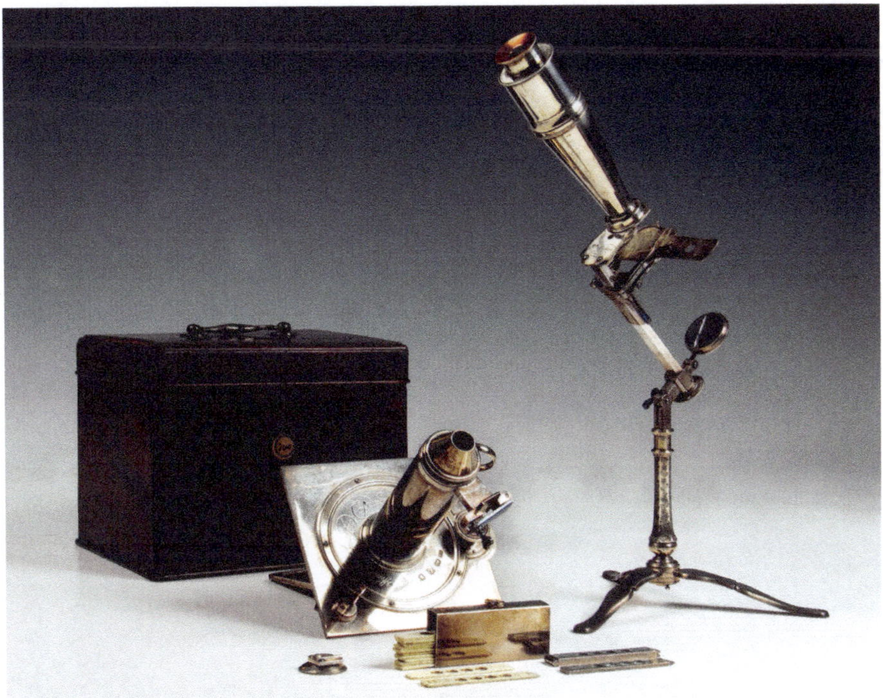

Plate 5 (above)
Compound, simple and solar microscope in silver, signed on the solar plate 'Fra. Watkins Charing Cross London', 1754. Museum of the History of Science, University of Oxford, inv. 53619

Plate 6 (left)
Compound and simple microscope to Francis Watkins's design, 1754. Museum of the History of Science, University of Oxford, inv. 47695

Plate 7 Portable Gregorian reflector telescope, signed 'Fra. Watkins Charing Cross London'. Museum of the History of Science, University of Oxford, inv. 69862

Plate 8 Hand-held refracting telescope by Francis Watkins. Private collection

Plate 9
Large refracting telescope on stand, by Francis Watkins. Private collection

Plate 10
Surveyor's wheel or waywiser, by Jeremiah and Walter Watkins. © National Museums of Scotland, T. 1912.53

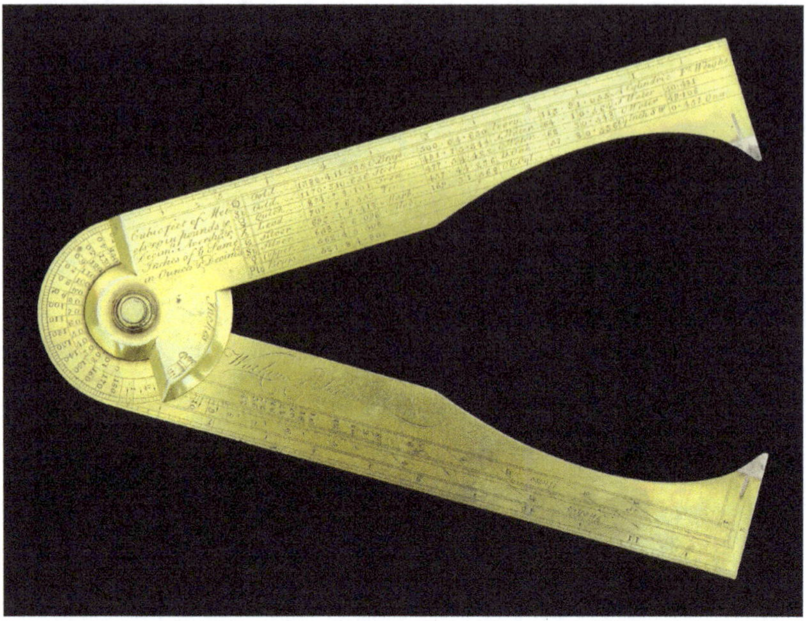

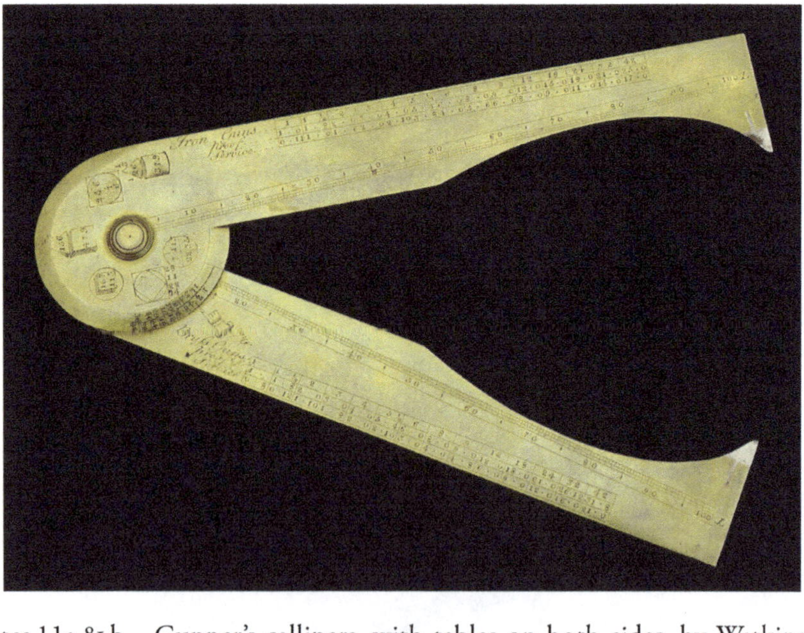

Plates 11a & b Gunner's callipers, with tables on both sides, by Watkins & Smith, London, c. 1770. Museum of the History of Science, University of Oxford, inv. 41742

Plate 12 Model of a beam engine by Watkins & Hill, c. 1850. Museum of the History of Science, University of Oxford, inv. 26418

Plate 13 Saxton magneto generator by Watkins & Hill, c. 1850. Museum of the History of Science, University of Oxford, inv. 94679

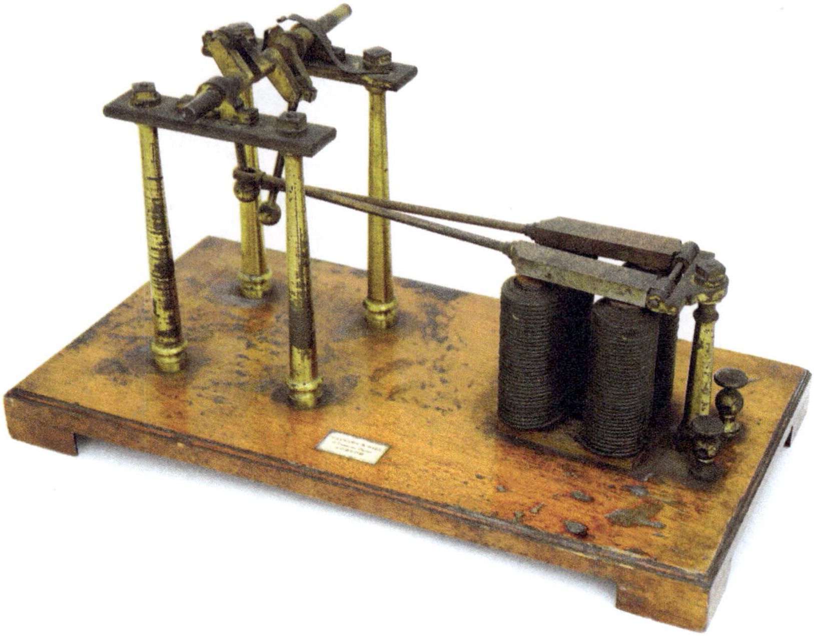

Plate 14 (above)
Reciprocating electric motor, by Watkins & Hill, London, c. 1850. Museum of the History of Science, University of Oxford, inv. 77942

Plate 15 (left)
Coulomb torsion balance electrometer by Watkins & Hill, London, c. 1850. Museum of the History of Science, University of Oxford, inv. 94227

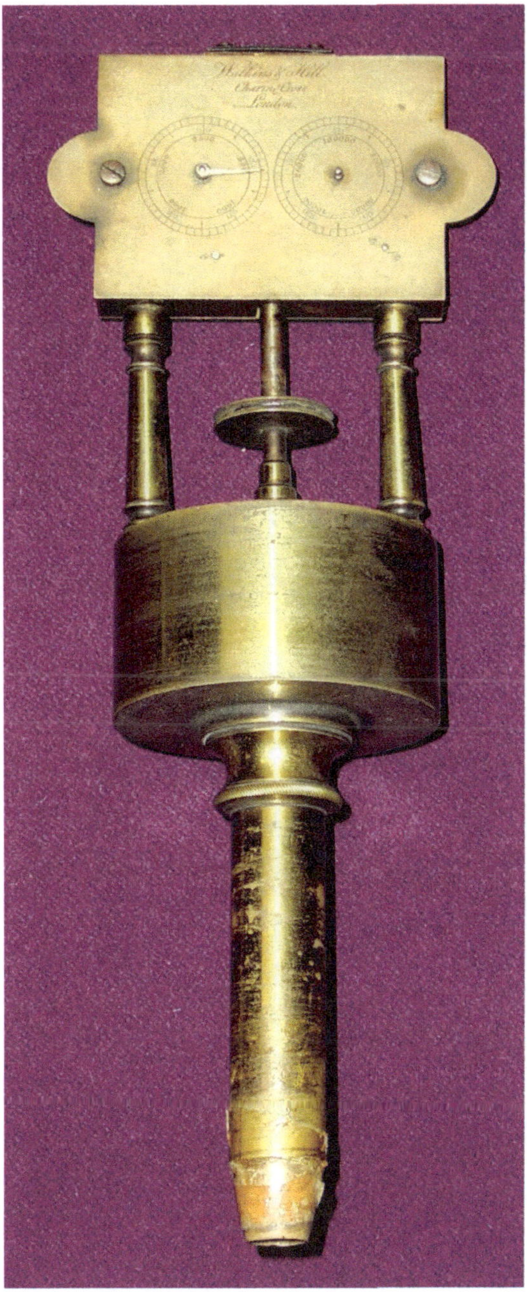

Plate 16 Twin dial Cagniard siren, by Watkins & Hill, Charing Cross, London, c. 1840. University of Aberdeen, Natural Philosophy Collection

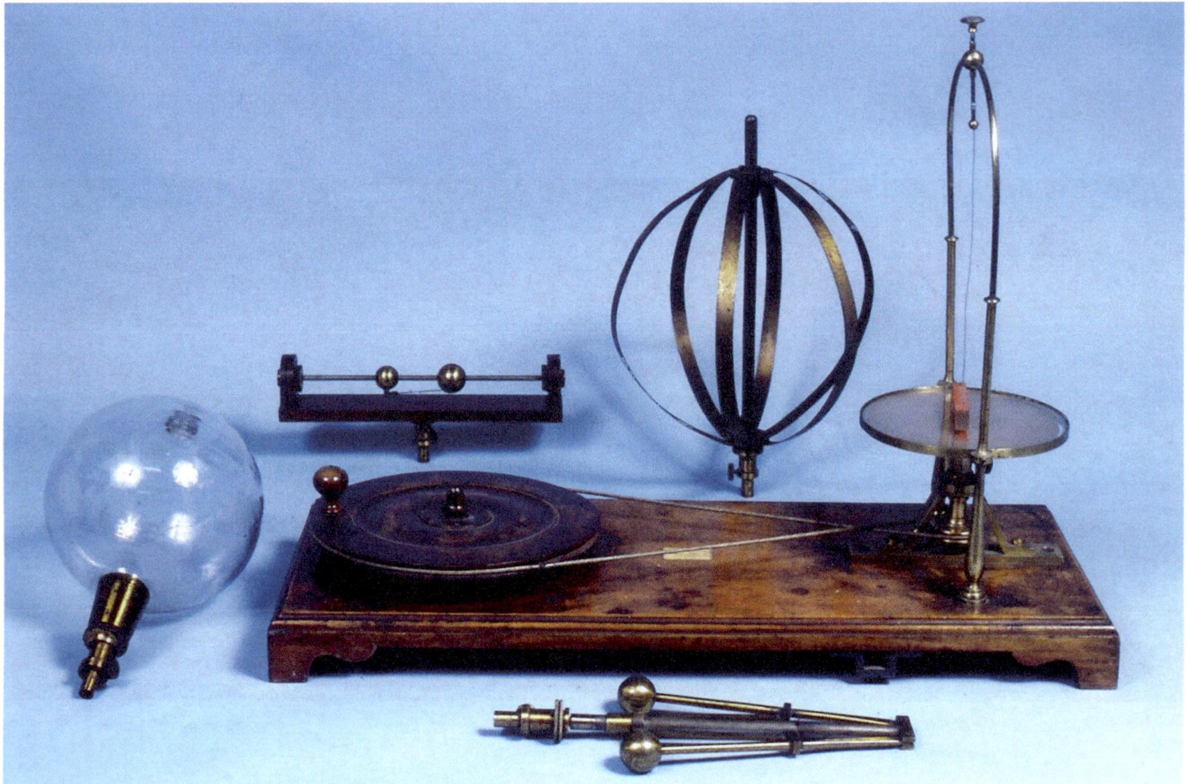

Plate 17 Whirling table, with accessories, by Watkins & Hill, London, c. 1840. University of Aberdeen, Natural Philosophy Collection

Plate 18 Device for demonstrating thermal expansion, by Watkins & Hill, 5 Charing Cross, London, c. 1840. University of Aberdeen, Natural Philosophy Collection

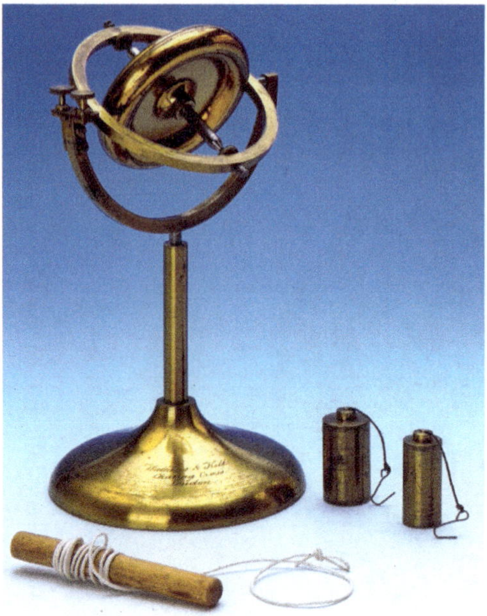

Plate 19
Gyroscope in gimbal mounting, by Watkins & Hill, Charing Cross, London, c. 1840. University of Aberdeen, Natural Philosophy Collection

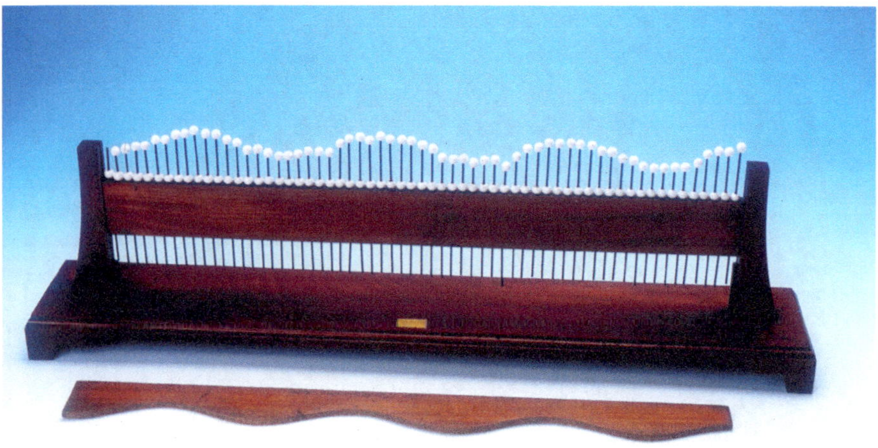

Plate 20 Wave model, by Watkins & Hill, 5 Charing Cross, London, c. 1840. University of Aberdeen, Natural Philosophy Collection

Chapter 8
New Conflicts within the Spectaclemakers' Company

Despite his acrimonious start in business, Peter Dollond went on to gain respect from within the instrument trade, and the firm flourished from about the time that he took John Berge as apprentice in 1765 and his own younger brother John in 1767. In December 1766 the business moved from their 'off Strand' shop to premises at the north side of St Paul's Churchyard – an instrument shop formerly belonging to the widow Mary Sterrop – and it was from this City location that the Dollond business developed a strong reputation as suppliers of a wide range of instruments including telescopes, microscopes, theodolites and sextants.[1] Many famous people visited the new premises, amongst whom were the Astronomer Royal Nevil Maskelyne, Captain James Cook (just before his famous expedition to the Pacific) and Captain George Vancouver (best remembered for his voyage in search of the Northwest Passage). Indeed, the name of 'P. & J. Dollond' became known throughout the world.

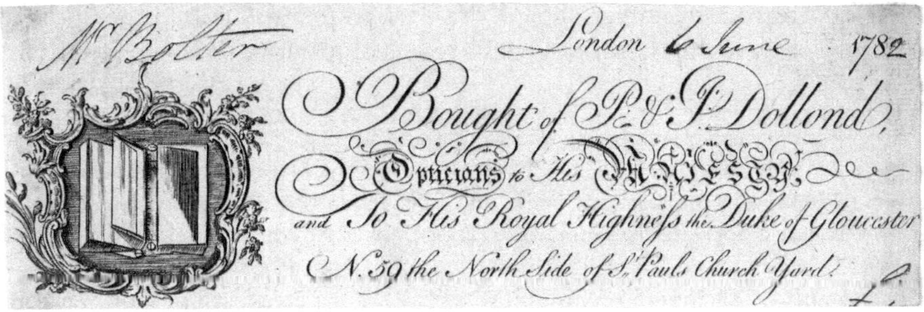

Figure 8.1 Bill head from P. & J. Dollond showing the achromatic triplet as their sign, 1782. © The Trustees of the British Museum, Heal, 105.38

[1] The 'Archimedes and Golden Spectacles' of Mary Sterrop was passed on to Peter Dollond for a consideration of £200, paid on 25 December 1766. 'Articles of Agreement between Mary Sterrop of St Paul's Ch Yd widow and Peter Dollond of the Strand'. GL: Dollond papers, CLC/444/MS14805/001 paper 4.

Over the same time span, Francis Watkins formed a partnership with his former apprentice, Addison Smith who operated independently from a shop at the western end of the Strand, opposite Northumberland Street.[2] These changes took place in 1763 at the time of Smith's freedom of the Spectaclemakers' Company and his marriage to Elizabeth Walder. According to directory sources, the 'Watkins and Smith' partnership ran for a decade or so, although transactions in Watkins's bank account suggest a working relationship until 1790, that is, the year before Francis's death. After about 1774, Smith traded independently as optician and, in 1783 he published a tract entitled *Visus illustratus; or the sight rendered clear ... the cause ... or defect of the present mode of constructing spectacles.* That same year he obtained a patent for a new method of constructing and grinding spectacle lenses.[3] The registered specification describes him as being 'of Richmond, in the county of Surry, gentleman', although according to rate records and other local searches, he does not appear to have owned any property in Richmond. However, about the time of submitting the patent, Francis and Clarinda Watkins intended to move to Richmond for their retirement and, towards this end, they purchased three properties, possibly hoping that the Smiths would become their neighbours. Instead, Addison and Elizabeth Smith retired to 79 Charlotte Street, Rathbone Place, a newly developed area in the parish of St Pancras.[4]

Watkins may not have secured the prestigious contracts to supply astronomical instruments to the newly emerging observatories as had the Dollonds, although he was in demand to supply instruments for some of the major expeditions. For instance, he supplied a flat-plate electric machine to Charles Green, astronomer aboard the *Endeavour* on Cook's expedition to Tahiti to observe the transit of Venus.[5] From comparative tests – logged in the

---

[2] With new street numbering, the West Strand address became '481 Strand'. Another address in St Martin's Lane, cited in Clifton (1995), p. 254, may refer to a temporary move around 1763. From the local rate books and Rocque's map it would appear that Smith's shop was nearly opposite the entrance to Craven Street. Smith used these premises until about 1783 when he retired to 79 Charlotte Street, Rathbone Place. See note 4.

[3] Smith (1783). Smith applied for letters patent for a design and method of making spectacles, described as 'an optical instrument of one or two sights, made in various formed frames, fitted to the head or nose or to hold in the hand.' TNA: Patent 1359 (7 July 1783).

[4] Smith's St Pancras address also appears on legal documents of the Lucy family of Charlecote, a prominent Warwickshire family. These papers concern an action by Smith (plaintiff) in 1788 following the Will of George Lucy over whom an issue had arisen concerning the deceased's sanity. It seems that Smith was owed money, that is, more than the £1,000 already received from this estate in 1787. WCRO: Lucy of Charlecote, *Legal papers* L6/1003 (Trinity 1788), L6/1004 (c, 1788); Settlement and Wills L6/1197 (20 December 1787).

[5] Fully logged details of the electrical experiments performed aboard the *Endeavour* appear in Beaglehole (1962), vol. 2, pp. 276–279.

ship's journal for 25 October 1768, five miles south of the equator – it appears that the Watkins machine performed far better than a similar machine made by Ramsden belonging to Joseph Banks. Sadly, the unfortunate young astronomer died during the voyage and no trace of what happened to his electric machine has been found. It was very fashionable to sell globes showing the latest updates of discoveries and observations made on the great expeditions at this time. In this respect, Watkins and Smith marketed globes made by Nathaniel Hill, who also supplied other prominent instrument shops.[6] Other connections are known, of which two examples will suffice: the Portuguese agent John Hyacinth Magellan placed orders with Watkins, so it is assumed that Watkins remained at the forefront of his trade for bespoke instruments;[7] and a fine parallactic telescope by Francis Watkins, now at the University of Coimbra, Portugal, indicates that he resumed work in achromatic optics after the Dollond patent expired.

As with the brothers P. & J. Dollond, Watkins responded to military demands by supplying gunnery instruments and field telescopes. Indeed, Horatio, Lord Nelson, reputedly owned a telescope and a microscope made by Watkins.[8] In spite of Watkins's earlier litigation with Peter Dollond, neither business faltered. Rather, each went from strength to strength, buoyed by military needs, rapid industrialisation and the emerging British Empire, thereby ensuring both establishments survived as prominent instrument houses well into the nineteenth century.[9] However, the conflict between the principals, Francis and Peter, was never fully laid to rest, nor was the issue concerning the priority over the invention of the achromatic object glass ever settled, at least in the mind of one influential contemporary. Echoes of the past surfaced during the last decade of the eighteenth century although, with the passage of time, the underpinning details became clouded and the matters of 'truth' and 'correctness' – over which battles had been fought through the courts – were hardly remembered, let alone comprehended, by the following generation of the scientific community who had scant evidence upon which to base any opinion. For his part, Peter Dollond maintained an effort to protect his father's good name because it was upon his crowning glory of the patent that the reputation of the Dollond enterprise was built. However, there were others – not the least of whom was Jesse Ramsden who had once worked very closely with Peter Dollond – who retained a view,

---

6 *Gazetteer and New Daily Advertiser*, 16 January 1765. Nathaniel Hill of the 'Globe and Sun', Chancery Lane, also supplied: Heath and Wing (Strand); John Bennett (Crown-Court, St Ann's, Soho), Dolone [*sic*] (Strand), Johnson (Ludgate-Street, near St Paul's), Sterrop (St Paul's Church-Yard), Ribright (Poultry), Pyefinch (Cornhill) and Stedman (Leadenhall-Street).

7 According to Anita McConnell's transcript of Magellan's account at Hoare's bank, he paid Watkins £27 6s on 23 April 1778 and £21 to Ribright in 6 May 1778.

8 NMSI Inv. no. 1869–2 (Parts 1 and 2).

9 This is apparent (though not expanded here) from the distribution of instruments worldwide (some still extant in Australia, North America and the Continent).

first argued by Watkins and Smith's counsel at King's Bench in 1763, that Chester Moor Hall must be held the true inventor of the achromatic object glass. This air of suspicion and doubt over John Dollond's actual contribution was thus perpetuated into the nineteenth century. Bearing this in mind, the present chapter opens by first considering the on-going tension between Francis Watkins and Peter Dollond in the wake of the litigation of the 1760s.

The Spectaclemakers' Court Minutes offer an insight into changes in the organisation of the Company's affairs. For instance, the forces of change are noticeable in dealing with the impact of expansion of the metropolis due to the rising population and developing industrialisation. These forces eroded the older ways of order and control because many of the trades and artisans began to operate beyond the customary bounds of a company's authority.

For a while, in the 1760s, the records of the Spectaclemakers' Company present – superficially at least – the same sense of orderliness that could be found 50 years earlier. Quarterly meetings – complete with reports and accounts dealing with apprentices, freemen and masters – were held with tedious but purposeful regularity. The numbers in attendance (assistants and Wardens) remained moderately constant at around a dozen or so. However, following Watkins's unsuccessful attempt to retrieve the status quo for the traditionalists through the petition to the king and after the Champneys case, attendance under James Tomlinson's Mastership in 1767 and 1768 dropped to five only. Ten years later, a minute for 27 June 1777 recorded – for the first time – a dire state of affairs: namely, 'The Business of the Company for some time postponed, a sufficient Number of Members not being present to compose a Court.'[10] Some of this slackening of attendance was also apparent in other City companies, but the minutes of the Spectaclemakers' Company also hint of tensions and potential dangers to the Company's survival from internal dissent and a certain lack of unity.

Only a few years earlier, in 1759, freemen and masters were issued with a printed sheet reminding them of their duties. This 'reminder' was an attempt to strengthen what was an already weaker and less cohesive company. It also refreshed the memories of those freemen who needed reminding of their oath to the Company. This was a laudatory creed extolling the virtues of 'the Fellowship of Spectacle-Makers' and their allegiance to 'the Sovereign Lord the King's Majesty'; it did not, however, necessarily engender the camaraderie that it was designed to promote. Moreover, the rule that 'you shall be ready at all Times to be at the Quarter Days, and every other Assembly, Matter or Cause that you shall be warned or called unto for the Affairs of this Fellowship, unless you shall have lawful and reasonable Excuse in that Behalf' may well have been treated with disdain if judged by the actual numbers present at meetings.

---

10 SMC Court Minutes 27 June 1777. GL: CLC/L/SH/B/002.

Attempts were made to modify the company's rules to make them more fitting to the new age. For example, the rise in the capital's population had inevitably led to an increased demand for apprenticeships, and so, on 3 October 1765, it was:

> ... *Ordered* That from after this day the anteint By Law relative to the restraining the Members of this Company from having more than one Apprentice at the same time till they Bear office in the Company shall be repealed, and that every Member shall be at liberty to take what number of Apprentices they please.[11]

Indeed, this improvement to the rules was long overdue because masters had, for some time, taken on additional apprentices even though, more properly, this was only allowed to those who had served on the court of assistants.[12]

Other corrections were needed: for example, a master might retain an apprentice well beyond the customary seven-year term, thus depriving the Company's coffers of its Quarterage, as in the case of Addison Smith, whose apprenticeship ran for 12 years. This often suited the master who wished to hang on to a good workman or potential partner, but it also showed disrespect for the system as a whole. Importantly, too, the Company lost funds that otherwise could have been raised through Quarterage and directed towards its higher ideals, such as providing support for orphan apprenticeships or assisting members in dire or extraordinary need. This lax approach caused the court of 28 March 1771 to take a firmer stance by compelling eight apprentices to take up their freedoms immediately because they had exceeded their seven-year terms.[13] Even more formality was brought to bear in 1773 when the court dropped the customary given christian names from the court minutes, substituting the formal title 'Mr'.

The Rising Mr Dollond

Further signs of the changing times may be detected through the relative positions of Francis Watkins and Peter Dollond. Two years after Watkins had served his term as Master, and despite his own non-attendance at the subsequent Quarter Day courts, there were still those who wished to see him return to a position of leadership. Maybe this hope was due to the respect they held for

---

11 SMC Court Minutes 3 October 1765.

12 SMC Ordinances. GL: CLC/L/SH/A/003/MS05215.

13 SMC Court Minutes 28 March 1771. Of course, there were often good reasons for not taking up a freedom, for instance, William Eastland Jr pleaded poverty and was thus excused his dues although the system caught up with him a couple of years later.

his integrity – as demonstrated through his proven correctness in book-keeping during his clash with Peter Dollond – because he was proposed auditor of the Company's accounts at the October meeting of 1766. As it happened, Watkins was absent when this proposal took place and consequently two older, but staunch, members (Bass and Cuff) filled the role jointly. On 4 October 1769, at the Michaelmas Day court, those faithful to Watkins advanced his name once more – along with proposals for Tomlinson and Nairne – for the position of Master. However, Peter Dollond – who had been a non-active 'foreign brother' since 1755 – had become an attending member of the Quarter Day courts, probably feeling justified in showing his face before those who had once acted against him in the petition to the king, but who now, in the year following Champneys's trial, could no longer assert any authority.[14] Thus, at the court of 30 March 1769, it was recorded:

> ... [that] Mr Peter Dollond who at last Court [December 1768] was chosen an Assistant of the Company, attended at this Court and paid his Fine and Fees; and was sworn in and took his place as an Assistant.[15]

It may also be significant that his younger brother, John, had also been sworn-in (by redemption) in the previous year.[16] Thus, with both Dollond brothers in attendance at the meeting of Michaelmas Day meeting (1769), it is not surprising that Watkins did not appear, even if it had been mooted that his name would be proposed for Mastership. It is, nevertheless, interesting to speculate whom among those present – Cuff, Howe, Ribright, Deane, Eglington, Tomlinson and Nairne – might have proposed his name. One possibility was Cuff, who had supported him in the petition although he would appear less likely than Thomas Ribright who had trading links with Watkins.[17] In the event, Tomlinson was favoured as the new Master.

Regardless of any split in allegiance between those who favoured Watkins and those who favoured Peter Dollond, it is also of interest to note that on the same day that Peter was sworn in, another significant election took place: namely, Addison Smith was chosen as an Assistant to the court.[18] Thus, with his

---

[14] Peter Dollond was sworn free of the Spectaclemakers' Company on 27 March 1755. SMC Court Minutes. GL: CLC/L/SH/B/003.

[15] SMC Court Minutes 30 March 1769.

[16] 'John Dollond was this day admitted by Redm & Sworn optician at Mr Dollonds St Pauls Ch Yd.' SMC Court Minutes 26 March 1767.

[17] From a bank ledger, it is known that Watkins dealt with Ribright. BAA: Biddulph and Cocks, fols 230–253 (current account).

[18] 'Mr Addison Smith who at last Court [December 1768] was chosen an Assistant of this Court attended this day and paid his Fine and Fees; and was sworn in and took his place as an Assistant.' SMC Court Minutes 30 March 1769.

partner as 'spy in the camp', Watkins probably felt it superfluous that he should attend court meetings and, more usefully, he could avoid all likely confrontation with the Dollonds. Indeed, it was an opportune time for him to divorce himself entirely from Company affairs. Maybe a further sign of his withdrawal from involvement in Spectaclemakers' Company affairs was his decision, in 1771, to turn over his third apprentice, George Temple, to John Glover, at a point half way through the boy's apprenticeship.[19]

Some jostling for elevation may have taken place in 1772 when Addison Smith was elected to serve as Warden alongside Peter Dollond. This move is particularly interesting, given the litigation that had taken place between the pair eight years previously, because these one-time opponents were now joint Wardens and so would be forced to co-operate. However, this was also the year that the Dollond patent expired and so there may have been some relaxation of old scores, although the Company's minutes do not reveal anything other than their respective roles being dutifully fulfilled. The shift in power is apparent 12 months later when Peter Dollond was made Upper Warden and placed in charge of the Company's accounts. From this senior position, it was not long before he was proposed as Master, on 6 October 1774. Thereafter, he served six years as Master: 1774–1776, 1778–1780 and 1782, also serving again as assistant during 1785 and Upper Warden in 1789.

The Dissenting Mr Watkins

During Peter Dollond's first term as Master, it was decided that Watkins, together with his Soho friend Edward Scarlett Jr, should be brought to account since both had declined to pay their one-shilling Quarterage for a number of years. The relevant minute of 31 March 1774 ran:

> Mr Scarlett & Mr Watkins not attending their summons It is Order'd that the Clerk do wait upon them with an Acct of their Arrear of Quarterage &c due to this Company [and further] Order'd that such persons who obstinately refused be Sumond to attend next Court to Show cause why they refuse to pay their Quarterage.[20]

Watkins refused to attend the next court; nor did he attend later courts to which he had also been summoned. In his case, the unpaid dues hung on, even until as late as 29 March 1781, when the Company ordered:

---

[19] George Temple (appr. 1767). Temple may have been turned over for a multiplicity of other reasons.

[20] SMC Court Minutes 31 March 1774.

> ... the Clerk consult Mr King of Cutler's Hall concerning the powers of this Court to oblige the said Mr. Watkins to attend; and what mode to adopt to compel him to pay the arrears.[21]

Twelve months further elapsed before Mr King – presumably a legal person or one with experience of a similar case – ruled that forfeits could be collected from a member living outside the City, which may have been the grounds on which Watkins refused payment. The dissident optician was then presented with a statement of arrears to 30 March 1782, by which time he owed fines amounting to £4 7s for the 17 and a half years between Michaelmas 1764 (when he ceased to be Master) and Lady Day 1782, together with Quarterage arrears for which a total of £2 18s was owing.[22] By this time, Watkins was nearing 60 years old, secure financially (from diverse activities that had nothing to do with the instrument trade), and with other plans afoot, he clearly did not intend to pay the final demand.

Francis Watkins and Peter Dollond were not only well separated from each other in their respective locations of business but Watkins appears to have been at pains to avoid his enemy at all times. Very possibly, the rift between them, together with any other unpleasantness caused by the patent, may have created a schism within the optical fraternity, with some supporters of the petition of 1764 remaining suspicious of the Dollonds and their assertive business style. By the mid-1780s, there were others outside the Spectaclemakers' Company and abroad who incorrectly – in Peter Dollond's esteem – belittled his father's achievement by laying priority elsewhere for the origin of the achromatic object glass. Thus, the account now shifts from the internal affairs of the Spectaclemakers' Company to the inner dealings of the Royal Society where Peter Dollond's discontent with overseas opinion was first articulated and where Jesse Ramsden argued in favour of Chester Moor Hall using supporting evidence supplied by Watkins and Smith.

Peter Dollond's Defence of John Dollond

Three very important communications were sent to the Royal Society in 1789: two from Peter Dollond and one from Jesse Ramsden. The authors, from their different standpoints, sought to clarify misunderstandings or misrepresentations and to establish the sequence of events relating to the invention of the achromatic lens on a surer footing. The first of Dollond's communications had been triggered by certain provocative remarks from abroad; Ramsden listened to the reading

21 Court and von Rohr (1930), p. 55 and p. 81, item 142.

22 Court and von Rohr (1930), p. 55.

of Peter's paper and then made haste to follow with his own version of the history, criticising Peter for his serious omission of Chester Moor Hall's name. Had Ramsden remained silent then, in all likelihood, Peter's paper would have been published in the *Philosophical Transactions*, as its author hoped. Dollond vehemently objected to Ramsden's interference and, as shown in a second communication, he struck out in an attempt to demolish Ramsden's collected evidence in favour of Hall. However, in consequence of this second paper, the Royal Society declined to give this matter any publicity whatsoever. Thus, the long-standing issue of whether or not Hall had any acknowledgeable place in the history of the achromatic lens remained a matter for debate, regardless of the judgement of Lord Mansfield against Watkins and Smith in which Hall's original invention was not denied but became an irrelevant claim due to the new way of defining originality.

John Dollond's early demise in 1761 had left his son, Peter, with the instrument business and the task of defending not only the family patent but also his father's good name as a contributor in the field of optics. The successful litigation against Watkins and Smith in 1763, the failure of the Spectaclemakers' petition to void the patent in 1764, and the crowning success of the test case against Champneys in 1766, each contributed to ensuring that the Dollond patent ran its full course until 1772. Abroad, in continental Europe, although the Dollond business went from strength to strength, there arose some irksome remarks that favoured the highly respected mathematician Leonhard Euler for the achievement, which, in Peter Dollond's estimation, was rightfully due to his father. Most of those abroad who knew of Dollond's correspondence with Euler in the early 1750s appear to have been unaware of Euler's later (1762) acknowledgement of Dollond's achievement; instead, they preferred to emphasise that since Euler had been the first to specify the parameters of a compound object glass, albeit with his theory untested, he should retain the credit.[23] Thus, well before Peter Dollond's communication to the Royal Society, there was already a tendency to support Euler's name

[23] Dollond's Royal Society paper of 1757 prompted Euler to review his own work on achromatism. In a paper read to the Berlin Academy on 16 September 1762 and published in 1769 entitled 'Considérations sur les nouvelles lunettes d'Angleterre de Mr.Dollond et sur le principe qui en est le fondement', Euler concluded that his own law of dispersion and that of Newton were distinct and that the sort of connection assumed earlier by Dollond did not exist. More simply put, Kelly (1808), p. 72, refers to Euler's retraction in the following way: 'In 1764 Euler says "I am not ashamed frankly to avow, that the first accounts, which were published of it, appeared so suspicious, and even so contrary to the best established principles, that I could not prevail upon myself to give credit to them ... I should never have submitted the proofs, which Mr Dollond produced to support this strange phenomenon, had not Mr Clairaut, who must at first have been equally surprized at it, had not most positively assured me, that Dollond's experiments were but too well founded."'

in the order of priority, and so to marginalise Dollond's name. By offering the scientific community a carefully constructed account of his father's work, Peter had hoped to correct such misunderstandings. It was, of course, understandable that he should seek to do this through the Royal Society because they had conferred honour upon his father precisely in recognition of that achievement.

The progress of science is rarely halted because of a single objection or contrary theory, however obstinate or influential the proposer of such may be. One view of scientific activity is that experimental work should always be treated rigorously and dispassionately and the results of experiments should be considered sceptically and tested ruthlessly and independently. Only then, by consensus, can a new idea – in this case one that entailed correcting the great Newton – be accepted by the scientific community at large, and only then can progress be said to have been made. In the case of Newton's experiment 8 in *Opticks*, the unhelpful conclusion – which seemed to assert that chromatic effects could never be eliminated – was not rigorously or sufficiently examined at the time, and so the advance of the refracting telescope was halted temporarily, although its progress did not stop altogether. It was inevitable that one day the experiment would be scrutinised; indeed, Euler did just that by considering the fallacious (but heuristically productive) analogy that the differing humours of the human eye were responsible for producing a colour-free image. Even earlier, the lawyer Chester Moor Hall, who was certainly fascinated by optics, may well have taken delight in detecting the error in Newton's law-like propositions for dispersion, although perhaps not daring or having the time to contradict the accepted viewpoint. Lamentably, any account that Hall may have confided to his personal notes had long-since vanished; in any case, no account of Hall's work appeared in print. So, from the Dollond viewpoint, no credit could be laid at his door, whatever his labours or success. Dealing with the legacy of the powerful Euler, however, was quite another matter.

Euler – whose prodigious and penetrating abilities were renowned in so many fields – tackled the dispersion laws to produce a mathematical model that cleverly retained Newton's experimental results although not his theorems. Of this intellectual exercise, the French mathematician Alexis Clairaut remarked that, without experimental confirmation, Euler's contribution was more ingenious than useful. Samuel Klingenstierna – who followed in Euler's path also without recourse to experiment – produced a variation in the dispersion theorem (dependent upon the trigonometrical sine rule) that could be applied to a compound object lens. Dollond denied employing any of these continental ideas and he certainly acknowledged neither Euler nor Klingenstierna in the all-important 1757 Royal Society paper describing his own refrangibility experiments. Yet, there were others who argued – with some justification – that it was precisely the modelling set out by these mathematicians (particularly that of Klingenstierna) that had caused Dollond to re-assess his own earlier imperfect experiments, and so bring about his own ultimate success.[24]

Unfortunately, John Dollond did not live long enough to defend his contribution to those critics abroad although he did have the support of one optical enthusiast: Alexis-Marie de Rochon. Long before his appointment as director of the Marine Observatory at Brest, he wrote a pamphlet in 1768 which drew attention to Dollond's achievement and Euler's failure. Recalling this detail in the 1790s, Rochon wrote:

> I advanced, in my mathematical printed pamphlet [of 1768], that Euler had been *the first who has wanted to correct the aberrations of refrangibility* by the use of different refractive materials. Maupertuis set himself the task to construct the [proposed] objective of Euler in Paris with water and glass, but this objective did not succeed, due to the relation now known between the refraction and the dispersion of plain glass and water. These relations are in the proportion as 155 to 133 for refraction, and for dispersion as 100 to 67. Now the large curvatures that must be given in order to destroy the chromatic aberration caused a very strong spherical aberration, so that which was gained on one hand was lost of the other. Newton shows this by giving the measurements of an objective of his composed invention; likewise, it is in the proof of Euler, for glass and of water, although unjustly correcting the aberrations due to the spherical face of glasses. *It was therefore necessary to fail Euler's project*, and, at the same time, to confirm Newton whose assertion seemed to remove all hope of attaining achromatism. From this time, one assigned to Jean Dollond the first understanding of the strong dispersion of a kind of glass called flint glass....[25] [Author's italics]

At this stage in Rochon's career his views carried little weight and, in any case, Peter Dollond was unaware of this when he wrote his first communication to the Royal Society.

What then were the concerns that led Peter Dollond to write formally to the Royal Society on 21 May 1789? To answer this, we must begin with Joseph Jérôme de Lalande of the Collège de France, for it was from his popular writings that repeated instances of mis-accreditation occurred. In 1764, Lalande had published a highly-regarded text on astronomy in which he drew attention to John Dollond's rejection of Euler's proposal of a new approach to the chromatic

---

[24] Brewster believed that Euler's 1747 paper was formative to Dollond's thinking for he writes: 'though the memoir of Euler is not entitled to the praise of originality [because Newton had preceded him], it had the great merit of exciting the inquiries of philosophers, respecting this long neglected subject.' Brewster (1830), I, pp. 104–105.

[25] Rochon (1801a), pp. 171–172, considers the availability of lead glass in England and how dispersion increased beyond a minimum according to the quantity of lead in the composition.

problem but how, after being convinced by Klingenstierna, he looked for a way to combine contrary refractions with flint and crown glasses to form achromatic object lenses.[26] There was nothing particularly inaccurate about Lalande's simple statements of fact and other authors copied his description of how refracting telescopes could be improved by this means. However, the weight of Lalande's description gave greater credence to Euler and Klingenstierna, thus Dollond's experimental endeavours were ranked, by implication, at a lesser level. Of course, this fitted the hierarchical conception of scientific knowledge in which theory always took primacy over experiment. For Peter, this was considered a snub of his father's achievement although the remark was allowed to pass, until 1783 when Nikola Fuss – a former student and colleague of Euler during the years he spent at the St Petersburg Imperial Academy of Sciences – composed a voluminous eulogy on Euler at his passing. Bluntly, this left the reader in no doubt as to where John Dollond should be placed in the order of credit.[27]

Fuss had contrasted Dollond's dependence upon Newton's authority in his naïve rejection of Euler's formulation with Euler's disregard of the Newtonian principles of dispersion from the very outset, concerning himself only with Newton's experimental results. Indeed, Fuss referred to some experiments carried out by Euler himself who had filled the meniscus between two lenses with different liquids to test whether refraction was possible without dispersion. Thus, Fuss concluded that Dollond should be remembered *only* for having:

> ... found two sorts of glass in the meanwhile which crowned at last, in 1757, the happy conjecture by Euler, by the invention of achromatic telescopes, thus forming a new epoch in astronomy and dioptrics.[28]

Of course, this remark was an oversimplification, albeit written to lionise the memory of the great mathematician. But Fuss had completely ignored Euler's later acknowledgement of the correctness of Dollond's experimental work which had led to his withdrawal of his original mathematical hypothesis of dispersion. Peter Dollond was even more indignant that Fuss had employed the phrases: 'Mr Dollond's success which *so substantially borrowed* from a theory that he had first criticised ...' and that suggesting his father had '*finally admitted* to Euler's conjecture in 1757 by the invention of achromatic glasses'. These remarks irritated Peter who claimed that his father's contribution had depended solely on his own individual work.

---

[26] Lalande, *Astronomie*, vol. 2 (1764), pp. 837–838; *Astronomie* 3rd edition, vol. 2 (1792), p. 574 (§ 2298).

[27] Fuss (1783).

[28] Quoted by Peter Dollond himself in Dollond (1789), p. 21. Dollond's own translation of Fuss has been compared here with Glaus's translation and so modified to make grammatical and intellectual sense.

The Lalande–Fuss viewpoint was reinforced in 1788 when Jean-Dominique Cassini, incumbent astronomer at the Paris observatory, although conceding that '[Dollond has] a just right to *partake* of the glory', went on to argue that this was only for having made use of Euler's ingenious idea of employing two different mediums. By now, Peter could no longer endure such 'false representation'. It was time to correct these inaccuracies for the sake of the history of the subject. It was on these grounds that Peter submitted his first paper to the Royal Society – 'Some account of the discovery, made by the late Mr. John Dollond, F.R.S. which led to the Grand Improvement of Refracting Telescopes, in order to Correct Some Misrepresentations in Foreign Publications' – where it received an airing on 21 May 1789.[29] The contents of this paper are of considerable significance, not because of Peter's honourable intention of defending his father's name, but because – as will be shown – it re-awakened the very issue that had underpinned the joint defence of Francis Watkins and Addison Smith in Peter's lawsuit against them in 1763, reiterated in the Spectaclemakers' petition of 1764, namely, that Chester Moor Hall should not be forgotten if the history of this optical development was to be debated.

Jesse Ramsden's Defence of Chester Moor Hall

Jesse Ramsden – or 'honest Jesse' as he later came to be called – was present at the Royal Society meeting on 21 May 1789 when Peter Dollond's paper was read. The mere mention, in the opening remarks, that it was now time for 'correction of any inaccuracies or false representations in the history of science' was sufficient to trigger the conscientious Ramsden into what he felt was his bounden duty. He felt he should ensure a place for Chester Moor Hall whose name was singularly lacking in Peter Dollond's account, just as it had been lacking in his father's paper of 1758. But it was now a lifetime – some 57 years – since Hall had made his first achromatic doublet and over 30 years since he had been subpoenaed to attend the court as a witness on behalf of Francis Watkins and Addison Smith in the action brought by Peter Dollond at King's Bench in 1763. Since that trial, Hall's place as originator of the flint–crown combination had not been in dispute; indeed, Lord Mansfield had denied this was an issue. But in terms of the historical record, the only published note of Hall having been engaged in an optical pursuit appeared as a vague aside in Caleb Smith's guide to the sea quadrant written about 1735. It was, therefore, important for Ramsden to gather whatever evidence there was to hand in order to complement Peter Dollond's concern for historical accuracy. Of course, Peter's concern did not stretch to allowing Hall any public acknowledgement because, in the eyes of those abroad who knew only of Lalande's version of events, it would have diminished his father's position even further.

29 RS: L&P decade IX, letter (P. Dollond), 131.

Figure 8.2 Jesse Ramsden, engraving by John Jones after the painting by Robert Home, 1790. © The Trustees of the British Museum, Ee, 1.24

Peter Dollond's initial communication of 21 May 1789 had been a reasonable defence against the continental authors Lalande, Fuss and Cassini whose mis-apportionment of credit would have been fixed for the rest of time and for that alone a correction was required. That his father's contribution only lay in finding the necessary proportions in the flint–crown glass combination to support Euler's hypothesis was an oversimplification, even wrong: indeed, John Dollond had done much more by confirming Newton's error and, through finely tuned experiments using flint concave and crown convex glasses, correcting the principal aberrations simultaneously. Moreover, his work was carried out independently of anything that Euler had proposed by way of his proposed combined object glass.

Beside the irksome remarks from abroad, there persisted, at home, among the optical practitioners, the suspicion that Dollond was only led to his success by having gleaned knowledge of Hall's work from within the trade. Hall had died in 1771; after so many years his part in events might have been entirely forgotten but for Ramsden's resurrection of the old controversial issue. Indeed, Ramsden probably felt it his duty – as a newly elected fellow – to inform those attending the next monthly meeting the truth (as he understood it) concerning Hall, thereby securing a place for him in the chronology of telescope development. Preparing evidence to this end, however, was not altogether a success: four of the crucial witnesses (Ayscough, Bass, Bird and Eastland) – who each must have held intimate, first-hand knowledge of Hall's efforts – were also dead. Fortunately, a few were still alive who could add weight to his cause.

The short timespan between Peter Dollond's initial communication and Jesse Ramsden's follow-up – a period of one month only – was sufficient for Ramsden to collect the salient evidence to form a supplement to the history of events outlined by Peter Dollond. This was compiled and presented by Ramsden himself on 18 June under the title 'Some observations on the invention of achromatic telescopes'. At the same meeting Nevil Maskelyne, in a semi-supportive way to the cause of Peter Dollond's claim for his father as inventor of the achromatic telescope, refuted Euler's hypothesis of 1748.[30] But, in Ramsden's paper, credit was given directly to Hall for starting this new line of optical investigation. In the absence of any published record, Ramsden's evidence was largely based on personal recollection and second-hand oral account. For example, one anecdotal fact – deemed 'the most Authentic evidence' – was no more than the recollection of a naval admiral who had recalled John Bird having made up achromatic telescopes for Hall. Another detail – from the depths of Ramsden's own 'perfect memory' (so he claimed) as a young man of 19 or so years – concerned Ayscough having told him personally of a telescope with a 'wonderful performance ... [based upon] ... a new invention'. Undoubtedly, such recollections contained an element of truth, as did John Dollond's own (alleged) admission to Ramsden that Hall *did have the invention before him*, a detail he claimed to have 'often been mention'd ... in conversation'.[31]

Something beyond these flimsy recollections was needed in his submission to the Royal Society and, to this end, Ramsden elicited written evidence from Addison Smith and Francis Watkins, both of whom had been at the nub of the Hall issue. Of these testimonies, Smith's is the most informative: indeed, it is from this source that the essential date and details of how Hall came to construct his first achromats is known. Thus, as vital and primary evidence, Smith's response to Ramsden's queries (dated 4 June 1789) was annexed to Ramsden's fair copy to the Royal Society. The following extract indicates how Hall kept the idea secret:

30 RS: MS L&P decade IX, letter (Maskelyne), 130.

31 RS: MS L&P decade IX, letter (Ramsden), 138.

> Attending his [Hall's] Experiments had fully convinced him that his Theory was well founded, in consequence of which he then applied to Mr Scarlett Optician in Soho (Father to the last Mr Scarlett) to grind him one glass of a certain diameter and spherical proportions, and afterwards applied to Mr Mann Optician in Ludgate Street for him or some of his workmen to grind him another glass, which was to answer in Diameter and Spherical figure to that Mr Scarlett had done for him, wch compound Object Glass made of Flint and Crown Glass ground with Concave and Convex surfaces he gave to me in the Year 1763 and told me it was the first that was made and to the best of his recollection was made about the year 1733, which of the two Glasses was worked by the two different Artists I know not, as I was not curious enough at that time to inquire, he told me his reason for employing different persons for grinding his Glasses was to prevent his Discovery being known ....[32]

As for Watkins's testimony – whose original letter to Ramsden is now lost – Ramsden's chosen extract is repeated here as a parting shot from the main optician who had felt so aggrieved by the whole issue. In one short passage, Watkins summarised exactly the outcome of Lord Mansfield's deliberations in 1763:

> It was clearly prov'd that Mr. Hall had several object Glasses ground by Mr. Bass and others of the trade, many Years before the late Mr. Dollond began to apply himself to grind any glass. Mr. Hall in court produced several object Glasses, and the court admitted him the prior Inventor, but Lord Mansfield argued very strongly that there had not been any sale, or the Invention made public & he then directed the jury to find for Mr. Dollond, – Mr Bird and myself went to Mr. Hall's house near Rochford in Essex. He shew'd us a very great variety of Glasses, and had some made as long ago as the Year 1732 or there abouts ....[33]

Peter Dollond's Response to Ramsden's Communication

Peter Dollond – although not a Fellow of the Royal Society – made it his business to attend three of the meetings following that at which Ramsden presented his case for Hall. As an outsider, his presence was by invitation: so, on 21 May and 25 June he appeared by invitation of Nevil Maskelyne and on 2 July by invitation of Edward Nairne. Maskelyne had been a good customer at the shop since his appointment as Astronomer Royal and Nairne knew Dollond as a fellow member of the Spectaclemakers' Company. Whether anything of relevance was said – informally or through discussion – at these gatherings can only be

---

32 RS: MS L&P decade IX, letter (Smith), 139.

33 Ramsden (1789).

imagined although it transpires that he was immensely annoyed by Ramsden's interference and was impelled to prepare a formal response. This came on 12 November 1789, after having had sufficient time to reflect upon what Ramsden had proposed in the form of 'the truth'.

Peter Dollond's response – 'An Answer to a Paper presented to the Royal Society by Mr. Jesse Ramsden F.R.S. relating to the Invention of the Achromatic Telescope' – opened with a summary of Ramsden's quibbles. Importantly, and for the first time, Peter admitted that his father had indeed gained a hint concerning the nature of flint glass during a visit to Bass's workshop, although he stressed that Bass was ignorant of the importance of flint glass in chromatic correction. From this visit he obtained a piece of plate glass – not crown glass, as stated by Ramsden – and a piece of flint glass. On returning home, his father showed him the two pieces and told him:

> ... if the difference of the dissipation of the different coloured rays could be easily perceived in looking through the two reading glasses in that way [that is, the glasses with different dispersive powers], he had no doubt but that the difference of the dispersion of the colours in the different kinds of glass would be found much greater than he had supposed it could be and on that account he should go on with his Experiments of trying all the different kinds of glass as fast as possible.[34]

In other words – as surmised in Chapter 5 – John Dollond had indeed perceived that flint glass would offer a solution to the 'stumbling block' in his experiments. Peter went on to explain that this revelation was not held as a secret because he had told a number of people of his intention to proceed, including Ramsden. Peter had then encouraged his father to proceed with speed.

Concerning the issue of Bass as informant, Peter explained:

> Mr Bass ... recollected to have ground glasses, something similar to those for which the patent was granted, many years before, for Mr Hall. He accordingly wrote to Mr Hall to inform him of what was done; and until this time, we had never heard of Mr Hall.

So it was Bass who was the informant and his letter caused Hall, on his next visit to London, to call upon Dollond at his home. As for the conversation that took place – re-told to Peter later by his father – Hall explained that he had attempted something along similar lines some 30 years previously but:

---

34 RS: MS L&P decade IX, letter (P. Dollond), 146.

> ... on account of the difficulties he had found in getting the glasses worked as he directed, and his being obliged, at that time, to apply himself to the study of the law, he had then laid aside his improvement of the telescope and had never taken it up since.

Peter added that 'he [Hall] was happy to find, that somebody had succeeded', and hoped that 'the invention was likely to become usefull'.

If Peter was truthful concerning these crucial remarks – that neither he nor his father had heard of Hall until the day that Hall turned up on their doorstep – then his father's position – of not having acknowledged any priority to Hall – may not have been so dishonourable. Conceivably, Dollond's visit to Ayscough – during which he informed the opticians (Eastland, Champneys, Stedman) that he had obtained the patent for Hall's invention – might then have taken place after Hall's visit to his home. However, no such support for this contention was given at the trial of Watkins and Smith in 1763, possibly because Lord Mansfield had quickly decided that Hall's position was irrelevant to what he had in mind by way of a new interpretation of the law concerning invention. Looking less kindly upon Peter Dollond, it is also conceivable that his account of Hall's encouragement to his father was fabricated. Leaving aside this question of authenticity and honesty, Ramsden's paper contained three further points that required specific attention: the evidence of Robert Rew, the evidence of Admiral John Campbell, and the written testimonies of Smith and Watkins. Each was dealt with in turn, and derisively.

Concerning Robert Rew, he was not an optician but a brass worker who worked for Francis Watkins. His name had appeared on the Spectaclemakers' petition, together with a sentence stating that it was he who had informed John Dollond, in 1755, of the process of making a combined object glass. Peter Dollond did not deny that such an encounter had taken place but rather admitted it. However, casting doubt on Rew's credibility, he suggested:

> To show, that one ought not to be too ready to give credit to the pretensions of others to any new invention, after it had been published, I shall mention another person, who was brought forward by Messrs Watkins & Smith. This was one Robert Rew, a workman of theirs in the brass way, who declared on oath in the Court of Common Pleas, before Lord Camden: that he [Hall] was the real inventor of the achromatic telescope; that on a certain day, he met Mr Dollond in Cheapside and went with him into the Fountain Public house and that there, over a tankard of porter, he communicated to him the whole principle of the invention. To this evidence some credit might have been given, had he appeared at the first trial; but, as there could be no cause shown, why he did not, his evidence falls to the ground.[35]

35 RS: MS L&P decade IX, letter (P. Dollond), 146.

'Common Pleas' referred to the trial against Champneys in 1766 while the 'first trial' meant that against Watkins and Smith at King's Bench in 1763. It is certainly curious that Rew had not been called to support Watkins and Smith at the first trial, although his absence might be explained in several ways: for instance, he may not have been called to bear witness because Watkins and Smith, or their attorney, may have considered him irrelevant to the argument in hand for which they had brought in Hall as their principal witness. But, if Rew's claim bears any truth at all then it must be understood that he simply knew of Hall's endeavour through gossip in the trade.

Rew's presence at Common Pleas in 1766 (by subpoena) was on behalf of Champneys, and so was irrelevant to the 'first trial'. However, since the verdict at King's Bench in 1763 did not favour Watkins and Smith, Rew may subsequently have been encouraged to add what he knew to the petition in 1764. Other questions arise: if Rew's volunteered information to Dollond on the achromatic telescope was of any value at all then it is curious that he did not volunteer the same know-how to Watkins who employed him. Indeed, Watkins did not admit any prior knowledge of the 'invention' at King's Bench in 1763, nor was Rew mentioned in Watkins's 'answers' to Dollond Jr's bill of complaint in 1765. Without further insight into Rew's credibility, the value of his evidence is marginal although the very fact that he met with John Dollond by arrangement does suggest that John Dollond was 'fishing' for information from within the trade. Finally, Peter Dollond's overly strong rejection of Rew as a credible witness might, otherwise, be construed to mean the opposite to what he had intended: in other words, that information received from Rew was very significant indeed. Concerning the evidence of Admiral John Campbell, RN, FRS – a known customer of Bird who recollected that he had made object glasses for Hall – Peter Dollond derisively remarked:

> [Bird] was a mathematical instrument maker, and perhaps did not understand much of optics [and, even if he had worked such achromats for Hall, he thought it] most likely they performed very badly; for it is well known that he had so great a prejudice against achromatic object-glasses that it was with the greatest difficulty he would be prevailed upon to apply them to the instruments which he made.[36]

Apparently, Bird was given a chance to speak at Champneys's trial, and this (according to Peter Dollond) was an opportunity lost had he anything of import to say. Despite Peter Dollond having cast aspersions on the quality of Bird's

36 Ibid. But Bird's prejudice was in applying achromatic glasses to the telescopes that he fitted to mural quadrants, finding that the extra weight of glass caused the telescope tube to flex when it was raised from the vertical. [Editors' note].

workmanship, it should be noted that Bird was heavily engaged by the Royal Observatory and therefore, for that very reason alone, must have been judged competent by those who employed him. Such disparaging remarks therefore smack of jealousies or fear of loss of credibility of his original argument in defending his father. Also, we should note that the very fact that Bird accompanied Watkins and Smith on their mission to Hall's home in Essex helps to confirm Bird's knowledge of Hall, possibly having made up flint–crown combinations for him. Thus, in court, Bird would have been able to corroborate the evidence presented by Watkins and Smith's attorney by authenticating Hall's object glasses.

Peter Dollond made no specific criticism about the written testimonies of Watkins and Smith. Instead, he simply remarked that he took both these opticians to court in order to prove the validity of the patent and, since this case was won, the presence of Hall and Bass as witnesses carried no weight in any subsequent argument or claim. Thus, having demolished Ramsden's evidence, Peter Dollond then punched home with the following memorable lines: [37]

> I hope that the philosophical truth of the invention of the achromatic lens by John Dollond will now appear to be as well established as the legal truth had already been – an ingenious distinction introduced into this subject by Mr. Ramsden, a distinction however not very evident to plain minds, as truth must be the same whether it be found in Westminster Hall or in the Royal Society.

This conclusion was plainly intended to quash for ever any hope of Chester Moor Hall being written into the history of the achromatic telescope. But, if Peter had imagined that his cleverly worded conclusion had cleared the way for his first communication, that of May 1789, to be published in the Society's *Transactions*, then he was very much mistaken.

Truth Will Out

Ramsden's attempt to rectify the history of the achromatic telescope by seeking some place for Chester Moor Hall had failed because of Peter Dollond's derogatory response. Recently discovered complementary evidence now shows that, at some point around August 1789, Peter had lobbied Alexander Aubert – a man of exceptional standing as the Royal Society's Secretary – by requesting him to allow a reading of a reply that he had prepared in response to Ramsden's paper.[38] Courteously, Aubert took Peter's paper and discussed it with eight other fellows at their dining club, held in a room above the George & Vulture (Figure 8.3).

---

37 RS: MS L&P decade IX, letter (P. Dollond), 146.

38 Aubert (1951–1952); Jungnickel and McCormmach (1996), p. 217.

Sir William Musgrave	Barrister
John Smeaton	Engineer
John Hunter	Anatomist
Henry Cavendish	Natural philosopher
Edward Nairne	Optician
John Lee	Physician
Sir Charles Blagden	Physician
Baron Mulgrave	Politician

Figure 8.3 Fellows of the Royal Society who convened at the George & Vulture to discuss Peter Dollond's objections to Ramsden's support for Chester Moor Hall

Together, they agreed that Peter's response was 'temperate' and 'discreet' and so they agreed to advise Joseph Banks, the President, that a reading of this 'innocent matter' might be allowed, pointing out to him that Dollond might otherwise construe a refusal to be a form of suppression. Banks concurred and, as we now know, the reading went ahead. As a safeguard, however, Aubert further recommended that 'all [the three papers] then be condemned to remain among the archives of y^{r} Society' and, just in case Banks disagreed with the dining club view, he added:

> ... pray pardon my troubling you thus & if you should think I am wrong in my judgement or in writing to you about it burn my letter ....[39]

Aubert's letter survives and, as revealed here, neither of Peter's communications, nor that from Ramsden's pen were published in the Society's *Transactions* and so the justifications and arguments that once seemed so vital have, in fact, lain unattended in the Society's archives for well over two centuries. Had the papers been published, then the names of Addison Smith and Francis Watkins would have come to the fore and so their statements would have been presented for posterity. At least, the case for Chester Moor Hall would have been presented openly.

Of course, the decision not to publish was very prudent; the Royal Society would not wish give space to a matter which, to outsiders, might appear to be no more than an internal family conflict due to Jesse Ramsden's separation from his wife, Sarah (Peter Dollond's sister); nor would the Royal Society wish to

[39] BL Add 33978 f. 251 and reprinted in Chambers (2007), 3, pp. 493–494.

lose favour with foreign members, some of whom Peter had criticised in his initial paper in defence of his father's priority. This outcome angered Peter so much that he immediately set about having the original communication of 21 May 1789 printed privately. This appeared with a fore date of 1 September 1789 and was distributed from the bookshop of J. Johnson, and probably also given gratuitously to all and sundry who visited the shop of P. & J. Dollond at St Paul's Churchyard. But such must have been the feeling of those who supported Ramsden's viewpoint that some anonymous person, aware of the archived contents of Ramsden's paper at the Royal Society, re-wrote all that had been said in favour of Chester Moor Hall and sent this to the *Gentleman's Magazine* under the pseudonym 'Veritas' by letter dated 30 September 1790. This outflow into the public domain was sufficient to create a ripple of half-truths among the scientific community, some of whom believed that Veritas was none other than Ramsden himself.

After Peter Dollond's private printing and the *Gentleman's Magazine* piece by Veritas, the facts became confused and the problem of attributing priority for the achromatic lens more a matter of choice and faith than due reference to primary sources. Nor was this matter of true record helped by a report appearing in the *Journal de Physique* which undoubtedly caused Peter Dollond further concern. Furthermore, the Veritas piece was partially reprinted in the *Philosophical Magazine* and this then reported in the *Journal de Physique*, which – as was the custom of the *Philosophical Magazine* – was re-translated back into English, and so the issue and confusion reverberated throughout the scientific community. Each of these communications (discussed below) tended to elicit sympathy for Hall, or at least strengthen his image as the true inventor of the achromatic lens, thus causing Peter Dollond further annoyance. Clearly, the necessary equivalence of 'philosophical truth' and 'legal truth' – as demanded by Peter in his reply to Ramsden – had not put the matter to rest. Snubbed by the Royal Society who would not publish his paper, and in fear of losing the fame upon which the reputation of the business was built, namely his father's 'invention', Peter then arranged for all his father's papers, including the unpublished communication to the Royal Society together with a relevant paper by Maskelyne, to be published privately. These appeared as an ensemble edited by John Kelly under the general heading *The Life of John Dollond, FRS*. However, this was not issued until safely after Ramsden's death.

Veritas Speaks

As an adult apprentice, Jesse Ramsden often visited the Dollonds who helped to improve his scientific understanding and in 1766 he married John Dollond's youngest daughter, Sarah. As Peter Dollond remarked of this time:

> [Jesse] was almost continually at our house getting all the information he could, both in the theory and practice of optics; for he had articled himself to a Mr Mark Burton, who lived very near to us; a very honest man and a maker of mathematical instruments but one under whom he could not improve himself in theory of any kind, or in the practice of optics .... [40]

It is almost certainly due to this intimacy with the Dollond family that no one invited him to sign the petition of 1764, but it is extremely unlikely – as sometimes reported – that he did not sign the petition because he had been promised a dowry for Sarah for her wedding from the patent proceeds.[41]

From about 1763 until 1766, Ramsden worked independently in the Strand, not far from the Dollonds; however, for the remainder of his working life he worked in the St James district of Westminster, from 1767 in the Haymarket and then from 1772 in Piccadilly. He and Sarah parted before 1780. Ramsden was, in every way, as reputable an instrument maker as his brother-in-law but it was his unusual talent for design and detail that eventually earned him his election to the Royal Society in January 1786. One of his most famous contributions was a dividing engine capable of an accuracy of 0.5 seconds of arc, a significant leap in accuracy over the best then available, which only offered a tolerance of 3 seconds of arc. In 1795 he was honoured with a Copley Medal for his 'various inventions and improvements to philosophical instruments'. By contrast, Peter Dollond, although recognised for his talent among the optical practitioners and astronomers, did not quite show the same all-round flair as Ramsden. Such differences in success, personality and temperament must surely have been relevant factors in the disagreement that developed between them.

As already stated, neither Dollond's nor Ramsden's communications to the Royal Society were passed for publication. Nevertheless, the contents of these unpublished papers revitalised the support for Chester Moor Hall following the leak by Veritas to the *Gentleman's Magazine* by letter dated 30 September 1790. The letter itself is curious and appears as a reaction to some published remarks elsewhere concerning the origin of the achromatic lens. Veritas began by suggesting:

> The writer of an Introduction to some Letters lately published on the Improvement of Ship-building seems to have been misled in saying, 'That great discovery in optics, the achromatic glasses, was entirely owing to three or four ingenious men attending at a public house in Spitalfields, to amuse themselves in friendly conversation upon mathematical and mechanical subjects.'[42]

---

40 RS: MS L&P decade IX, letter (P. Dollond), 146.

41 McConnell (2007), p. 17.

42 Veritas (1790), p. 890. The source of this quotation by Veritas remains unidentified.

Clearly, the author of the article on 'ship building' had known of Dollond's Spitalfields origin and the mathematical society there, but Veritas was keen to correct the content regarding the origin of the achromatic lens:

> ... As the invention [of the achromatic lens] has been claimed by M. Euler, M. Klinginstierna, and some other foreigners, we ought, *for the honour of England*, to assert our right, and give the merit of the discovery to whom it is due; and therefore, without farther preface, I shall inform the author of the above quotation, that the inventor was CHESTER MORE HALL, Esq. of *More-hall*, in *Essex*, who, about 1729, *as appears by his papers*, considering the different humours of the eye, imagined they were placed so as to correct the different refrangibility of light. He then conceived, that if he could find substances having such properties as he supposed these humours might possess, he should be enabled to construct an object-glass that would shew objects colourless. After many experiments, he had the good fortune to find these properties in two different forms of glass; and by forming lenses made with such glass, and making them disperse the rays of light in contrary directions, he succeeded. About 1733 he completed several achromatic object-glasses (though he did not give them this name), that bore an aperture of more than 2½ inches, though the focal length did not exceed 20 inches; one of which is now in the possession of the Rev. Mr. Smith, of Charlotte Street, Rathbone Place....[43] [Author's italics]

Two points now arise: firstly, that the achromatic lens was certainly *not* due to John Dollond or the Spitalfields clique, and secondly, that Chester Moor Hall had pre-empted all others (if any others at all) in the invention of an achromatic object glass. There is the assertion that foreigners (and probably implying those of Huguenot parentage) should not be allowed to claim a discovery which was due to an English gentleman. This remark has a similar overtone to that recorded by Benjamin Martin's earlier claim for Caleb Smith. Particularly interesting in this quotation is the remark about 'his papers', which suggests that at the time of the Watkins–Smith–Bird's visit to Hall's home in Essex there existed some written account of Hall's hypotheses and experiments.

Conclusions

In this chapter we have examined the changing allegiance in the membership of the Spectaclemakers' Company which followed in the wake of the petition and litigation of the 1760s. It is noteworthy that Francis Watkins withdrew from active participation at the same time that Peter Dollond made his presence felt.

---

43 Veritas (1790), p. 890.

More intriguing still is the apparent cover-up by the Royal Society in 1789 which did not wish to become embroiled in the issue surrounding the disagreement between Ramsden and Peter Dollond. Peter's father had caused sufficient embarrassment with foreign members in 1761 and it would appear that those who discussed Peter's objections to Ramsden's attempt to find a recognised place for Chester Moor Hall feared further complication and embarrassment should their respective papers appear in print. Thus, apart from a simple reading, the issue was allowed to slide into oblivion. This caused immense disappointment for Peter although he arranged for his first paper of 21 May 1789 to be printed privately. The issue concerning Chester Moor Hall would not go away and, as we have seen, the 'truth' came out in the following year through subterfuge and anonymity by one Veritas. This tactic only added to the mystery, and so, led to one of the longest running sagas in the history of science.

We now understand that Mansfield's revolutionary interpretation of the concept of 'invention' had the immediate effect of denying any prior invention that had not been presented in the pursuit of commerce. Hall did not do this and so the path was open for Dollond. Mansfield's clever switch of meaning of the term 'invention' was very much in keeping with the envisioned demands of patent law and, indeed, within the bounds of the 'spirit of the law' which needed to evolve to cope with the growing industrial trend. So, from the time of Peter Dollond's success over James Champneys, it was deemed that 'the publisher is the true and first inventor' and not – as in the case of Hall – 'the one who confined something to the closet'.[44] Watkins himself, in writing to Ramsden in 1789, reminded him of how strongly Mansfield had argued this point at the King's Bench trial in 1763. Thus, the legal twist had generated a new outlook and, of course, as a law – tried and tested again in 1766 in the case of *Peter Dollond* v. *Champneys* – it had to be upheld. Of course, what motivated Watkins to defend his position initially stemmed from the devious manner in which John Dollond had acted, but that was a 'wrong' of quite a different nature.

Some 35 years later, Ramsden still could not accept that Hall had no place in the scheme of recognition because he knew the details of the method of combining flint and crown glasses had been divulged to John Bird, then Ayscough, and onwards to Eastland and Champneys who did make modest, commercial headway. John Dollond, too, was aware of the practice within the trade using crown and flint glasses prior to submitting an application for patent through Chancery. As we now can see (from the corollary above), the patent – while it lasted – did reap an enormous advantage for the Dollond business. Other instrument makers, more attuned to an older style of trading depending on customer loyalty, remained disconsolate and readily put their names to the Spectaclemakers' petition in the hope of stopping Peter Dollond in his tracks.

44 Webster (1844), p. 43.

Chapter 9
The Unexpected Longevity of Chester Moor Hall

Over the years, several opinions have been expressed about Chester Moor Hall's place in the history of the refracting telescope, mostly depending upon the evidence provided by Veritas in 1790, and none leading to any definite conclusion. The confusion apparent at the end of the eighteenth century remained throughout the nineteenth century, and so some support for Hall's name as inventor of the achromatic lens remained, despite Lord Mansfield's ruling that an invention must be 'published' commercially through a patent. Such a ruling did not fit comfortably with the longer tradition whereby the first to establish a new device was honoured as inventor. The quest to identify Veritas continues to reappear from time to time. He was surely known among the close circle of scientific persons of the 1790s, particularly Jesse Ramsden whose views he represented. Because neither the court actions, nor Ramsden's original unpublished paper to the Royal Society, were known to those striving to establish the origin of the achromatic telescope, Veritas offered the only published evidence in Hall's favour. Peter Dollond had dismissed Hall's name, and moreover, it was completely ignored by the family in a subsequent publication on John Dollond's life. The present chapter, therefore, relates how these strongly opposing views persisted throughout the following two centuries.

Residual Accounts at the Turn of the Century

Alexis-Marie de Rochon and the Institut de France

The first person abroad to draw attention to the unhappy situation between Peter Dollond and Jesse Ramsden – nominally brothers-in-law, but hardly on friendly terms since Sarah's parting from Jesse to return to the family fold – was Alexis-Marie de Rochon (1741–1817).[1] Seldom mentioned in the literature, it is worthwhile pausing to consider why this obscure Breton should have had more than a passing interest in both Dollonds (father and eldest son) and their achromatic object glasses.

---

[1] Fauque (1985); Dezos de la Roquette, article 'Alexis de Rochon', *Biographie universelle ancienne et moderne (*reprint 1995), vol. 36, pp. 258–263.

In fact, Rochon had much to offer the new French Government in the 1790s through his wealth of experience. His talent for experiment, possibly unrecognised during his youth, was nearly lost forever when his father, an *aide-major* at the Chateau of Brest, placed him as intern at the monastery of St Martin-la-Garenne (Nantes). He left the monastic life a few years later to return to Brest where, in 1765, he became librarian and astronomer at the marine academy. Following his overriding passion for practical optics, it was not long before he submitted a paper – '*Sur les moyens de perfectionner les instruments dioptrique*' [Concerning methods of improving optical instruments] – to the Académie royale des sciences in 1766. In April 1767, he was able to join the ship *Union* – conveying the consul and ambassador to Morocco – thereby gaining invaluable navigational experience. Thereafter, he made further voyages to Madagascar, Ascension Island and the East Indies.

It was through concerns for improving the telescope at sea that Rochon first became aware of John Dollond's patent and Peter Dollond's much improved version, the triple achromat. These developments inspired him to suggest to the Académie royale des sciences in February 1774 that he might improve further upon the performance of double or triple achromats by interposing a fluid (diaphane or other oils) in the spaces between the crown and flint glass components.[2] Initiatives such as this led to his appointment as curator of the royal cabinet of physical and optical instruments at the Chateau of La Muette at Passy. Here he was allowed to pursue his scientific interests, working alongside Jean-Baptiste Le Roy.[3] In 1787, an opportunity arose to return to his native Brittany, this time to take up the post formerly held by Roger Boscovich SJ, the first director of the marine academy at Brest. Thus, it was at Brest, far from the dangers of revolutionary Paris, that he provided guidance in marine science. Still within the reach of the new Government, however, the Count de Montmorin, minister of foreign affairs, commanded him to act as envoy to London on matters requiring a scientific understanding.

---

2 The results of the ensuing investigations were met with moderate success. Borda, Le Gentil and Cassini each tested this novel idea using different fluids. Resumés of their findings appear in Rochon (1801a and 1801b).

3 Rochon's various improvements included an attempt to minimise the defects in glass production using platinum-lined crucibles to cast flint at the Saint-Gobain glassworks; improvement of Bouguer's heliometer, an engraving machine; an improved Hadley's reflecting quadrant using achromatic prisms instead of mirrors; a prismatic micrometer using the double refraction of quartz to produce two images that were made to coincide by moving one optical element along a scale (effectively, a rangefinder); efforts to improve reflectors and achromatise refractors; invention of the diasporameter to measure the refractive index of glass using a variable angle prism; and, somewhat ahead of his time, application of an objective prism to observe the spectra of a field of stars although, in this case, his interpretations were inaccurate.

Rochon made that visit in 1790 – shortly after the fall of the Bastille – ostensibly to investigate the British system of weights and measures with a view to adopting something similar in France. This was not the only change afoot for the citizens of the New Republic: an entirely new calendar was already begun and changes to the monetary system were also well underway. Among those that Rochon met in the capital was Jesse Ramsden who expressed nothing but praise for Peter Dollond but only where praise was felt due. However, it does not appear that he encountered Peter Dollond although he did learn of the recently voiced differences of opinion at the Royal Society regarding the invention of the achromatic telescope objective.[4] Not until June/July 1796 – Messidor, Year IV according to the post-Revolutionary calendar – did Rochon's notes concerning the dispute between the two English instrument makers over the invention of the achromatic telescope become available in France, and even then the paper in which he published the remarks did not become generally available until 1798 when it appeared as an aside in a report to the Institut de France published in the *Journal de Physique* concerning some related work on the measurement of small angles carried out by Lexell, Ramouski and Fuss at St Petersburg.[5] From this account we may note that despite his knowledge of Peter Dollond's attempt to correct foreign misunderstanding regarding his father's priority, Rochon remained resolute that 'the immortal Euler' should not be lost from the French record although he advocated:

> *La nature des rescherches de ce savant opticien* [John Dollond], *& le parti qu'il a fu en tirer, lui ont mérité l'honneur de partager avec Euler la gloire de cette belle & importante découverte.*[6]

In other words, Rochon conceded that Dollond merited the honour of *sharing* the discovery with Euler.

The report to the Institut also referred to the trial of *Peter Dollond* v. *Watkins* at King's Bench in 1763 and, from what he had gleaned by word of mouth (probably from Ramsden), he also felt obliged to offer some remarks concerning Chester Moor Hall. This account was partially translated into English for the benefit of readers of the *Philosophical Magazine* but the editor, Alexander Tilloch, became concerned that Rochon's account, as it stood, would have perpetuated errors of fact. For convenience, therefore, he replaced a substantial part of what Rochon was attempting to say with a shortened version of the

---

4 Rochon (1801a), p. 173. Rochon alludes to his own *Recueil de Memoires sur la Mechanique et la Physique* (1783) as his source.

5 Rochon (1798a) and (1798b). During the Revolution all 'academies' were abolished. In 1795, however, the Académie des sciences was re-established as part of the Institut de France.

6 Rochon (1798a), pp. 3–15.

Veritas article.[7] To justify this switch, Ramsden was consulted to ensure that the facts in Veritas were accurate.[8] On noting this editorial 'adjustment', Rochon, once again, revised his account for the Institut de France, this time taking his corrections from the substituted facts contained in the Veritas extract.

The final French version of Rochon's account appeared in an 1801 issue of the *Journal de Physique*, antedated for Fructidor, Year IX (August/September 1798).[9] In summary, it emphasised the following points: that Hall ought to hold priority, having made several achromatic doublets by 1733; that a 'tribunal' had taken place at 'Westminster' (meaning King's Bench, Westminster Hall) under Lord Mansfield, who ruled that the benefit [of patent] was not necessarily due to the one who had the scientific idea but the one who brought it to the public; that Mansfield had acknowledged Hall as the inventor but due to his position and fortune the latter had disdained of receiving 'for his beautiful discovery' any pecuniary reward; that 'Ayscout' (meaning Ayscough) of Ludgate Hill had, in '1756' (*sic*, probably earlier), owned one of Hall's objectives and this was deemed to be 'very authentic'; that Dollond offered flint–crown achromatic objectives to the public in '1755' (*sic*, not before May 1758) and that, in the 1757 *Philosophical Transactions* paper, he stated he could destroy the refrangible aberrations simultaneously with those of spherical aberration; that, in France, it was generally accepted that Dollond was the inventor of these achromatic lenses until 'Walsius' (*sic*, meaning Watkins) had doubted the validity of the patent due to the revelation about Hall; but that Lord Mansfield decided who should have the priority, based on who had brought it forward commercially (as Hall had not done); and, finally, that 'This interesting discussion was renewed lately (*sic*, within the past ten years), between Dollond the son and his brother-in-law Ramsden; they addressed each other through memoirs to the Royal Society, but this scholarly Society thought it must "nip in the bud" what might follow from this process which divided the two celebrated artists'.[10]

Despite the vagaries of dates and spelling, Rochon did, indeed, present the events and their outcomes correctly, if somewhat inadequately expressed. Unfortunately, Watkins's position – *very* apparent in Ramsden's unpublished communication to the Royal Society – continued to remain obscure due to Rochon's misspelling, nor was Addison Smith mentioned at all. With Ramsden

---

[7] Two trivial errors appearing in Rochon (1798a) are: that '*M. Boss*' (meaning George Bass) lived in '*Brideville*' (meaning Bridewell), and, more seriously, that the 'tribunal' under Lord Mansfield was '*contre Jean Dollond*' whereas, strictly, it was Dollond's action *against* Watkins and Smith, not *vice versa*.

[8] Rochon (1798b), p. 177.

[9] Rochon (1801a). A note of this paper appeared in *Phil. Mag.* Rochon (1801b). One purpose of this paper was to announce an angle measuring instrument (using variable prisms) to determine the distance of ships superior to one by Ramsden.

[10] Rochon (1801a), p. 175 [in translation].

dead (just a few months prior to this translation into English), no one, on either side of the Channel, would have any inkling of who 'Walsius' could have been although they would gain a general sense of the order of historic events, so reinforcing the opinion that the unassuming Hall may have deserved some recognition.

Jean Montucla and the Histoire des Mathematiques

Rochon's account to the Institut de France does not appear to have been heeded by Jean Etienne Montucla who, instead, derived his information directly from Lalande. The relevant passage from his *Histoire des Mathematiques* (1802) ran as follows:

> *Ce fut* Chestermonhall [sic] *qui, vers 1750* [sic], *eut l'idée des lunettes achromatiques. Il s'adressoit à Ayscough, qui faisoit travailler Bass.*
> *Dollond ayant eu besoin de Bass pour un verre que demandoit le duc de Yorck* [sic], *Bass lui fit voir du crown-glass et du flint-glass.*
> *Hall donna une lunette à Ayscough, qui la montra à plusieurs personnes; il en donna la construction à Bird, qui n'en tint pas compte. Dollond en profitra.*
> *Dans le procès qu'il y eut entre Dollond et Watkin* [sic] *au banc du roi [King's Bench], cela fut prouvé; mais Dollond gagna, parce qu'il étoit le premier qui eût fait connoitre les lunettes achromatiques.* [11]

From the evidence already described concerning Addison Smith, we can see that the first line of Montucla's entry is not correct. Chester Moor Hall gave James *Mann* (Ayscough's master) one of the lenses (flint or crown) which he then subcontracted to George Bass; the other lens being submitted to Bass via the elder Scarlett. The second detail concerning Dollond's visit to Bass's workshop is correct. The third detail is probably correct for Ayscough did take up the flint–crown formula, which he subsequently passed on to Eastland. Whether it was Ayscough who passed on the same formula to John Bird is not known although Bird was certainly involved. However, it was too simplistic to report that Dollond profited only because Bird did not pursue the achromatic formula.

Interestingly, the remark concerning Peter Dollond's success against Watkins at King's Bench was known and recognised for its significance. However, Montucla conflates father and son in referring simply to Dollond. Settling the order of priority mattered as much to the French as the British and, as this extract shows, the debate was still very much to the fore when Montucla's text appeared in 1802.

---

[11] Montucla (1802), vol. 3, p. 448 (note 3). Montucla also provides a full account of the achromatic lens, including the work of Clairaut, Euler, Dollond, d'Alembert, the abbé Bouriot and Rochon (vol. 3, pp. 447–495).

John Kelly and his 'Life of Dollond'

The combination of the reprinted Veritas piece in 1798 and Rochon's second paper, 1801, in English, must have been the last straw for Peter Dollond, who, 12 years after submitting his paper to the Royal Society in 1789, was still pestered with 'mistaken' opinions favouring Hall. Some attempt to bolster the name of the founder of the Dollond business was made in 1804 by John Kelly, who married Peter's first daughter, Louisa. Kelly was rector of Copford and Ardleigh in Essex and very much a man of letters, and published a short article on John Dollond's life.[12] With Ramsden having died in 1800, the ageing Peter Dollond must also have considered it opportune to offer a far more substantial testimony to his father by reprinting Kelly's article together with other papers and letters by his father, and those relevant contributions of Euler and Maskelyne which pointed to his father's recognition as inventor. Again, the task of compiling these was allotted to Kelly. This work – *The Life of John Dollond. F.R.S.: Inventor of the Achromatic Telescope* – appeared in 1804, with a reprint in 1808.[13] In effect, it was Peter Dollond's last attempt to promote an irreproachable statement of his father's achievement and offer an outward impression of the honourable basis for the family business, then having traded over 50 years. Probably in accord with Peter's wishes, Kelly avoided any mention of Hall although – in one easily missed passage – he did refer to those 'who endeavoured to find a prior inventor', possibly alluding to Jesse Ramsden who did, indeed, seek to establish Chester Moor Hall as the one and only true inventor of the achromatic telescope objective.[14]

It is interesting to observe how details of historic importance are often lost or muddled by the mists of time, only later to be resurrected to support some new view or purpose. During the nineteenth century, the Veritas article (the only printed source of evidence available in favour of Chester Moor Hall), was taken up by individuals whose motives were more extreme than those of its author. One writer wished to demonstrate that British achievement was understated, particularly in the light of developments in achromatic telescope making elsewhere; another used the Veritas piece as evidence to expose the family feud between the brothers-in-law, Ramsden and Dollond Jr; while others, more altruistically, took it upon themselves to make a case for the long-deceased Hall who was portrayed as an unsung hero, overshadowed by the mighty Dollonds, and therefore much deserving of support and attention.

---

[12] Kelly (1804).

[13] The 1808 edition is generally referred to as the *third* edition, presumably because it contained the third printing of Peter Dollond's (unpublished) 1789 paper to the Royal Society.

[14] Kelly (1808), p. 13.

At various times throughout the nineteenth century the names of John Dollond and Chester Moor Hall were lauded as 'inventor of the achromatic telescope'. The promulgation of John Dollond's name as 'inventor' was, in no small measure, due to vested interest in that instrument by Peter Dollond. Even on the death of his mother, Elizabeth Dollond, Peter took this opportunity to remind the public of his deceased father's importance. The obituary announcement placed in the *Whitehall Evening Post* for 15 August 1782 for his mother ran: '[Died Thursday 8 August], Same day, at Brompton, Mrs Dollond, relict of Mr J. Dollond, *the ingenious optician that invented the achromatic telescope*.'[15] [Author's italics]

Questions raised through the *Philosophical Magazine*

The issue was still very much to the fore in 1829 when a correspondent to the *Philosophical Magazine* inquired how it could be that Hall's name appeared in a chronological listing of astronomical instruments in a French almanac, as inventor of the achromatic telescope in 1750, some eight years before Dollond's published account.[16] The year 1750 seems arbitrary but, more importantly, the author of the query (signed 'R.B.') wished to know the current whereabouts of Hall's achromatic telescope, as a mark of respect and for the sake of priority for British science. After all, he lamented, Galileo's telescope was preserved by the Florentine authorities, and the British Museum was the guardian of Newton's reflecting telescope. Indeed, if Hall was to be accorded any credit at all then certain questions had to be answered, namely:

> Are any of Mr Hall's achromatic telescopes now in existence?
> Is any correct information now to be obtained respecting the performance of Mr Hall's telescopes?
> Are any of Mr Hall's papers containing the principles of his discovery extant?
> Is any information to be obtained from records of the trial in Westminster Hall, respecting the original discovery of achromatic telescopes?
> Is anything further known respecting the philosophical labours or the life of Mr Hall?

'These enquiries,' wrote R.B., 'are ... not undeserving of the attention of the Astronomical Society.'[17] Certainly, the quest was laudable enough and a

15 *Whitehall Evening Post*, 15 August 1782.

16 Statements occurring in the *Annuaire présenté au Roi par le Bureau des Longitudes* [date unknown but before the Revolution] ran as follows: 'Hall construit une lunette achromatique—1750.' and 'Dollond publie la découverte des lunetttes achromatique—1758.' [R.B.] (1829).

17 [R.B.] (1829), note 16.

Figure 9.1 James Ferguson, FRS, mezzotint by Robert Stewart, after a painting by John Townsend, 1776. © The Trustees of the British Museum, 1902,1011.5189

challenge for this recently founded group of like-minded enthusiasts of optical and mathematical persuasion, although in-depth inquiries were not pursued until 50 years later.

James Ferguson and David Brewster

The Scottish lecturer James Ferguson FRS did not discuss the new achromatic telescope in his influential text *Lectures on Select Subjects* which came out in nine editions before 1799. By 1805, Ferguson was long dead, and his new editor, David Brewster, was adding fresh information that 'this instrument was first invented and constructed about the year 1758 by our countryman John Dollond'. In his second edition, in 1823, Brewster embellished this further:

Figure 9.2 Sir David Brewster, LL.D., FRS, engraving by William Holl, after a painting by Sir Henry Raeburn, R.A., 1832. © National Maritime Museum, Greenwich, London, PAD3677

> ... we have the most unquestionable evidence that this instrument [the achromatic telescope] was invented in 1729, and constructed about the year 1738, by our countryman Chester More Hall, and afterwards in 1758 by Mr Dollond, who seems to have been unacquainted with the labours of Mr Hall.[18]

It was also Brewster who declared that it was John Bevis who proposed to distinguish telescopes not affected by prismatic colours by the name 'achromatic' because it was 'an appellation which they have hitherto retained, though some have erroneously stated that it was first given them by the astronomer Lalande.' A second printing of this text appeared in 1823 when David Brewster (as editor) added a personal note, directing readers to the

[18] Ferguson's text was edited by Brewster in 1805 and 1823. The quotation appears in Ferguson (1823), vol. 2, p. 309. This work was valuable to lens makers for its tables of curvatures to which flint- and crown-glasses should be ground for double- and triple-achromats.

Edinburgh Encyclopaedia in which he had laid plain the little that was known about the origin of the achromatic lens.

Brewster based his opinions on the Veritas article, but went one step further by pointing out that, even if the facts could be verified, one should be cautious of publicising Hall's achievement solely due to the Veritas source: '... because it is evidently intended to insinuate, that Mr Dollond had availed himself of information obtained, either directly or indirectly, from Mr Bass.' Prudently, he warned:

> If the claims of Mr Hall are substantiated by unexceptional evidence, *we cannot withhold from him the merit of being the original inventor*. But while we allow that he may have anticipated Dollond in the discovery, we must, at the same time, guard against every insinuation that has the least tendency to tarnish the fame of that ingenious optician [John Dollond].[19] [Author's italics]

Brewster went on to refer to the indisputable influences of Euler and Klingenstierna on Dollond's thinking, but it is also clear that he had no knowledge of Watkins and Smith, nor any inkling of Hall's presence as witness at their trial in 1763. Not surprisingly, he was unaware of the papers by Peter Dollond and Ramsden which in 1789 had been banished to the Royal Society's archives. In short, Brewster had no basis upon which to assess Bass's involvement, although his impartial remarks may have succeeded in cautioning those who would otherwise argue simplistically in Hall's favour, thereby unwarrantably diminishing John Dollond's standing.

John Farey and a Parliamentary Inquiry

In the new industrial age, with a considerable increase in the number of patents, there also arose an increased number of disputes. In the late 1820s a parliamentary Select Committee had been set up to investigate the principal decisions arising from cases under the old Monopolies Act of 1624 as reported in authoritative casebooks.[20] To this end, John Farey, an independent civil and mechanical engineer, was commissioned to draw up a 'List of the most important cases of Trials at Law respecting Patent Rights'.[21] The very first case to appear on

---

19 Brewster, *Edinburgh Encyclopaedia* (1830), vol. 1, p. 106, and repeated vol. 14, pp. 607–608. Brewster (1855), vol. 1, pp 111–123, again refers to Chester Moor Hall with information from Veritas.

20 It is likely that Farey obtained information regarding the case of *Boulton and Watt* v. *Bull* from Blackstone (1827).

21 Woolrich (1997); Farey, (1829), Appendix B: 'A list of the most important cases of trials made at law ...', pp. 182–215.

Farey's list was that of *Boulton and Watt* v. *Bull* in 1795 because, from it, an important principle was derived: namely, that a patent could be granted for a foreign invention, even if it was practised 'beyond the sea' although, of course, it had to be a 'new manufacture within this realm'.[22]

More importantly in the present context, was the emphasis given by Farey to the *Dollond* v. *Champneys* case, for which he described the achromatic lens as being 'a most important invention'. Indeed, as he pointed out, this case had 'great relevance' in determining the outcome of another issue in the case of *Bolton and Watt* v. *Bull*. On delivering his report to the Select Committee on 8 June 1829, Farey considered the defendant's (Champneys) principal objections: namely, that the patent was issued for a '*mere method*' and that could not be sustained, and that the patent described a '*mere principle*', but not the mechanical construction. Farey summarised the patent itself and noted that Chester Moor Hall had made such glasses as early as '1720' (*sic*, probably a typesetting error, 1730 would be more in keeping with the year stated by Addison Smith in his letter to Ramsden) although 'he [Hall] had never disclosed the secret'.[23]

In fact, three judges had sat at the *Bolton and Watt* v. *Bull* hearing and it was from their separate reports, that Farey extracted these remarks:

> *Mr Justice Buller*: 'As Dollond first made it public, he was held to be considered as the first inventor. Dollond's telescopes are certainly a manufacture within the statute (21st Jas. I) [i.e., the Monopolies Act as it was held at the time]: they consist of principles reduced into form and practice; and the patent is for glasses completely formed, not for mere principles. The specification describes the manner in which the invention is to be carried into effect, with all the perspicuity of which the thing is capable.'
>
> *Mr Justice Heath*: 'I consider Dollond's patent to be substantially for an improved machine. A patent for an improvement of a refracting telescope, and a patent for an improved refracting telescope, are in substance the same. The same specification would serve for a patent taken in either of those terms.'
>
> *Chief Justice Eyre*: 'Dollond's patent was perhaps objectionable, being for the method of producing a new object glass, instead of being for the new object glass produced; because the mechanism or process by which it was produced, though perhaps new, will only be useful as producing the new article.'[24]

---

22 Farey (1829), p. 182.

23 Farey (1829), Appendix B: 'A list of the most important cases of trials made at law ...', pp. 182–183.

24 Ibid. The *Dollond* v. *Champneys* case was reported by Carpmael (1843) and (1851), and again by Thomas Webster (1844), 1, pp. 43–44, p. 50, and p. 126, case 13. Apart from quoting the three judges at the *Boulton and Watt* v. *Bull* case, Webster also provides proper

None of these remarks suggests that the Dollond specification – first queried by Watkins and Smith at King's Bench in 1763, and doubted again in the 'answers' to Chancery by the five deponents to Peter Dollond's bill of complaint of 1765, and most likely (but not known due to missing folios) raised again by Champneys's attorney at his trial at Common Pleas in1766 – was ever scrutinised for its *sufficiency*.[25] But the sufficiency or adequacy of a specification, and the need to make it plain and indisputable in a technical sense, was a topic that plagued patent presentation until reform, when the development of a technical (as well as legal) role for patent agents became necessary. However, the Dollond patent had stood up to the test of Champneys's trial and so – with the precedent already set – there was no need to re-examine the Dollond patent for its sufficiency in 1795.

Ranyard and the Royal Astronomical Society

There had been no call for any further investigation of the question posed in 1829 of who should be credited with the invention of the achromatic telescope until 9 May 1881 when Richard Bissell Prosser, a functionary of the Patent Office, presented a facsimile of a legal document bearing the signature of Chester Moor Hall to the Royal Astronomical Society.[26] This specimen signature added nothing to telescope history beyond the simple fact that, long ago, there had once lived a man called Chester Moor Hall who had served as a justice of the peace.

Nevertheless, it was a sign that there remained a certain curiosity about Hall and whatever he may have contributed to the invention of the achromatic lens. For the time being, the committee meeting of the Society simply closed with the bland remark: 'It is very curious the conflicting evidence we have to reconcile, but I think the balance of evidence is in favour of there having been a prior invention

references, suggesting perhaps that he consulted the actual judgements at Common Pleas for the Champneys case. These older texts are now replaced by an 11-volume (reprinted) set entitled: *Hayward's Patent Cases 1600–1883* (1988) in which Webster's account and that from the *Parliamentary Papers* (Farey) are combined. Hayward (1987), 11, p. 172.

25 Given Hulme's privileged position and his special interest in the Champneys case, it is possible that he was allowed to transfer the relevant Champneys folios H.6 Geo III, 626 Middx from the Public Record Office at Chancery Lane, after which point they became mislaid or were never returned.

26 Ranyard (1881). Signature of C.M. Hall: RAS: MS Add 98. Richard Prosser was a Birmingham engineer and a patent reform campaigner who worked with Bennet Woodcroft in compiling retrospective lists of specifications; his son (R.B.P.) worked at the Patent Office from 1856 until 1888 and did much to preserve interesting material. Hewish (1980), p. 26.

of achromatic object-glasses before the date of Dollond's patent.'[27] Others too were interested in Hall, as may be witnessed by a short article by W.G. Lettsom who inconclusively pointed out two minor errors in the handed down story: that Arthur Ranyard, Secretary of the Royal Astronomical Society, had spelt 'More Hall' wrongly (a detail perpetuated from the Veritas piece); and that Sir John Herschel, in his article on 'Light' in the *Encyclopaedia Metropolitana*, had stated that Hall came from Worcestershire (rather than Essex).[28] Nonetheless, the mere resurrection of his name in 1881 once more caused those of historical bent to bear in mind the possibility of a nearly forgotten inventor who, perhaps, deserved recognition. It may, however, have worried the directors of the House of Dollond who continued to publicise their founder's reputation as the inventor of the achromatic lens.[29]

At another meeting of the Royal Astronomical Society, held on 9 April 1886, James Glaisher, as President, began by thanking Richard Prosser for his donation of Chester Moor Hall's 'autograph' (presumably referring to the facsimile signature that he had located five years previously) and then continued by calling upon Arthur Ranyard, the Society's Secretary, to elucidate.[30] Ranyard obliged in a cursory fashion, but with nothing new to report other than a promise that the facsimile signature would be framed and hung in the Council Room, no doubt as a salutary reminder that the history of that most important instrument of astronomers, the achromatic telescope, still awaited attention. Ranyard – clearly a very conscientious man who had already seen the brief report of the Champneys case in the *Gentleman's Magazine* – persevered in the quest to find out as much as he could about Hall although his report merely reflected upon his frustrating and unsuccessful searches for the 1766 Common Pleas judgement rolls for the case of *Dollond* v. *Champneys*.[31] So, nearly a century after Ramsden's attempt to secure credit for Hall, the quest was no further advanced; moreover, a crucial piece of legal evidence which might have helped – the very document that Ranyard sought in the Record Office – had vanished. Fortunately, the practical expertise behind the method of construction of the achromatic lens had survived, and to good effect too, with further improvement to the telescope objective and, from the 1820s onwards, application to the microscope.

---

27 Reports of meeting of 9 May 1881 by Ranyard (1881) and (1886a), and quoted by Humphreys (1914), King (1881) and Ranyard (1886a). Ranyard was an officer of the council of the RAS.

28 Lettsom (1881).

29 This claim was still apparent at the start of the new century. See 'Dollond discovers the achromatic lens', in Dollond's *Book of Eyesight* (c.1905), pp. 27–28.

30 Report of a meeting of the RAS held on 9 April 1886; Ranyard (1886b) pp. 177–178.

31 Ranyard (1886b), pp. 460–461. Ranyard may have suspected that Hall had been present at the Champneys case (by subpoena). It is due to Ranyard's painstaking searches that existence of the case of *Dollond* v. *Pyefinch*, retailed in Chapter 7, is known.

Agnes Clerke and the *Dictionary of National Biography*

During the nineteenth century vague references to Chester Moor Hall often appeared in various publications. In the first decade of the century, Thomas Young included Hall among his 'Historical Portraits', lectures before invited audiences at the Royal Institution, probably because the memory of the Veritas controversy was not too distant. But later, when popular texts became abundant, Simon Newcomb avoided the obscure Hall altogether in his *Popular Astronomy* (1878), leaving only a brief mention of Euler and Dollond. More formal textbook references to the invention of the achromatic lens in the nineteenth century were even less informative than Lalande's *Astronomie* of the previous century. Of course, allotting honour and recognition retrospectively was a laudable goal although no one appears to have appreciated the immensity of the task in providing satisfactory evidence in favour of Hall. By the last decade of the nineteenth century, even the names of those optical practitioners who had known Hall – such as Bass, Watkins, Ayscough, Eastland and Bird – were largely forgotten, and unknown to the community of astronomers. Little did Hall himself imagine that, after his death, there would be such strenuous efforts to immortalise his name, merely for something he had tackled out of curiosity but abandoned without formal record or secured recognition.[32] So, when Agnes Mary Clerke, the historian of astronomy, took on the task of preparing an entry for John Dollond for the first *Dictionary of National Biography*, she could hardly avoid noting the recently discussed, but ill-understood, possibility that Hall might be deserving of some stake in the history of the achromatic telescope.[33]

---

[32] Perhaps the only contemporary testimony to Hall's life and interests were those recorded by Martha Hall, his elder sister, who ensured a monument to her brother in the church at Sutton, Essex. On this is inscribed that he was 'a judicious lawyer, an able mathematician, a polite scholar, a sincere friend, and a magistrate of the strictest integrity'. Humphreys (1914), p. 313.

[33] From the references cited by Clerke in her article on Hall (*DNB*, 1890, vol. 24, p. 62), Ranyard's account from the *Astronomical Register* (1881) appears as her principal source of biography but her mention of Brewster, Barlow and Herschel added nothing new regarding Hall's position as inventor. Clerke's view of Hall's invention being 'fully proved' is incorrect although her reference to Lord Mansfield is correct. It is clear that she had no proper understanding that Mansfield's judgement referred to the double case of *Dollond* v. *Watkins* and *Smith* in 1763, and not the *Dollond* v. *Champneys* trial of 1766 to which she referred. Therefore, the use of Ranyard's article caused her, unwittingly, to conflate the results of these two trials. Agnes Clerke was well known among members of the Royal Astronomical Society and a personal friend of Arthur Ranyard. It is, therefore, understandable how she would have been strongly influenced by Ranyard's researches when she wrote her article on Hall. Brück (2002).

Figure 9.3 Agnes Clerke (1842–1907), photographer unknown. Royal Astronomical Society/Science Photo Library

Agnes Clerke was a remarkable woman by any yardstick. Born in County Cork, Ireland, but spending her early life in Dublin, receiving a good private education, then sojourning in Florence with her sister for seven years, and following her father's passion for astronomy, it is easy to appreciate how she was confident to take on the task of academic writing. However, it was not until 1884 – when the Clerkes moved to Redcliffe Square, Kensington – that Agnes properly began her writing career. Her reputation grew out of the hundreds of articles written for the *Edinburgh Review* and *Encyclopaedia Britannica*, many on scientific subjects but it was her intense interest in astronomy – witness her *Popular History of Astronomy during the Nineteenth century* (1885), a work of extraordinary thoroughness – that led to her acceptance at the Royal Astronomical Society. With Arthur Ranyard's friendship, the rule forbidding women access to the library was relaxed in order that she could use its facilities to prepare entries for the *Dictionary of National Biography*. Ranyard was then Secretary of the Society and keen to allot some recognition to Hall. It is not surprising, therefore, that Clerke added Hall's name to her task list of nearly 150 potential articles.

However, the inclusion of Hall was based less on critical research and more on the material already compiled by Ranyard. Raised to a point beyond that which he truly deserved, Hall's name has been written indelibly into the history of the telescope.

The Ranyard–Clerke account in the *Dictionary of National Biography* effectively told all that could be said although it was not truly a definitive account. Generally, wise authors of twentieth-century astronomical texts and other recent historical contributors are guarded in attributing any priority to Hall, other than acknowledging him as an independent, amateur inventor. Without the knowledge of Watkins and Smith and the litigation of the 1760s, others have questioned Hall's place in the history of the telescope. For instance, Thomas Court and Moritz von Rohr in 1929 championed Hall and Henry King in 1955 drew attention to Peter Dollond's lack of acknowledgement to Hall.[34] But although these authors were thorough in their researched remarks, they caught only a glimpse of the arguments through Veritas and absolutely nothing of the unpublished communication from Ramsden which lay dormant in the archives of the Royal Society. The wider view – as told in the present study incorporating an account of the origin of the patent – remains. The extra notes, offered here, show that the twentieth century began and ended with much the same search for answers to questions that had been sought 100 years previously.

Correspondence in *Notes and Queries*

The issue concerning Chester Moor Hall arose again in 1914 – this time from the other side of the globe – when Cecil Owen, a teacher at the High School, Perth, Australia, asked readers of *Notes and Queries* for details of the history of glassmaking, particularly of lenses for achromatic work. Briefly, Owen wished to learn more about Hall and also the effect of the British glass tax.[35] Of the two replies, one was immediate: A.L. Humphreys recited the discussions that had taken place at the Astronomical Society in 1886, related above. The second, more interesting, reply appeared tardily (probably due to wartime) in 1917 in which 'R.B.P.' (Richard Bissell Prosser) chose to describe – as an example of John Dollond being credited with the invention rather than Hall – the case of *Boulton and Watt* v. *Bull* of 1795 in which the *Dollond* v. *Champneys* case was cited as precedent. Prosser quoted Justice Buller who had observed:

---

34 Court and von Rohr (1929b), p. 228; King (1955), pp. 154–155.

35 Owen (1914a), p. 230.

> The objection to Dollond's patent was that he was not the inventor of the new method of making object glasses, but that Dr Hall had made the same discovery before him. But it was holden, that as Dr Hall had confined it to his closet and the public were not acquainted with it, Dollond was to be considered as the inventor.[36]

Although it is apparent that Prosser himself knew little more at this stage than he had in 1886 at the meeting of the Royal Astronomical Society it would, nevertheless, appear that some headway was made into the subject by his mentor at the Patent Office, the librarian Edward Wyndham Hulme.

Today, Hulme is recognised as a respected historian of patent law.[37] Indeed, it is very much due to his efforts that the document of the Spectaclemakers' Company's petition resurfaced, thus re-awakening an historical interest and bringing to the fore the names of long-forgotten instrument makers of which Prosser had cited 'G. Ribright, J. Troughton, J. Cuff and George Bast [*sic*]' in his reply to Owen. It was also because of his familiarity with patent law, that Hulme became aware of the concurrent cases of *Dollond* v. *Watkins* and *Dollond* v. *Smith* under Lord Mansfield at King's Bench although, according to Prosser, Hulme did not explore the implications of these cases because his concern lay only in searching for the test case of *Dollond* v. *Champneys*. In other words, the double case at King's Bench against Watkins and Smith under Mansfield in 1763 was passed by and so the golden opportunity to understand the position of Chester Moor Hall as original inventor had been missed.[38]

Thomas Court, Moritz von Rohr, and the Optical Society; N.V.R. Nordenmark and J. Nordström, and the Royal Swedish Academy

A new quest to identify Veritas was initiated by historian Thomas Court and Moritz von Rohr, an optical scientist working for Carl Zeiss at Jena. While writing up their history of the telescope in 1929, they championed Chester Moor Hall for his achromats while cautiously pointing out that the anonymous

---

36 Quoted in Hayward (1987), 1, p. 470. Also see Webster (1844), 1, p. 43 and repeated by Prosser (1917), p. 336. Also see Carpmael (1843), and (1851); Farey (1829), p. 182. Strictly speaking, Hall was not a doctor; this error was perpetuated in Hayward.

37 Hulme (1902) and (1917).

38 The fruits of Hulme's researches on the Spectaclemakers' petition and the *Dollond* v. *Champneys* case are noted (without historical interpretation) in Hayward (1987), 11, p. 626. Prosser (who appears to have had access to Hulme's notes) quoted the full title of the Champneys case in his reply to *Notes and Queries* (1917). The title – *An action for infringement of his Patent of 1758, for a new method of making the object glass of Refracting Telescopes* – is repeated here because it is just possible that these missing folios, once in the hands of Hulme, may one day resurface.

Veritas must have held a great deal of exact knowledge, and must have been familiar with Bass and Ayscough.[39] The latter detail implies that whoever fitted the identity would have been of a sufficient age and familiarity with the optical trade of the 1750s and 1760s. These authors also implied that Veritas was none other than Ramsden himself although, on closer inspection of their text, they *do not* state this identity in absolute terms, but only that: '... it is *not impossible* to ascribe the article [by Veritas] to Jesse Ramsden [Author's italics].'[40] Being 'not impossible' is not to say that it was the case.

In their analysis, Court and von Rohr noted certain variations of spelling made by Veritas for the names Kling*e*nstierna (using *i* rather than *e*, as in Kling*i*nstierna) and Chester *Moor* Hall (using instead Chester *More* Hall); they also noted that Veritas referred to Lord *Mansfield* rather than Lord *Camden*.[41] The first of these details helps little because anyone reading Peter Dollond's handwritten communication of June 1789 would most likely have copied the spelling of the unusual name 'Kling*i*nstierna' (using *i* as in the original unpublished communication); the spelling 'Chester *More* Hall', however, may hold a clue to the author's identity because that was the spelling adopted by Addison Smith in his letter to Ramsden. However, the mention of Lord Mansfield by Veritas was *not* incorrect (as suggested by Court and Rohr) because he was the Chief Justice who presided at the King's Bench trial of *Dollond* v. *Watkins* and *Dollond* v. *Smith* (Camden was judge only in the case of *Dollond* v. *Champneys* and so was irrelevant in the context of the Veritas article). This would be in keeping with Addison Smith as author. We should note also that Hall donated at least one of the early achromatic doublets to Watkins and Smith (possibly for exhibition in court) and it was Smith who offered Ramsden and his friends the opportunity to examine this early example at his house in Charlotte Street. Indeed, Smith appears strongly in support of Ramsden's cause, and his evidence was sufficient for Ramsden to attach Smith's letter to his own communication to the Royal Society. It may, therefore, be more fitting to suggest Addison Smith as a candidate for the identity of the mysterious Veritas. Certainly, he admired Chester Moor Hall and spoke of him as 'a Gentleman of great Candor and scientific knowledge'.[42] Furthermore, the address stated in the Veritas article

---

39 Court and von Rohr (1928–1929b), p. 234.

40 Court and von Rohr (1928–1929b), pp. 233–234 were the first to point out the dispute between Ramsden and Dollond as raised by Rochon (1801a or 1801b). King (1955, p. 155) detected minor mistakes in Rochon and so (unfortunately) dismissed these accounts on the grounds they were simply 'inaccurate narratives.' Despite the inaccuracies in Rochon's papers of 1798 and 1801, they influenced French readers of the *Journal de Physique* and, hence via translation in the *Philosophical Magazine*, to English readers.

41 Court and von Rohr (1928–1929b), p. 229.

42 Addison Smith to Jesse Ramsden (4 June 1789). RS:MS L & P decade IX, letter (Smith), 138.

(Charlotte Street, Rathbone Place) is correct for Addison Smith. However, there remains an anomaly because the Veritas article refers to Smith as a reverend gentleman and, to date, no record has yet been found to confirm this status.[43] So, if Veritas was neither Smith (from whom the essential evidence first arose) nor Ramsden (whose reputation for honesty and openness was renowned), then the individual behind the pseudonym must have been someone else privy to the same information from within the Royal Society.

An important study of John Dollond with regard to his claims to priority and, more particularly, the contribution of Samuel Klingenstierna, was undertaken by N.V.R. Nordenmark and J. Nordström in 1938 and 1939 for the Royal Swedish Academy.[44] The first part of their study comprised a discussion of Newton's discovery of the different refrangibilities of the colours comprising white light and the dispersion law that led him to conclude that colour aberration could not be corrected; a chapter on Chester Moor Hall, discussing the extent to which John Dollond might have understood his prior contribution; followed by a chapter dealing with the Dollond–Euler–Klingenstierna issue. The second part of this study concentrated on the lack of formulation, by Dollond, of a proper mathematical theory and, hence, the motivation for Klingenstierna to produce such a theory. Nordenmark and Nordström published their work in *Lychnos* (the transactions of the Royal Swedish Academy) in Swedish with a summary in English. Some of this excellent scholarship was considered by members of the Antique Telescope Society who adopted the Dollond–Hall priority issue as a suitable focus for their annual convention in October 1999. The scholarship of Nordenmark and Nordström has also been invaluable in this work in tracing the Swedish connection behind John Dollond's actions in the 1750s.

The Antique Telescope Society

At their 1999 meeting in Victoria, British Columbia, the Antique Telescope Society reconsidered the age-old question: 'Who invented the achromatic telescope?'[45] The conference conveners, in their report, appear confident that Hall had indeed preceded Dollond although it does not appear that they

---

43 A search of the few surviving records relating to Addison Smith's life (legal documents, Marylebone Poor Rates, and so on) lend no support to the notion that Smith had become 'reverend'. He does not appear in the records of the Church of England (www.theclergyonline.com accessed 1 July 2011), although he could conceivably have been active as a nonconformist. He was not a Fellow of the Royal Society, though he may have attended meetings.

44 Nordenmark and Nordström, part I (1938) and part II (1939).

45 The Antique Telescope Society discussion panel was organised by Eugene Rudd and Peter Abrahams and discussions were based on papers by Richard Sorrenson and

debated the crucial issue of the term 'inventor', either in the old sense (the '*first and original*', meaning by chronology) or the new sense given to it by Lord Mansfield (meaning by '*publication and commercial pursuit*'). Other issues that arose included the importance of the testimonies of Rew and Bass in the Spectaclemakers' petition, the influence of Klingenstierna (old ground), and the unpublished memoirs of Peter Dollond and Ramsden in 1789 (new ground). With regard to Francis Watkins, the only new detail to appear concerned his obtaining the patent on behalf of John Dollond, but, unfortunately, this was conflated with the ownership of the patent, which Watkins *did not share* with Dollond. Mention was also made concerning the extract from Watkins's letter to Ramsden in 1789, although no suggestion was offered as to why this testimony should be considered important, that is, the delegates appeared unaware of the quintessential legal proceedings of 1763, and no account was made of why the unfortunate reporting by Veritas has persisted to the present day. Among the more basic questions asked, one such – 'How successful were Hall's lenses?' – echoes similar queries raised back in 1829, quoted at the beginning of this section. Of course, no clear answers were available simply because no identifiable Hall achromat, such as that once in the possession of Addison Smith, was known at that time; nor is any known today. From the findings and reconstruction of the context described in the present work, we might once again be tempted to ask the same questions about the achromats of William Eastland or James Champneys, but this too would be to no avail. Another conference under the Society's auspices, entitled 'The History of the Telescope', held at New York University on 16–17 April 2010, gave no further attention to the Hall–Dollond controversy.

Popular articles have sometimes touched upon Hall's name, and in so doing have once again perpetuated a degree of confusion. Usually these accounts are based on secondary sources although one of the better researched articles of this genre – submitted by George Reed for the Hughes *Griffith Observer* essay contest in 1984 – presented a very readable, though simple, account of the known events.[46] Today – from the additional material revealed in the present work – it might now be safer to credit John Dollond *only* for what he is assuredly due, namely, the conception and execution of the beautiful series of experiments – although economically described – by which he confirmed the fault in Newton's experiment 8 of *Opticks* and formulated the correct curvatures of the flint and crown glasses in combination for colour-free images. As stated elsewhere, his

Rolf Willach, plus some material from Robischon's thesis (1983), supplemented by a valuable addition from Duane Jaecks. Rudd et al. (2000).

[46] Reed (1984). Reed was a professor at the West Chester State College, Pennsylvania, and popularised science with articles such as 'The man who was murdered by Newton'. The Hughes Essay Competition was sponsored by the Hughes Aircraft Company.

work represents a determining step towards the widespread adoption of the achromatic object glasses; he was not, however, the 'first or original inventor' – in the sense commonly understood and implicit in the original Monopolies Act of 1624 – that is, as understood and believed by instrument makers generally and which Francis Watkins, Addison Smith and Jesse Ramsden held as the only truth in spite of Lord Mansfield's shift of definition in 1763. For them, an invention was the tangible result of intellectual effort and so, pragmatically, under any circumstances, those who had it first in whatever form – as Chester Moor Hall presumably had it – should be admitted into the recognised chronology.

It is quite apparent, from the details presented in this chapter, that much of the evidence surrounding the crucial events of the mid-decades of the eighteenth century was poorly understood and mostly forgotten. Understandably, the newly founded Royal Astronomical Society sought to investigate the origins of the achromatic refracting telescope. Inevitably, this led to an investigation into the position of Chester Moor Hall, although all such inquiry tended to fail. Nevertheless, Arthur Ranyard, an official of the Society, pursued the history and what little was gleaned was set in print by Agnes Clerke. Thus the mystery was frozen in time and perpetuated rather than solved.

Even 100 and more years onward, in 1999, the conclusion issued on the same mystery by the Antique Telescope Society ran:

> ... it is unlikely that any *new revelations* will change the fact that, as with many important advances in science and technology, the development of the achromatic telescope depended on the ideas and efforts of more than one person.[47]

Now although this remark is not incorrect, such a wide sweep discourages further advance in our understanding. It does, however, encompass the pity that Chester Moor Hall had not prepared a public statement of his work; it also reinforces the greater pity that John Dollond did not acknowledge those upon whose shoulders he had stood while pursuing his quest.

Conclusions

The issue today surrounding whether or not Chester Moor Hall has any rightful place in the history of the achromatic telescope is no longer one of interpreting priorities since that was answered long ago in Lord Mansfield's court. Rather, the greater dilemma concerning 'philosophical truth' *vis-à-vis* 'legal truth' – as posed by Ramsden in response to Peter Dollond's omission of Chester Moor Hall from his initial account to the Royal Society in 1789 – remains to this day.

---

47 Rudd et al. (2000), p. 11.

Situations can arise in scientific endeavour where these distinct truths coincide, although whether they should do so *of necessity* (as stipulated by Peter Dollond) can still be a subject of debate. As we have seen with practitioners of high moral integrity such as Francis Watkins, situations can arise where 'philosophical truth' (as distinguished by Ramsden) and 'legal truth' (according to Lord Mansfield's definition) are at odds, and so judgement should be weighed with great care, even if the final judgement at law does not reflect the level of moral correctness that might be expected from those who practise in the name of science.

Chapter 10
One Hundred Years at Charing Cross

No prices of the Dollond–Watkins telescopes have been noted from advertising for the period between the acquisition of the patent and the death of John Dollond, that is, 1758–1761.[1] It is known from the 'terms of agreement' between John Dollond and Francis Watkins that telescopes of this period were priced at 12, 16, 21 and 32 shillings for tubes of 2, 3, 4 and 5 feet, but it is not clear whether these were the prices at which the finished telescopes were to be retailed. From the modest balancing payments made by Watkins to Dollond through Drummond's Bank it might be supposed that the sales of new telescopes were sluggish at the start. By the time that Peter Dollond was minded to dissolve the partnership with Watkins, sales had begun to take off. For the year 1763–1764, according to an estimate – probably much exaggerated – by Watkins, Peter Dollond's profit in this field was of the order of £800 per annum.[2] None of this lucrative trade was available to Watkins and Smith, Eastland or Champneys after July 1767 because of the injunction placed on them by the Court of Chancery.[3] Indeed, by July 1764 (at the time of the Spectaclemakers' petition), only Adams, Ayscough, Cole, Gilbert, Gregory, Johnson, Lincoln, Nairne, Pyefinch, Scatliff and Sterrop were permitted to trade in achromatic telescopes containing the patented combined object glass.[4] Nevertheless, it may be assumed that these links would have provided the Dollonds with a more than modest income.

Of these principal makers, we have evidence of prices from three well-known retailers of the Dollond telescopes: namely, Benjamin Martin, George Adams Sr and Henry Pyefinch. Martin had only been trading in London about one year when, in 1757, he advertised 'Dalond's' 4, 3 and 2-foot refractors but with no price given.[5] The curious spelling would appear, at first, as a simple misprint, therefore suggesting that Martin retailed Dollond's

1 Bengt Ferrner met John Dollond in October 1759. He ordered two achromatic telescopes: one for the Royal Swedish Academy costing seven guineas; the other for Klingenstierna (price unknown). Sorrenson (2001), p. 51, note 16.

2 'The Several Answers of Francis Watkins one of the Defendants to the Bill of Complaint of Peter Dollond, Complainant, 30th November 1765'. TNA: C 12/1956/ rolls 19.

3 TNA C 33/427 (17 July 1766). 'C 33' deals with Chancery entry books, decrees and orders (1544–1875); '/427' covers the year 1766.

4 *Gazetteer and New Daily Advertiser*, 18 July 1764. PC: McConnell (15 June 2008).

5 Millburn (1986b), p. 33.

ordinary refracting telescopes although this may now be considered the 'artful' advertising to which the Dollonds alluded. With the Dollond–Watkins partnership in place from May 1758, Martin would not have been able to retail achromatic telescopes, nor would he have wished to do so because, overtly, at that time he listed ten reasons against the use of crown–flint combinations.[6] However, in 1762, he included achromatic telescopes in his price list which superficially appear to be from Dollond due to the then agreed pricing at one guinea a foot although, curiously, Martin was not on Dollond's list of approved sellers two years later in 1764. The Adamses, senior and junior, held off selling achromatic telescopes although (as witnessed by their 1766 price list) they could not afford to ignore the newer directions in telescope development. Pyefinch, as described in Chapter 7, struck up an agreement with Peter Dollond although this did not last. The prices charged to the end customer, according to the catalogues of Martin and Adams and, for Pyefinch, according to an advertisement seen by the Swiss savant Jean-III Bernoulli, are given in Table 10.1.[7] All these refracting achromatic telescopes were sold at 'a guinea a

Table 10.1 Prices of achromatic telescopes compared

Focal length	Martin 1762	Adams Sr 1766	Pyefinch 1768	Adams Sr 1771	Adams Jr 1777	Martin 1780
20 inch			£1 11s 6d		16s 0d	
2 feet	£2 2s 0d	£2 2s 0d	£2 2s 0d	£2 2s 0d	£1 12s 0d	£1 1s 0d
3 feet	£3 3s 0d	£3 3s 0d	£3 3s 0d	£3 3s 0d	£2 8s 0d	£2 2s 0d
4 feet	£4 4s 0d	£4 4s 0d		£4 4s 0d	£3 4s 0d	£3 3s 0d
		pro rata at one guinea a foot	pro rata at one guinea a foot	pro rata at one guinea a foot	pro rata at 18s a foot	pro rata to 21 guineas

Sources: Martin 1762: Millburn (1976), p. 220; Adams sen. 1766: Millburn (2000), pp. 364–365; Pyefinch 1768: Bernoulli (1771), p. 71; Adams sen. 1771: Adams (1747), pp. 8–9; Adams jun. 1777: Adams (1777), catalogue p. 10; Martin 1780: Millburn (1986b), p. 33.

6 Martin (1759), pp. 85–87, and Appendix 6 in this volume.

7 Pyefinch's prices are given in the letter dated 20 January 1769, Bernoulli (1771), pp. 70–71.

foot' for those upward of 2 feet. Thus, from the simple comparison of prices, it is clear that Peter Dollond's threats and actions at court gave him complete control over the market. High prices were maintained until the expiry of the patent in April 1772; thereafter prices dropped markedly.

The original design of the achromatic objective had worked well enough at the start although John Dollond was keen to produce object glasses of shorter focal length to be used in conjunction with a concave eyepiece. Since the field of view using a concave eyeglass depends upon the aperture of the objective (combining crown convex and flint concave) the first attempts were unsatisfactory. The effect of the single crown convex glass was then divided by sandwiching the flint glass between two convex glasses. This arrangement was successful when used in conjunction with a concave eyeglass and the aberrations were thereby reduced. However, no success was achieved when this triple glass arrangement was used to achieve longer focal length telescopes using a convex eyepiece.

After his father's death, Peter Dollond returned to this perplexing problem because he was convinced that the previous arrangement using two crown convex and one flint concave should have worked for convex eyeglasses too. With perseverance, he achieved success, and so the first commercial 'triple achromat' was born. A prototype of focal length 3½-foot and aperture 3¾ inch was offered to James Short for trial during the latter months of 1764. Peter Dollond then announced this latest improvement of the refracting telescope to the Royal Society via Short, dated 7 February 1765. On relaying this letter to Thomas Birch, Secretary of the Royal Society, Short added: 'I have tried it [the new triple achromatic telescope] with a magnifying power of 150 times, and I found the image distinct, bright, and free from colours.'[8] Peter then sold or loaned this instrument to Nevil Maskelyne, newly appointed as Astronomer Royal at Greenwich. Maskelyne's records show that for the night of 25 September 1765, Jupiter's satellites and the immersion of the star δ-Capricorni into the Moon's dark limb were successfully observed.[9] It was noted that the achromatic triplet improved the resolution of double stars;[10] also the new arrangement with the extra lens reduced the effect of the secondary spectrum without compromising the correction of spherical aberration.

Impressed, Maskelyne then allowed Dollond to replace the objectives in other instruments with state-of-the-art achromatic triplets. Gradually, the shorter and lighter achromatic instruments replaced the longer focal length refractors. So pleased was Maskelyne with this development that, in April 1767, he wrote (in Latin) to the astronomer Roger Joseph Boscovich of Pavia observatory explaining his success with Dollond's 3½-foot triple achromatic telescope.

8 Dollond (1765), p. 54.

9 Howse (1989a), p. 70.

10 King (1955), p. 160.

Figure 10.1 Nevil Maskelyne, chalk drawing by John Russell, R.A., c.1776. © National Maritime Museum, Greenwich, London, presented by the descendants of Nevil Maskelyne, ZBA4305

This recommendation was sufficient for Boscovich to order one immediately. Around this time Maskelyne also informed Jean-III Bernoulli (astronomer to Frederick the Great of Prussia) and Benjamin Franklin (just then advising on an observatory for the Philosophical Society at Philadelphia) both of whom took up this idea. Among Dollond's other important customers was the Duke of Orléans whose achromatic telescope cost 1,400 livres (about £70);[11] another recorded purchaser in the 1770s was the astronomer Antoine Darquier de Pellepoix, of Toulouse whose instrument was of 3¾ inch aperture with a 42 inch focus.[12] Thus, initially by word of mouth and correspondence, news of Dollond's triple achromatic object glass quickly spread.[13] In a remarkably short period of time, Esprit Pézenas in France included a particular example of the 'théorie de la lunette de M. *Dollond* qui grossit 150 fois' for students studying the newly translated (1767) French version of Robert Smith's *Optics*.[14]

For Peter Dollond, the investment of allowing Maskelyne to undertake a trial had paid off. By 1793 – according to his reply to an inquiry from Maskelyne on 8 July – his quotation for a 10-foot achromatic telescope was '150 gns – it would be more to anyone else'.[15] For this price, the instrument would have been bespoke. As for Dollond's prices generally, a simple ready-made and fully mounted achromatic refractor could be obtained for as little as two and a half guineas (£2 12s 6d). Lalande – in the revised 1792 edition of his *Astronomie* – relates that both Dollond and Ramsden sold achromatic objectives at competitive prices: a 3-foot objective cost 3 guineas (£3 3s); 9 feet, 8 guineas (£8 8s); 12 feet, 10 guineas (£10 10s); and 18 feet, 15 guineas (£15 15s). However, a good 3½ feet (focal length), 3½ inch (aperture) achromatic refractor which used to cost 26 louis plus one louis for carriage (£24 5s + 18s 6d) at Dollonds was difficult to obtain although the Parisian makers Etienne-Antoine Putois and Noël-Jean Lerebours sold both at 25 louis (£23 7s) for the objective and 5 louis (£4 13s) for the mounting.[16] Prices varied also according to aperture and, of course, the quality of the optics and complexity of the mounting.

By 1793, the manufacture of achromatic telescopes by the brothers P. and J. Dollond was so proficient that no ordinary optical instrument maker in London – except perhaps Jesse Ramsden – could meet their standards and expertise. The Dollonds commanded large prices although there was just sufficient competition from Ramsden and other makers abroad to moderate their profits. After 1772, when the Dollond patent expired, the City three (Eastland, Stedman and Champneys) were at liberty to resume production of achromatic telescopes although it would be interesting to learn if they did. Watkins (but perhaps not Smith) certainly did. In retrospect, Dollond's litigation had paid

11 Daumas (1989), p. 145.

12 King (1955), pp. 159–160.

13 A scrapbook relating to the Dollond family business includes a printed handbill (instruction sheet) headed 'The Use of the Three Combined Prisms ... to shew the Principles of the Achromatic Object-Glass' [n.d.] stating (at end) 'as demonstrated ... by the late Mr John Dollond.' MHSO: Dollond Archives, Blundell, item 23.

14 Pézenas (1767), p. 482.

15 RS: MS Gh.130.

16 Howse (1989b), p. 9.

off for, over that very critical period, the window of opportunity was of just sufficient duration to establish a very dominant position among the company of spectacle makers and a very lucrative position in the niche market as suppliers to the world's astronomical observatories.

As late as 1782, the Spectaclemakers' Company was still pressing Watkins for back payment of the Quarterage dues that had accumulated over the years. By that time, however, he was approaching 60 years old and had reached the time when he should look to the immediate future, particularly in the making of provisions for the continuance of his business. With no sons – and his daughters, Frances and Clarinda, already married to persons with no interest in the business – it might be supposed that Addison Smith would have succeeded to the old firm but this was not the case. Smith was only nine years younger than Watkins and he appears to have managed successfully on his own account. Clarinda's son by her first marriage, Henry-II Walder, was already deceased; there were no sons from Addison Smith's marriage to Elizabeth Walder. At one time, there might have been a possibility of Francis Watkins's nephew, Francis-II, son of William Watkins, entering the business because he had started out as an apprentice to Addison Smith. However, his apprenticeship was not completed and it now seems possible that he moved to Hereford with his father.

Handing over the Business

The solution of what was to be done with the business was settled along the following lines. Jeremiah-III Watkins (eldest son of Francis's eldest brother Jeremiah-II) – possibly seeking permanent work in the capital where two families of aunts and uncles now resided – became the next in line. His background is unclear although it is assumed that he had gained experience elsewhere since he was nearing 30 years old by the time he joined his uncle Francis at Charing Cross. Meanwhile, Jeremiah-III's younger brother, Walter, had already moved from Herefordshire to London in 1774 to take up an apprenticeship in printing under the supervision of his uncle (William Owen) at Homer's Head in Fleet Street, where it is presumed that he lived with his uncle and aunt (Mary Owen, née Watkins) and where also he remained until joining his elder brother in partnership at Charing Cross. The transition of the business – from Francis to these nephews – appears to have been carried out in 1785 although, as mentioned already, Jeremiah and Walter did not properly accede to the leasehold of the shop until Francis's death in 1791.[17] It therefore is probable that Francis

[17] Death notice (1791). According to the Sun Insurance policies, the shop at Charing Cross was insured for £200 (house) and £1,600 (utensils and stock) in 1780 (policy 427163); not until 1785 did the joint names of Jeremiah and Walter Watkins appear (policies 476969

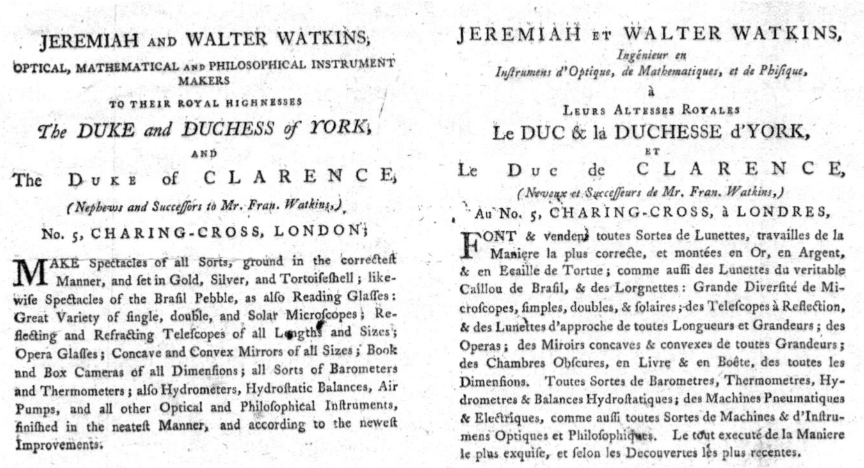

JEREMIAH AND WALTER WATKINS,

OPTICAL, MATHEMATICAL AND PHILOSOPHICAL INSTRUMENT MAKERS

TO THEIR ROYAL HIGHNESSES

The DUKE and DUCHESS of YORK,

AND

The DUKE of CLARENCE,

(Nephews and Succeſſors to Mr. Fran. Watkins,)

No. 5, CHARING-CROSS, LONDON;

MAKE Spectacles of all Sorts, ground in the correcteſt Manner, and ſet in Gold, Silver, and Tortoiſeſhell; likewiſe Spectacles of the Braſil Pebble, as alſo Reading Glaſſes: Great Variety of ſingle, double, and Solar Microſcopes; Reflecting and Refracting Teleſcopes of all Lengths and Sizes; Opera Glaſſes; Concave and Convex Mirrors of all Sizes; Book and Box Cameras of all Dimenſions; all Sorts of Barometers and Thermometers; alſo Hydrometers, Hydroſtatic Balances, Air Pumps, and all other Optical and Philoſophical Inſtruments, finiſhed in the neateſt Manner, and according to the neweſt Improvements.

JEREMIAH ET WALTER WATKINS,

Ingénieur en

Inſtrumens d'Optique, de Mathematiques, et de Phiſique,

à

LEURS ALTESSES ROYALES

Le DUC & la DUCHESSE d'YORK,

ET

Le DUC de CLARENCE,

(Neveux et Succeſſeurs de Mr. Fran. Watkins,)

Au No. 5, CHARING-CROSS, à LONDRES,

FONT & vendent toutes Sortes de Lunettes, travailles de la Maniere la plus correcte, et montées en Or, en Argent, & en Ecaille de Tortue; comme auſſi des Lunettes du veritable Caillou de Braſil, & des Lorgnettes: Grande Diverſité de Microſcopes, ſimples, doubles, & ſolaires; des Teleſcopes à Reflection, & des Lunettes d'approche de toutes Longueurs et Grandeurs; des Operas; des Miroirs concaves & convexes de toutes Grandeurs; des Chambres Obſcures, en Livre & en Boête, des toutes les Dimenſions. Toutes Sortes de Barometres, Thermometres, Hydrometres & Balances Hydroſtatiques; des Machines Pneumatiques & Electriques, comme auſſi toutes Sortes de Machines & d'Inſtrumens Optiques et Philoſophiques. Le tout executé de la Maniere le plus exquiſe, et ſelon les Decouvertes les plus recentes.

Figure 10.2 Jeremiah and Walter Watkins's trade card, 1793. © The Trustees of the British Museum, Heal, 105.109

retained the apartment at 5 Charing Cross in order that he could give assistance and guidance to his nephews in the early years of their partnership. Walter was taken ill at Charing Cross in the house of a neighbour and he died of apoplexy in 1798.[18] Jeremiah-III worked alone until his death on 3 February 1810.[19]

From his various incomes from property in the London area and Radnorshire acquired over the years, Francis-I Watkins was able to prepare for retirement in grand style by purchasing substantial property at Richmond. Not even in retirement, however, could he resist the temptation to capitalise. It so happened that in 1789 an opportunity arose to purchase the lease of the remainder of Johnson's Court (later re-named Trinity Place) immediately to the rear of the Charing Cross shop (probably allowing for workshop expansion) together with the leases of the six front houses, which included the well-established instrument shop.[20] Yet, despite all the opportunities that must have arisen over the years to move the business to a more central location, the situation at 5 Charing Cross was eminently well-situated for the affluent inhabitants of Westminster and the military personnel who passed by on their way to the Army Office at

---

and 507824). GL: Sun Life 11936/313 and 330. Tardily, the Poor Rate entry in 1786 still listed 'Francis Watkins' as the owner although his name had been deleted in favour of 'Jeremiah and Walter'. WA St Martin-in-the-Fields, F 588.

18 Death notice (1798). The circumstances of Walter's death were discussed at length in *Gentleman's Magazine*.

19 Death notice (1810).

20 London County Council (1937), vol. 18, p. 7.

Horse Guards or the Admiralty Office in Whitehall. Indeed, the position of the shop – being situated at the western edge of the commercial metropolis – was ideal and Watkins appears to have been as fond of the location as was his better-remembered contemporary, James Boswell, who, on quoting Dr Johnson for 2 April 1775, famously remarked: 'I think the full tide of human existence is at Charing Cross.'[21]

Property and Real Estate

It is clear that Francis-I Watkins was intellectually as competent as any of the mathematical instrument makers and opticians of the first rank, and it may also be said that his contribution in diffusing philosophical science to the wider public was of moderate significance. Unlike his contemporary practitioners John Cuff and James Ayscough, Watkins was not always to be found at his workbench or even in attendance at the shop because, beyond the base trade as optician and instrument maker, he developed a separate profitable line through the acquisition of property. It is for this reason that he always remained solvent, possibly enviably so by Cuff and Ayscough who each suffered acute financial problems during their careers. By his mid-60s, after 38 years of conscientious effort and varied trade, Watkins was able to retire, a privilege not open to all in the eighteenth century. Most important of all, he was able to leave a fully functioning and profitable enterprise to his nephews who perpetuated the family name as instrument makers well into the nineteenth century.

In parallel with his instrument-making life, Watkins derived a separate income from the rent and lease of property (Table 10.2). This knowledge of property and estate management was not something learned under the Scarletts where he had spent the more mature years of his apprenticeship, but rather appears to have been acquired from within his own family who were well-versed in such matters. His father and two eldest brothers (Jeremiah and John) had prospered in this way among the minor gentry of Radnorshire and Herefordshire and Francis, too, made headway in the same manner by acquiring land or leases, first in Herefordshire and later in the capital.

An extensive search for information relating to Watkins's wider interests, both in metropolitan London–Westminster and in rural Radnorshire–Herefordshire, has revealed documents bearing his signature. His involvement in real estate had begun in Herefordshire when, on 21 June 1749, he took on a lease of possession for land at Pembridge in the Court of Noke.[22] In this transaction,

21 Boswell (1953), p. 608.

22 Land at Pembridge in the Court of Noke, HRO Document AK49/44*ff* (conveyed 21 June 1749).

Watkins obtained Kirkwood's Farm in common with the 'Chamberlain ladies', the daughters and co-heirs of Thomas Chamberlain of Ocle Pichard, gent. Such an opportunity may have been guided by his father who, in turn, had married into the local landed gentry of the Whitney family. A second opportunity arose in late 1752 when the Pontvane estate, a farm in the parish of Newchurch where Watkins spent his youth, became available. In this instance, it was very likely that Francis would have known Henry Carnop, the previous tenant. The land comprised an old Welsh longhouse (which stands to this day) together with 130 acres of land, partitioned into fields and woods which have changed little since it was first surveyed in 1754.[23] Watkins's name is visible through his signature on the deeds of conveyance, under the style 'Francis Watkins of Charing Cross liberties of [the] city of Westminster, co. M'sex, optician'.[24] It is from this document that his London solicitor (John Caldecott of Serjeants Inn, Chancery Lane) is known. A second dorse for the same estate (dated August 1753) shows that he shared the tenancy with his brother, John.[25] As partners, John appears to have administered the estate while Francis was a sleeping partner.

Watkins's first major acquisition in Westminster took place in 1757 when he purchased the freehold of three 'messuages' (old brick houses) together with the site of Mary Kirke's house at Charing Cross.[26] These properties were situated across the road from his own shop: two houses stood next to Spring Gardens and the other – commonly referred to as 'Oliver Cromwell's house' – was let to the landscape painter, John Harris.

---

23 NLW (Maps and Prints) Map 7605, 134/2/10: 'An exact survey of Pontvane (Wales)' (1754).

24 PCA Dorse R/D/GNX/153–154. *Note:* Thirteen deeds of conveyance of the Pontvane estate passed through the solicitors Green and Nixson of Presteigne and Knighton between 1714 and 1820. Four names are cited on DNX/156 (No. 10 and No. 11): (i) Francis Watkins of Charing Cross, liberties of the city of Westminster, co. M'sex, optician; (ii) Thomas Jones, late of town of Presteigne, now p. Eardisley, co. Hereford, gent; Ph[oe]be his wife; (iii) John Watkins of town of Kington, co. Hereford, gent; and (iv) John Caldecott of Serjeants Inn, Chancery Lane, London; David Davies of Kington, gents. Thos. Jones paid John Watkins £420 and Fr. Watkins £250.

25 PCA Dorse of R/D/GNX/156. By April 1766, there were eight parties to this estate, again including Francis's brother John (and his son, also named John), who then lived at Tiduer in the parish of Lugwardine. By this time, the consideration had risen to £4,900. Curiously, a conveyance of 17–18 August 1753 details that one of the party owners, Thomas Jones, was indebted to John Watkins in the matter of £250, resulting from a Judgement in favour of Watkins in King's Bench (Michaelmas 1744). Clearly, the family were well-versed in the law. Dorse of DNX/156 (No. 10 and No. 11).

26 The 'three messuages late in the tenure of Mary Kirke' were bequeathed in 1753 to John Lambert (Mary Kirke's upper servant) and Elizabeth Cox. LCC, *Survey of London* (1937) vol. 18, p. 85.

Table 10.2 Properties acquired by Francis Watkins

Year	Property and/or Estate
1749	Lease of possession of land at Pembridge in the Court of Noke, Herefordshire (acquired June 1749).
1750	Lease of the Manor of Ocle Pichard, Herefordshire (acquired January 1750).
1752	Party (initially with (i) David Morgan; then (ii) his brother, John) in the lease of Pontvane farm and land at Newchurch, Radnorshire (acquired (i) 11–12 December 1752; (ii) 17–18 August 1753).
1757	Property bought at Charing Cross comprising three 'messuages': two next to Spring Gardens and one (Cromwell's House) protruding into the street. Included in the purchase was the site of Kirke House.
1771	Compensation received for loss of land due to Westminster Bridge road-widening project.
1771	A 'new brick messuage' built on the site of Kirke House; rented to Biddulph & Cocks, bankers.
1782	Four adjoining houses in Barbican Court, City of London (probably leased).
1782	City residential property at 77 Aldersgate Street (probably leased).
1782	Obtains lease for some property in Johnson's Court, Charing Cross (behind his shop).
1782	Subleases property in Johnson's Court.
1785	Three properties in Spring Gardens.
c.1785	Purchases 'Ivy House' at Richmond and two adjacent residences.
1789	Obtains lease for remaining properties in Johnson's Court, Charing Cross.
1789	Obtains lease for 'front houses' at Charing Cross (that is, numbers 4–8 Charing Cross).

This purchase was opportunistic because, not long after making the transaction, Watkins was approached by the commissioners for the Westminster Bridge who wished to remove the protruding Oliver Cromwell house in order to widen the street and so improve the access to the new bridge over the Thames. Watkins was not the only property owner to be affected by the road-widening scheme when demolition began in 1757–1758;[27] the Earl of Dysart was similarly approached but, with him, the bridge commissioners met their match.[28]

[27] An announcement by Tucker & Bateman, woollen drapers, opposite Craigs-court, Charing Cross, concerned their need for removal 'on Account of the Houses on that side of the Way being pulled down to widen the Street'. *Daily Advertiser*, 30 March 1757.

[28] The Earl of Dysart's property was rented to John Lambert (note 26) of whom a descendant (William Lambert) acted as executor to the will of William Hill (of Watkins and Hill) in 1847.

From the wrangling which ensued over this road-widening scheme, Watkins must have learned a great deal. In short, Dysart, the fourth earl (and great-grandson of Lady Dysart of the influential and aristocratic Lauderdales of Ham House, Richmond), was unwilling to relinquish land without proper compensation. A lawsuit ensued in which it was argued that 'if his Freehold Estate ... is to be taken away by Force ... he [Dysart] shall look upon such Proceedings as an Encroachment on his Property'.[29] Although the demolition of the Oliver Cromwell house and the reduction of Dysart's freehold went ahead in 1758, both Watkins and Dysart refused to accept either the commissioners' offer or the verdicts of two juries at King's Bench.[30] The matter was not simple: Watkins joined with Dysart in petitioning the House of Commons. A counter petition was taken by the commissioners but eventually, in 1765, recompense was made: Dysart received £1,670 for his two front houses that had been pulled down and Watkins received £2,000, although he had to wait until 1771 before payment.[31] For all the wrangling, the street was only widened by 5 feet to allow a uniform width of 35 feet.[32] Thereafter, a house and tavern (The Ship) were built on Dysart's plot and Watkins erected a 'new brick messuage' on the site of Mary Kirke's house which became the tenure of the bankers, Messrs Biddulph and Cox.[33]

Retirement

In the mid-1780s Francis and Clarinda Watkins decided to retire to Richmond, perhaps having already followed the many aristocratic and affluent Londoners – including the more successful shop owners – who took weekend recreation there, travelling by carriage or boat along the River Thames. At this time Richmond was an extremely fashionable resort which held an atmosphere similar to that to be found at Versailles before the Revolution. There were balls, masquerades, regattas and displays of philosophical fireworks and, of course, there was always a chance of rubbing shoulders with better class customers.[34] Francis and Clarinda would have noted the urban development that had already begun to make its mark, particularly along Richmond Hill, which climbed from Petersham near the Thames to the gates of Richmond Park. It was thus to Ivy House, just off

[29] LCC, *Survey of London* (1937), vol. 18, p. 90.

[30] TNA: C 54/6165, No. 11.

[31] Indenture (22 May 1771) between Francis Watkins and Clarinda, his wife, and Samuel Seddon and John Simpson. TNA: C 54/6314, No. 2 cited in LCC, *Survey of London* (1935), vol. 16, p. 86 and pp. 243–244.

[32] TNA: Works 6/35, pp. 243–244.

[33] LCC, *Survey of London* (1935), vol. 16, pp. 85–86.

[34] Thorne (1876), pp. 488–505.

this road and amidst other important and historic mansions, that they retired in the mid-1780s. But this property was not Watkins's sole purchase for he also bought the two adjacent houses in the terrace although these remained vacant during the time of his residence there.[35] For Ivy House (the largest of the three) he paid an annual Rent (meaning rates) of £30, together with a Poor Rate set at £3 15s and Highways Rate of £1 10s, thus totalling £35 5s per annum. For the three properties (taken together) his annual outlay was just short of £100 but this would be of little consequence to the 'comfortably off' Watkins who had planned for an annuity of £150 to commence in his sixty-fifth year.[36]

The retired couple had lived only four years in their new surroundings when Clarinda died on 19 January 1789 in her seventy-third year.[37] She was interred in a vault along the south wall of St Peter's, their local church in Petersham.[38] Francis then turned once more to the purchase of property, this time taking the opportunity to acquire the lease for the block of shops and houses incorporating the old instrument shop. It is possible that he was then living in the separate apartment that he had retained at Charing Cross for, in November 1790, it was reported that 'A house at Richmond Hill, belonging to Mr. Watkins, of Charing Cross, was broke open on Monday night, and a quantity of plate and other valuable articles stolen', perhaps suggesting that the Richmond house was unoccupied on this occasion.[39] This loss, coming on top of the loss of his wife, may have taken its toll on his health because, a year later, on 17 November, in his sixty-ninth year, he died.[40] His funeral took place at St Peter's on 24 November 1791 where, according to a marginal note in the burial register, he was duly interred 'In the vault with Mrs Clarinda Watkins.'[41] Neither frugal, nor extravagant, Francis stipulated in his Will that only one mourning coach should be hired for the occasion of his funeral.

---

[35] These properties were later numbered 1–3 Mansfield Place (with Ivy House being no. 3); today they are numbered 144, 146 and 148 Richmond Hill. RCRL: from 'Extract of rates paid by residents of Richmond Hill'.

[36] Francis Watkins had paid James John and Benjamin Vassar of Winchendon and William Lee of Hartwell (Bucks) £1,050 for an annuity of £150. Quarterly receipts commenced on 25 December 1787 and ran for one year only. BCC: D-LE/15/81. It appears that Watkins saw that he could manage without the annuity because he then sold it on to one Thomas Ryder of Lincoln's Inn for a consideration of £1,000. BCC: D-LE/15/82a.

[37] Death announcement of Mrs Watkins, 'wife of Francis Watkins, Esq.'. *Whitehall Evening Post*, 20 January 1789.

[38] SHC: Petersham Burial Registers: ref. P48/1/12. Many famous people lie at this church, including Captain George Vancouver, mentioned in Chapter 8. Thorne (1876) p. 468.

[39] [39] *London Chronicle*, 9 November 1790, and other newspapers.

[40] *Whitehall Evening Post*, 17 November 1791; *Morning Chronicle*, 18 November 1791; *The World*, 19 November 1791; and *Public Advertiser*, 21 November 1791.

[41] Petersham Burial Registers, SHC: P48/1/12.

At his passing, Francis Watkins's business formally fell to his nephews. He did, however, make provision for his two married daughters – Frances (Mrs Samuel Bourn) and Clarinda (Mrs John Woods) – by way of annuities of £50.[42] By a separate trust of £100 yearly, he also provided for his 'good friend', the widow Sarah Langley who kept his house at Richmond, but conditional upon her not marrying. Thus, on his death, Sarah came into his household goods, plate and furniture and had the free use of his apartment at Charing Cross for as long as she needed it. The Smith family with whom he had shared so much of his life was also remembered. As an 'esteemed friend' Addison received a token of £100 and Elizabeth-II (Addison and Elizabeth's daughter) received the remarkable sum of £500. His nephews, Jeremiah and Walter, executors to the Will, had the onerous task of arranging these bequests, trusts and annuities, although they were amply recompensed by their inheritance which, for the most part, comprised freeholds and leaseholds. Thus, their start in business had been facilitated in a way not experienced by Francis at the start of his own working life. Time would tell whether this inheritance would be put to good effect in the turbulent years ahead, although it is sufficient here to remark that the business did survive a further 65 years.

Instruments

A glance inside the shop at Charing Cross at the time when young Francis was apprenticed to Nathaniel Adams has been outlined in Chapter 3. Spectacles and spectacle frames were available there in plenty, although no lens-grinding machinery was found when the inventory was drawn up on Adams's death. It is therefore suggested that it may have been the practice of opticians such as Watkins to buy in prepared lenses from specialist grinders and polishers with proven skills, such as Oliver Combs or George Bass, or from the Dollonds who supplied Watkins with patented compound objectives for the 'new-invented telescope'. Other saleable goods – walking canes, spectacle cases, and opera glasses – were bought in, as were books, prints and maps. Stocking up in this manner allowed Watkins and his apprentices to concentrate their practical skills on making bespoke instruments, or dealing with on-going maintenance and repairs. Watkins was both highly skilled and inventive and, given his all-round capabilities, he was evidently very much in demand. His entry to the trade had taken place during a period of moderate economic stability, and given his good health and supportive wife, it would appear that he was destined to succeed from the very start (see Table 10.3).

[42] Will of Francis Watkins (signed 7 November 1791). TNA: PROB 11/1211, sig. 548. Proved 22 November 1791.

Table 10.3 Instruments signed by Francis Watkins, in public collections in the United Kingdom

Instrument	Signature	Location
Mercury barometer	F. Watkins/London	NMSI 1920-54
Solar and compound microscopical compendium	F. Watkins/London	NMSI 1993-1118
Single microscope	F. Watkins/fecit	NMSI 1906-59
Single microscope	Unsigned, but by Francis Watkins	NMSI 1928-821
Universal microscope	F. Watkins/London	NMSI 1928-825
Universal microscope	Watkins/Charing Cross/London	NMSI 1928-794
Compound microscope	F. Watkins/Charing Cross/London	NMSI A56327
Compound and solar microscope	F. Watkins/London	Museum of the History of Science, Oxford 43487
Compound and simple microscope	F. Watkins/Charing Cross/London	Museum of the History of Science, Oxford 47695
Compound microscope	F. Watkins/London	Museum of the History of Science, Oxford 39147
Compound, simple and solar microscope in silver	Fra. Watkins Charing Cross London	Museum of the History of Science, Oxford 53619
Simple microscope	Fra. Watkins/London	Museum of the History of Science, Oxford 63814
Simple microscope	Unsigned, but by Francis Watkins	Museum of the History of Science, Oxford 23243
Gregorian reflecting telescope, brass, on pillar and claw stand; lacks eyepiece	Fra. Watkins/Charing Cross/London	Castle Museum, York
Gregorian reflecting telescope, brass, on pillar and claw stand	Fra. Watkins/Charing Cross/London	National Maritime Museum, Greenwich AST0959
Gregorian reflecting telescope, brass, on pillar and claw stand	Francis Watkins/ Charing Cross/London	Whipple Museum for the History of Science, Cambridge Wh:296

Instrument	Signature	Location
Gregorian reflecting telescope, brass, on pillar and claw stand	Francis Watkins/ Charing Cross/London	Whipple Museum for the History of Science, Cambridge Wh:1278
Gregorian reflecting telescope, brass, on pillar and claw stand	Francis Watkins/ Charing Cross/London	Museum for the History of Science, Oxford 69862
Monocular opera glass	Watkins/London	Museum for the History of Science, Oxford 11346
Galilean spy glass telescope	Watkins/London	National Maritime Museum, Greenwich NAV1535
Galilean spy glass telescope	Watkins/London	National Maritime Museum, Greenwich NAV1539
Solar and scroll microscope in mahogany box	F. Watkins/London	Kedleston Hall, Derby
Drawing instruments in ray skin box: signed on protractor and diagonal scale	F. Watkins/London	Ipswich Museum, Suffolk
6 inch sector in ivory and silver	F. Watkins/London	Tyne & Wear Archives & Museums, Newcastle TWCMS: J4866
Globe electric machine; vertical globe and table clamp	Fra. Watkins/London	Royal Institution, London
14 inch radius ebony frame octant	F. Francis Watkins/ Charing Cross/London	Castle Museum, York
Waywiser with iron bound wheel (dia. 31¾ in), brass dial on handle and divided to measure up to 12 miles	Fra. Watkins/London	Bruce Castle Museum, Tottenham, London
Brass telescope with wood cased outer tube	Watkins/Charing Cross/London	Royal Naval Museum, Portsmouth
Mounted angle barometer and thermometer with hygrometer and perpetual calendar	Watkins/fecit	Blair Castle, Blair Atholl
Mounted angle barometer and thermometer with hygrometer and perpetual calendar	F. Watkins/London	Museum for the History of Science, Oxford 24756

Numerous examples of Watkins instruments may be found in the holdings of many institutions, both at home and abroad. As an optical and philosophical instrument maker, Watkins responded mostly to the demand for microscopes and telescopes, although he was equally comfortable with electric machines and pumping engines. In the 1760s and 1770s, Watkins – along with his partner Addison Smith – branched into mathematical instruments such as surveying and gunnery, as well as coin scales; they also dealt in barometers and thermometers. Occasionally, surviving instruments turn up abroad: for example, a stick barometer can be found at the Colonial Williamsburg Museum (USA), an achromatic telescope is in the collections at the University of Coimbra (Portugal), another at Schloss Arolsen (Germany), and a universal microscope at the Ernst–Abbe–Stiftung Optisches Museum in Jena (Germany). Unlike his well-known contemporaries Benjamin Martin or George Adams Jr, whose price lists are known, no price list has yet been found for Watkins. Despite this lack of information, the diversity of Watkins's output is apparent, including objects such as a universal ring sundial, a circumferentor, a wheel barometer, an armillary sphere, and a brass cannon scribe, which, all too briefly, appeared at public auctions and then quickly disappeared from public gaze into private collections.[43]

Watkins's reputation remained with him to the end. He is remembered today for his instruments and for his published works on the electric machine (1747) and microscope (1754); it is also hoped that he will now be recognised for his place in the chronicle of actions and petition to the Crown by which he tried to defend members of the Spectaclemakers' Company from the monopoly of a patent held by the Dollonds. At the time of his death, his name was of sufficient merit to warrant a brief announcement in the *Gentleman's Magazine*, which simply recorded that he died 'At his house at Richmond, Surrey, in his 69th year, Francis Watkins, esq., formerly an eminent optician at Charing Cross.'[44] That 'eminence' was completely lost on Victorian society but he has been memorialised in the recent *Oxford Dictionary of National Biography*.[45]

[43] For example, see the catalogues of Sotheby's (5 December 1975) and Christie's (15 May 1996), the *Newsletter* of Trevor Philip 5130 (issue 5), and *Tesseract* No. 54 (Fall 1996). It is not the purpose here to attempt a catalogue of all known Watkins instruments, even those available in public collections.

[44] Death notice (1791).

[45] Francis Watkins (b.1723–1791), www.oxforddnb.com/view/article/51718 accessed 2 July 2011.

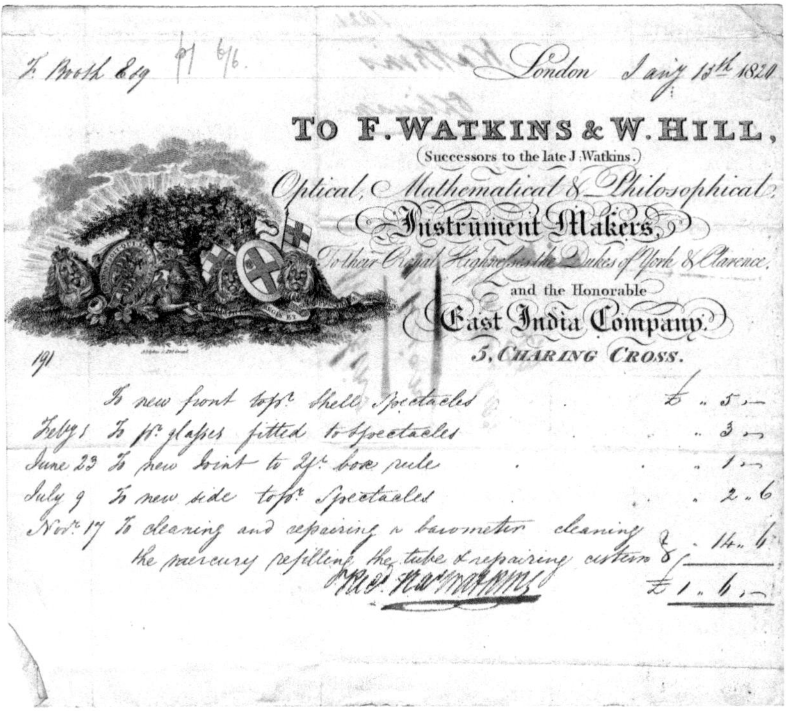
London Jany 13th 1820

TO F. WATKINS & W. HILL,
(Successors to the late J. Watkins.)
Optical, Mathematical & Philosophical
Instrument Makers,
To their Royal Highnesses the Dukes of York & Clarence,
and the Honorable
East India Company.
5, CHARING CROSS.

	To new front topp. Shell Spectacles	£ .. 5 .. –
Feby 1	To pr. glasses fitted to Spectacles	.. 3 .. –
June 23	To new Joint to 2ft. box rule	.. 1 .. –
July 9	To new side topp. Spectacles	.. 2 .. 6
Novr. 17	To cleaning and repairing a barometer cleaning the mercury refilling the tube & repairing cistern &c.	.. 14 .. 6
		£ 1 .. 6 .. –

Figure 10.3 Francis Watkins-III and William Hill's billhead, 1820.
© The Trustees of the British Museum, 1958,1006.2794

Watkins & Hill

In 1811 the business featured in Holden's *Directory* as C. Watkins, probably Jeremiah-III's widow Charlotte, still trading as optical, mathematical and philosophical instrument makers with appointments to their Royal Highnesses the Dukes of York and Clarence and to the East India Company. In 1816 and 1817 the firm is listed as J[ohn] Watkins. From his death until Francis-III Watkins was of age, the business was run by William Hill (d. 1847), an employee of long standing who had been with the firm since at least 1798. By 1826 it is listed as Watkins & Hill, a partnership of Francis-III and Walter Watkins with William Hill.

This chapter now relates principally to the business of Watkins & Hill, and more specifically, the presence of Francis-III Watkins in the context of the development and diffusion of science in the second quarter of the nineteenth century. Some of the archival material, together with other details from the literature and trade ephemera, highlights his contribution, at a time when a utilitarian philosophy of 'science for all' prevailed. This period was also an era of a developing interest in current electricity and electromagnetism, which followed in the wake of advances by Alessandro Volta, Hans Christian Oersted, André-Marie Ampère, Jean-Baptiste Biot and Michael Faraday. Watkins was enthusiastic about the various electrical developments and very keen to disseminate what he knew to those of a practical turn of mind. This chapter is thus centred on the mechanics' movement, that is, a phase in the educational development of the nation before state intervention.

The University of London was also set up at this time and it is now clear that Watkins & Hill were much involved in providing scientific apparatus during its early years. Much of the company's efforts in the 1820s are reflected in the detailed trade catalogues which appeared in the 1830s. Watkins & Hill were not, of course, alone as suppliers to the rising educational market. Despite their early prominence and their success as leaders in the instrument trade, they were faced with the rise of other competitors, particularly J.J. Griffin and Elliott Brothers, both before and after the Great Exhibition, as well as other factors which led to the eventual closure of this century-old business house.

Science for the Working Classes

The selection of instruments and accessories advertised in the Watkins & Hill catalogues of the 1830s was enormous in comparison with earlier publications of a similar nature. This was not entirely due to purchases made by university professors of experimental philosophy and chemistry at Oxford, Cambridge and London, nor by the local lecturers at the City Philosophical Society or the Russell Institution or other local literary and philosophical societies, nor, indeed, by schoolmasters. In fact, only a privileged minority attended university, formal societies were few, and demands for apparatus by schoolmasters in public schools were rare. The situation in parish or dame schools was even worse, and teachers inexperienced in science could hardly be expected to promote the subject, other than by silent reading of popular works such as Jeremiah Joyce's *Scientific Dialogues*, or John Paris's *Philosophy in Sport made Science in Earnest; being an attempt to implant in the young mind the first principles of Natural Philosophy by the aid of Popular Toys and Sports of Youth*. Occasionally, teachers would insist that pupils recited aloud from Samuel Parkes's *Chemical Catechism*, although Michael Faraday in his youth preferred to read Jane Marcet's *Conversations in Chemistry*. As Francis-III Watkins was of a similar age to Faraday he could also have attended the same courses at the City Philosophical Society in his youth, perhaps even coming across Faraday. However, the size and scope of the Watkins & Hill instrument catalogues reflect a time when a new adult education movement had begun to take off among the working classes.

The first sign of interest in the scientific education of the working classes is usually credited to John Anderson, professor of natural philosophy at Glasgow, in the closing decades of the eighteenth century. By inviting ordinary workmen to his classes, Anderson perceived the benefit that a basic scientific education could bring, and so willed some of his personal fortune to further his cause for the working classes. Although this did not realise all that he might have wished, there was sufficient to fund a chair of physics, whose first incumbent was George Birkbeck, hitherto trained in pharmacy and medicine.[46] Putting some of Anderson's ideals into practice between 1800 and 1803, Birkbeck too was struck by the curiosity shown by workers, who clearly had the capacity to benefit from instruction. Tentatively, like Anderson before him, Birkbeck began a regular course of free lectures, based on the scientific principles governing their daily work, presenting these by demonstration and in simple language. His motives were not entirely altruistic for he realised that an educated worker might benefit society in terms of increased productivity. Large numbers attended his lectures and, from this, it was decided that a 'mechanics class' should become a permanent feature of the institution. After Birkbeck's departure in 1804, the mechanics' class came under the direction of Andrew Ure, and it was to his chemistry class that John Joseph Griffin (of whom more later) came. It was around this time in the early 1820s, that the class members showed a desire to break free from the founding institution, which led to the formation of one of the first Mechanics' Institutes in 1823.

In the same year that the Glasgow Mechanics' Institute (GMI) opened, J.C. Robertson, in London, launched a new publication – the *Mechanics' Magazine* – aimed to popularise science and widen its applications. This venture was backed by Birkbeck who – aside from medical practice – was also associated with the London Institution, a body set up in 1805 to diffuse science and the arts through public lectures. He contributed an introductory article describing his own experience at Glasgow. These moves were quickly followed by a public meeting on 11 November 1823 at which, reputedly, some 2,000 persons attended – men who, according to Robertson, resolved both 'to think and act for themselves.'[47] With a little encouragement from Birkbeck, a committee of traders – comprising a printer, an engineer, a tailor, a shoemaker, an oilman, a coachmaker, a painter, a glazier and two students of law – was formed and, within three months, the London Mechanics' Institution (LMI) was born.

It was here, at the LMI, that Francis-III Watkins appears to have studied among his fellow artisans. Courses were given on chemistry, mechanics, electricity, heat and the steam-engine and all were illustrated by models and apparatus. Watkins's own involvement is plain, for on 12 November 1827 it was

---

[46] Kelly (1957).

[47] 'Public meeting' (1823), p. 177.

minuted that 'Mr Watkins of Charing Cross had present an instrument of his construction for producing the rotation of electrical wires by electromagnetism'.[48] In fact, the rapid success of the LMI spawned similar groups throughout Britain, so that within two years there were over 80 such institutions. This movement had the support of Henry Brougham, a barrister and politician who had received his scientific education under John Playfair at Edinburgh and who, at this time, exerted considerable influence. In 1825, he published *Practical Observations upon the Education of the People, addressed to the Working Classes and their Employers*, a tract dedicated to Birkbeck in recognition of his valiant efforts to channel the energies of those who, indeed, could 'think and act for themselves'. Much to Brougham's relief, this adult education movement had begun without any need for government intervention, although – despite the initial wave of enthusiasm – there were, plainly, many obstacles to be overcome, not the least of which were those associated with time and money.

In reality, the working classes had little time to spare for study and they could hardly afford to pay for expensive courses or books. As Brougham argued, these fears could partly be allayed by providing a more 'summary and compendious method of instruction', and by providing 'cheap and elementary works, concentrating more on practical applications'. As for practical demonstrations – such as those given by Watkins of his rotatory electromagnetic devices – these would certainly inspire, and explanations could easily be given in terms of rules and effects rather than abstruse theory. Concerning texts, Brougham proposed the formation of a publishing venture – the Society for the Diffusion of Useful Knowledge (SDUK) – which, in 1826, began to prepare texts for use among the mechanics.

The march of the mechanics' movement in the 1820s moved step-by-step alongside the development of the new electromagnetic science in a way that is forgotten today. Humphry Davy had announced Oersted's famous discovery at the Royal Institution and, soon afterwards, Faraday discovered a remarkable effect, namely, that a tangential force existed on a freely suspended current carrying wire about a fixed magnetic pole (and vice versa), thus leading to the notion of a means by which rotatory movement could be achieved. Peter Barlow at the Royal Military Academy at Woolwich also followed the path opened up by André-Marie Ampère at the École Polytechnique, Paris, and quickly evolved a course comprising 20 electromagnetic demonstrations. Two others improved upon these early demonstrations: James Marsh, assistant to Barlow and chemist at the Woolwich Arsenal, and William Sturgeon, tutor at Addiscombe College for the East India Company. Marsh's contribution was an improved battery of cells to power the demonstrations (1823); Sturgeon's efforts

[48] Minutes of the Committee of the LMI (12 November 1827). Birkbeck College Archives: LMI Minutes.

were directed towards amplifying the visual electromagnetic effects by applying the electromagnet principle using a soft-iron core (1824). Both contributors were rewarded by prize medals from the Society of Arts.[49] Watkins, too, seized upon the subject, for which he developed a commercial range of demonstration pieces.[50] These he showed before attendees at the LMI, and also before members of the newly formed Institution of Civil Engineers, a group with whom he was associated. By 1832 some of this range had been boxed and made portable for the peripatetic lecturer.[51]

The subject itself was intricate and without theoretical precedence. Moreover, investigations had to proceed according to new rules, never before experienced in experimental philosophy. Suspended wires, floating helices or solenoids (potentially poisonous due to the splashing of mercury in contact cups), and batteries of cells (not without their inherent dangers of inflammability due to the production of hydrogen and other noxious gases), were the order of the day. Ampère, with his instrument maker, the elder Hippolyte Pixii, was at the forefront of these investigations. By June 1822, he had modelled an expression for the action-at-a-distance between two current-carrying elements of a conducting wire and, within a year, Jean Firmin Démonferrand had explained the basis of Ampère's experiments in *Manuel d'Electricité* (1823).[52] Meanwhile, Peter Barlow had published a second edition of his *An Essay on Magnetic Attraction* (1823) which contained a theoretical view of electrical current based on a fluid conception, rather than Ampère's 'molecular' (material) standpoint. Although an early account of Ampère's work had appeared in the Royal Institution's *Quarterly Journal of Science, Literature and the Arts* in 1822, the subject did not appear in comprehensively until about 1826–1827, in the *Philosophical Magazine* (*PM*) and the *Mechanics' Magazine* (*MM*), by which time James Cumming at Cambridge had published *Manual of Electrodynamics* (1827), which was a translation of Démonferrand's text.

Important though these scholastic presentations were, they were unlikely to be easily understood by those who attended the Mechanics' Institutes of which, in the London region alone, there were some ten institutions by 1828.[53] Mechanics thrived better on visible effects rather than abstractions and mathematical analysis. It was for this reason that Watkins resolved to present another version of the subject, written in such a style that would have direct appeal to those like himself. As he explained:

---

49 Morus (2005).

50 Watkins and Hill (1828). For a fuller description of Watkins's apparatus for electromagnetic rotations, see Gee (1989), pp. 51–55.

51 Watkins and Hill (1832), p. 61.

52 Ampère (1823); more fully in: Ampère (1826). See also Caneva (1980).

53 Kelly (1957), appendix IV.

> But as many persons may wish to be acquainted with the subject, who, from not being familiar with mathematical expressions and formulae, would have a difficulty in pursuing Professor Cumming's work, [for them] this Popular Sketch has been compiled; and to assist to the same class of inquirers, and in the hope of spreading over a wider field, a more general knowledge of the parent sciences, Electricity and Magnetism,—an outline of these sciences has been carefully traced.[54]

Thus it was that Watkins's *A Popular Sketch of Electro-Magnetism or Electro-Dynamics* made its appearance in July 1828. Completely avoiding the 'ingenious and plausible [but] far from satisfactory [theories]' by concentrating on showing the ways and means of making things go, Watkins began the task of interpreting electricity and magnetism analogously as the parents of electromagnetism. 'All that is intended,' he wrote, 'is to lay the facts before the student in such a language as will make the matter clear to him; avoiding theoretical disquisitions...'[55] As the first text of its kind, it attracted long reviews in the *PM* and the *MM*.[56] Both reviewers sang praises for the painstaking way in which the author had culled the literature: that of *PM* – being somewhat more generous in his remarks – offered commendation for Watkins's undertaking to 'make this beautiful and interesting department of knowledge, a branch of *popular* science' while *MM*, more grudgingly, held the effort to be 'unambitious but very meritorious'. Both cited the value of the text for experimenters with its 'clear and connected detail of the principal observations and results of experiment ... and directions for their performance' (*PM*) and the 'many useful hints' (*MM*). There was also something to be said about the way in which Watkins had treated the history with *PM* seeing merit in Watkins naming the 'original and first-rate authorities ... instead of referring only to compendiums and compilations' although *MM*, still less complimentary, complained that 'He never quotes his authorities (book, chapter, or verse, we mean) ... and never makes any use of letters of reference'. That last detail, however, would hardly have mattered to the mechanics; it was, after all, a work of simple diffusion, not scholarship.

When it came to the way by which Watkins had modelled the subject, in the form of 'parent sciences', he was, in effect, perpetuating the older (pre-Oerstedian) perception of electricity and magnetism as distinct or unrelated phenomena. This presented certain difficulties for *MM*, who complained: 'We know of no authority, however, which Mr. W. has for such a classification.' Elaborating further, the critic continued:

---

54 Watkins (1828), pp. iii–iv.

55 Watkins (1828), p. iv.

56 'Notice' (1828); 'New Publications' (1829).

> *Galvanism* (which he includes under electricity) is at least as much a separate science as *magnetism*; and *electro-dynamics* is no more the offspring of the one than the other; but the conclusion to which philosophers have now generally come, is, that *electricity, galvanism, magnetism, electro-magnetism* and *electro-dynamics,* are only so many different names for different modes in which the electric fluid exhibits its effects to our observation.[57]

The complaint was that Watkins had ploughed through the history of the subject 'without catching more than once or twice a glimpse of that polar star which has guided the researches of Volta, Oersted, Ampère, Farriday (sic) and others, – namely the universality of electricity, as a principle of nature'. Adding his own solution to the new vision of oneness in nature, the reviewer offered:

> A term that will include the whole of this class of phenomena is still wanting; unless, indeed, men of science could be persuaded of the propriety of reverting to the use of the word *electricity* in the universal sense in which it was first employed, and which it is still as capable of bearing as ever.[58]

Watkins's 'parent sciences' and the reviewer's 'universal electricity' are the differing frameworks for all subsequent criticism. In the introduction to the *Popular Sketch,* Watkins had portrayed an already obsolescent view although it was conceded that the 'child' electromagnetism might well provide insight into whether the 'parents' are 'either distinct and separate principles, or that they are one and the same principle differently modified'.[59] But the choice of model was considered awkward and damning. In citing the various sources of electricity, for example, Watkins had placed that produced by friction, contact and pressure, in the parental section under 'Electricity', while that originating from a 'disturbance of equilibrium' (later termed thermoelectricity) was allotted to the parent 'Magnetism'. 'Need we seek a stronger exemplification of the impropriety of Mr. W's classification?' asked the reviewer rhetorically. 'The moment he attempts to follow it out, it involves him in this practical absurdity, – that either there is no truth in what he affirms of the modes by which electricity is developed, or that electricity and magnetism, which he classes as two distinct things, are in reality one and the same.'[60]

Having pounced on this weakness the reviewer then felt at liberty to quibble over other theoretical matters which had not been a part of Watkins's plan. For example, Watkins had offered know-how on which acids and

57 'New Publications' (1829), p. 29.

58 Ibid.

59 Watkins (1828), p. iii.

60 'New Publications' (1829), p. 30.

saline solutions worked best in batteries made to deliver 'quantity' but *MM* complained that no insight was given as to why such solutions are employed. Of course, the burning issue of the time lay in understanding whether these aqueous solutions act as conductors or whether they are the primary sources of electricity by oxidation with the metallic plates, the latter being the opinion of 'the late lamented Dr Wollaston'. 'Mr. W. might, at least, have mentioned that such a difference of opinion on the subject existed, and have stated such ascertained facts as have a bearing upon it' quibbled *MM*.[61] But this was seeking theory which, as stated in the prefatorial remarks, Watkins was not prepared to debate. The reviewer did, however, concede the text to be 'an extremely useful performance'.

Both reviewers anticipated that the text would be reprinted and so gave suggestions for further improvement. However, it was not to appear again at Watkins's instigation. A somewhat more rigorous text did appear in 1832 when Peter Mark Roget, then Secretary to the Royal Society, published *Electricity, Galvanism and Electro-Magnetism* under the auspices of the SDUK. This was a work of considerable clarity, containing some recognition of Watkins's electromagnetic apparatus as asides. Watkins's contribution, however, had served a double function: namely, it launched the new subject among mechanics and their teachers while, at the same time, it helped to promote Watkins & Hill as leaders in an emerging and perceptibly important field.

Curatorship at the New University

Completing the quest to provide 'science for all' in London also required provision to be made for the more privileged who could afford to pay for their education, and who had the time to attend courses. For them, an important opportunity arose when the new university was set up in 1826/7 (to be known from 1836 as University College).[62] This was the first university to appear in England since the medieval foundation of the universities of Oxford and Cambridge. Its origin sprang from the same Whig–Radical and nonconformist baseline as had the recently established Mechanics' Institutes of George Birkbeck, and was driven forward by the same utilitarian ideals of Henry Brougham who was on the Provisional Council when it was first considered in December 1825. Another like-minded Scot on the Council – who also features in correspondence with Francis-III Watkins – was Leonard Horner, a man of considerable drive who had been a leading spirit in the foundation of the Edinburgh School of Arts in 1821. Birkbeck, too, was

61 'New Publications' (1829), p. 31.

62 Harte and North (1991).

brought in by Brougham along with other supporters of the mechanics' movement.[63] Indeed, all on this Council had no hesitation in placing science high on the agenda of the intended curriculum and straightway a post was advertised in the autumn of 1826 for someone with strong utilitarian ideals to take up the chair of natural philosophy and astronomy.

Among the interviewees to report before the Council in the following April was the somewhat flamboyant Reverend doctor Dionysius Lardner, an articled lawyer of about 35 years of age who spoke convincingly of the utility of science with the kind of sentiment which Brougham wished to hear. He was well-versed in steam technology, a subject on which he had delivered lectures at the Royal Dublin Society and was already known in London through a début in the theatre of the Royal Institution on the same subject. The outcome of his interview was probably a foregone conclusion and the chair was offered to commence on 12 July 1827.[64] Envisaging fees of five or six guineas per pupil and other extra-mural honoraria for popular courses to fashionable circles, Lardner lost no time in persuading the Education Committee – apparently without reference to the Council – to advance him £200 towards apparatus for his grand plan for a 'museum of physical apparatus'.

Most of the apparatus and instruments were ordered from Watkins & Hill. But with the buildings barely completed and apparatus about to flow in, Lardner had to badger the Council to rent a house in Percy Street for temporary storage of the apparatus. With preparations at the new university in a state of flux, Watkins – possibly at Lardner's suggestion – took the opportunity to suggest to the Council that he might fulfil some useful role. Only three years earlier, the instrument maker John Newman had applied to the managers of the Royal Institution for recognition and, after acceptance, he proudly displayed this patronage on printed material.[65] Such an advantage had not escaped Francis Watkins when, on 2 November 1827, he wrote to Leonard Horner, Warden and Secretary, proposing that he could fulfil gratis a role as superintendent of the laboratory assistants and supervise the care of equipment in return for permission to call himself 'curator of philosophical instruments and apparatus'. Within three days, the Committee of Management ratified Watkins's proposal and the Warden was directed to notify him that his title would be permitted 'without salary or other direct pecuniary remunerations in the construction of apparatus'.[66]

---

63 Bellot (1929), Chapter 2; Kelly (1957), p. 153.

64 Dionysius Lardner, www.oxforddnb.com/view/article/16068 accessed 2 July 2011.

65 Minute of 7 April 1823. RI: Archives, *Managers Minutes* VI, p. 384; Dawes (1996); Gee (1996); Clifton and Gee (1997).

66 Committee of Management. Letter to Watkins (5 November 1827). UCL: Records.

An extant Apparatus Book, dating from 26 November 1827 (detailing maker, date of delivery and cost) with entries in Watkins's handwriting bears testimony to his carrying out proper duties.[67] This same source also reveals how ineffective the management was in preventing Watkins from having anything but a favoured position. By a simple analysis, the number of items purchased costing more than £1 ran as follows: Watkins & Hill (140 items), Cuthbert (25), W. & S. Jones (16), Fraser (5), Stiles (2), Dollond (1), Dunn (1), Lynch of Dublin (1), Bramah (1).[68] Lardner's free-spending reign came quickly to an end in 1831 when he left to seek other remuneration as author and editor of the SDUK. At the same time, the Committee of Management urged the Clerk 'to ascertain how the services of the Curator of Philosophical Apparatus could be dispensed with without injury to the establishment'.[69] Despite Watkins's subsequent attempts to retain his privilege under the new incumbent professor, William Ritchie, he was snubbed by the authorities and ignored by Ritchie whose meagre requests for new equipment could hardly warrant retaining Watkins and who, in any case, expressed his satisfaction to make do with the services of Francis Kirby, the laboratory steward of three years standing.

Trade Catalogues

From the results of a project to locate and identify instrument makers' catalogues, and from an examination of those by contemporaries of Watkins & Hill, it would appear that very few metropolitan suppliers were actually involved directly in the supply of instruments and apparatus to the various metropolitan institutions.[70] However, during the second quarter of the century Watkins & Hill was one of the most prominent suppliers. They were not the first to publish priced lists although they were among the first to include a separate division for chemistry instruments. Whereas most earlier and contemporary catalogues comprised only a few pages, the one-shilling Watkins & Hill catalogues of the 1830s ran to 84 pages and more in later editions, crammed with over 2,000 priced items of various designs and dimensions, and all well-suited to any enlightened institution or enthusiastic lecturer aspiring to promote science by demonstration or individual practical work.

Today, these catalogues – more than those by other contemporary suppliers such as Newman or Cary – provide valuable reference to the sorts of instruments, apparatus, machines, models and ancillary materials then in use. Importantly,

67 UCL: Archives, MS Add. 31 and Add. 32.

68 Ibid.

69 Committee of Management Minutes (16 August 1831). UCL: Records.

70 Anderson, Burnett and Gee (1990).

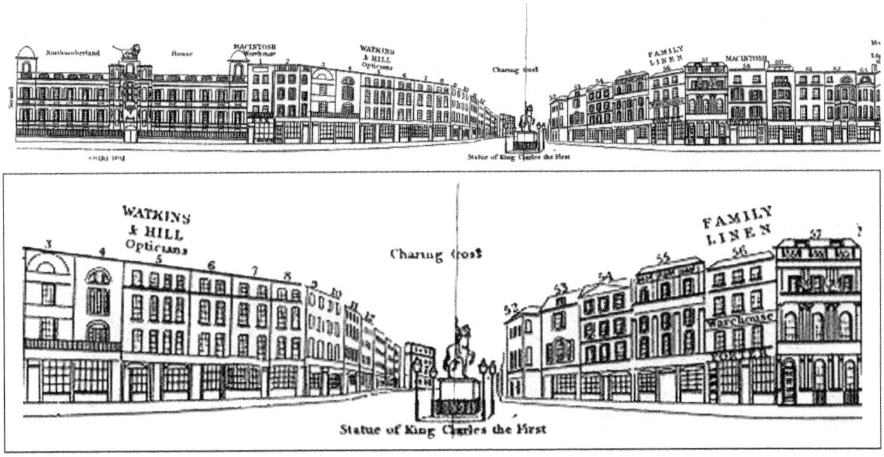

Figure 10.4 View of Charing Cross, with Watkins & Hill's premises, from John Tallis's *London Street Views 1838–1840* (with an enlarged view of the central road). City of London, London Metropolitan Archives

they offer a snapshot of the kind of scientific experience that was possible by demonstration in the lecture room, or gained by individual practice in the laboratory. This was, in fact, a very important period when science was perceived to be an essential ingredient in the education of the male labouring classes.[71]

As each new scientific development came along, Watkins worked furiously to perfect some way of utilising or presenting this advance. It was precisely such distinguishing qualities that had been recognised by Robert Hare, professor of chemistry at Philadelphia, who met Watkins in 1836. Writing to Joseph Henry of Princeton College, New Jersey, just before his expedition to England in 1837, Hare advised:

> In London you will find a great number of shops in which philosophical instruments are sold. The persons so engaged are generally scientific men and I found no difficulty in making myself acquainted with them. Watkins of Charing Cross I found more intelligent than the majority of them. I do not like Newman, No. 122 Regent St. He is however much employed.[72]

[71] In comparison with the situation in earlier times and elsewhere, Millburn (1986b) compares three of Benjamin Martin's catalogues; de Clercq (1997), pp. 65–72 has looked at the catalogues of Musschenbroek of Leiden, and Morrison-Low (2007), pp. 164–174 has analysed those of W. & S. Jones.

[72] Letter of Hare to Henry (13 February 1837). Reingold (1979), vol. 3, pp. 150–151.

Henry took Hare's advice and, within a day of his arrival in London, he sought out Watkins and the two met several times thereafter. Indeed, such was their friendship and parallel interests that both agreed to correspond.[73] This they did on equal terms, with Watkins eagerly awaiting Henry's discoveries and in return relaying to him gossip from within the Royal Society, although not as frequently as had been proposed during their meeting. Later, Watkins was to remind Henry:

> ... my position in society is as you well know of no very high degree in my Aristocratical Society.[74]

Here, then, is a salutary reminder of the nature of the society in which the instrument maker worked. But regardless of his rank, Watkins was much appreciated by those whose leanings were more scientific than aristocratic. On one occasion, John G. Children, Secretary of the Royal Society, had been so impressed by Watkins's investigations that he encouraged him to submit a report for the *Philosophical Transactions* on the strengths and retention of magnetism using pre-electromagnetised soft iron cores. Children read Watkins's very worthy paper to the Society on 16 April 1833 and straightway he was invited to perform his magnetic experiments before the President, the Duke of Sussex, at a 'conversation party' held at Kensington Palace on 27 April 1833.[75] This recognition and encouragement was formative to later research which led him to design two distinct forms of electromotive engine.[76] So, despite his marginalised position in the capital's highly structured society, Watkins's visit to Kensington Palace was in itself an accolade in recognition of his scientific standing. Indeed, his presence is also noted at the Royal Institution where, among the many subscribers to Faraday's lectures, he inevitably rubbed shoulders with the capital's population of aristocrats.[77]

But it was mostly among like-minded scientific individuals that Watkins probably felt most at ease. In addition to Joseph Henry, Watkins's reputation in electrical matters attracted many visitors to the shop including Gerrit Moll from Utrecht, Auguste de la Rive from Geneva, Stephano Netschayef from Russia and Wilhelm Eisenlohr from Karlsruhe.[78] Some sought Watkins's advice and some bought materials with which to pursue their own experiments. For example, Moll, in 1828, became inspired to improve the lifting of the electromagnet after being shown by Watkins just what a nine pound horseshoe could achieve. Of his

[73] Letter from Henry to Watkins (19 June 1839). Reingold (1981), vol. 4, p. 240.

[74] Letter from Watkins to Henry (March 1845 and 16 December 1845). Rothenberg (1992), vol. 6, pp. 264–267; quotation p. 265.

[75] Watkins (1833).

[76] Gee (1991), pp. 60–63.

[77] Subscription list to Faraday's lectures (1838). RI: Archives M9 fol. 44 (4 June 1838).

[78] Watkins (1838).

visit, Moll wrote: 'I immediately determined to try the effect of a larger galvanic apparatus ...', and indeed, at home he did so, using an even larger soft-iron horseshoe from Watkins which he carted back to Utrecht.[79] Possibly it was through Moll that his fellow countryman J.G.S. van Breda, then working at Teyler's museum in Haarlem, also made his various purchases from Watkins & Hill.[80]

In addition to those institutions mentioned earlier in this chapter: the London Mechanics' Institute, the Institution of Civil Engineers, the Royal Society, the University of London, Watkins's name can be found elsewhere in other contexts. For example, early on in his career, in 1827, he was chosen by George Birkbeck to serve on a Committee of Inspection for the Society of Arts exposition of 1828 at the National Repository held in the former Royal Mews.[81] Since Birkbeck's committee also included T. Bramah, B. Donkin, C. Holtzapffel, M.I. Brunel, T. Winkworth and a dozen other engineers and academics of some standing, we can see that Francis Watkins held a good reputation. His name was also pencilled into Faraday's juvenile chemistry lecture notes, thereby implying a role as demonstrator or source of some borrowed item of apparatus for the Christmas lectures.[82] Another venue at which Watkins appeared was the Adelaide Gallery for Practical Science in the Lowther Arcade off the Strand. This institution was founded in the 1832 by some members of the Institution of Civil Engineers, a group with which Watkins was associated.

At the Adelaide Gallery, on 14 November 1833, Watkins convened a meeting of some of London's most important scientific persons – Daniell, Lardner, Pepys, Phillips, Turner, Moseley, and, most important of all, 'Dr Faraday' – to an event dealing with a subject very close to his particular interests in electricity. This was a competition to determine the relative merits of two recently invented, but differently designed, quite extraordinary machines for drawing electricity from magnets. These were the magneto-electric machines of Joseph Saxton and Hippolyte Pixii. The reported issues surrounding this important soirée are both lengthy and contentious and have been described elsewhere.[83] Briefly, Pixii and Saxton, quite independently, had succeeded in obtaining a spark from a steel magnet. Saxton, in fact, had achieved this before Faraday himself who also lured a spark from a lodestone, although Watkins praised Saxton as being 'the first in England' in this achievement. Watkins had, in fact, befriended Saxton since his arrival from Philadelphia in the winter of 1832, possibly because of Saxton's reputation in making permanent steel magnets.

---

79 Moll (1830), p. 329.

80 See Bruijn (1979).

81 *Catalogue of the specimens of new and improved productions of the Artisan and Manufactures of the United Kingdom* ... (1828). RSA: Archives, National Repository, item 83638.

82 Faraday, 'Juvenile Lecture Notes'. RI: Archives, Faraday F4C (5).

83 *'Arts and Sciences'* (1833), p. 730; Gee (1993b), pp. 101–133.

Figure 10.5 'Long Room, Gallery of Practical Science, Adelaide Street, Strand', drawn and etched by Thomas Kearnan, 1834–1840. © The Trustees of the British Museum, 1954, 1103.12

Indeed, Saxton was invited by the engineering proprietors of the Adelaide Gallery to manufacture a steel magnet of enormous strength as a show piece for the gala opening and it was this magnet that was converted into the magneto-electric machine. With regard to the evening of the competition, Saxton presented his own machine while Pixii's machine (a machine borrowed for the occasion by Watkins from the Count di Predivalli) was demonstrated by that E.M. Clarke who, at this point, was in Watkins's employ. Each machine was subjected to various tests (ability to decompose water, strength of the 'shock' and so on) and, in discussing their individual merits, each received praise although the Saxton machine received the greater credit. Thereafter, Watkins retailed Saxton's machine for 12 guineas and, shortly afterwards, seemingly on realising that considerable profit could be made by this new source of current electricity, Clarke forsook Watkins in order to set up independently as 'magnetician' (a specialist in magnetoelectricity), an entirely new term in the lexicography of the instrument maker.[84]

[84] Clarke (1836–1837); (1838).

Watkins's presence may not only be found in the literature of the second quarter of the nineteenth century but also he appears in occasional correspondence with contemporaries such as Moseley, Babbage, Daniell and Wheatstone. From the literature, it is clear that he was not to be fooled in those areas in which he had carried out his own experimental research and formed his own opinions. One instance of standing his ground came in 1835 when a squabble broke out between Watkins and William Sturgeon, an infantry man returned from the Napoleonic wars who, in an amateurish way, had turned to instrument making and teaching at Addiscombe College. The issue over which they disagreed concerned whether 'true polar decomposition' was possible using a magneto-electric machine for the electrolysis of water. Unfortunately, their disagreement, though restrained at the start, led to derogatory remarks and inflammatory comments. After a lengthy exchange – with Sturgeon objecting to Watkins's 'unwarrantable attack' and Watkins describing Sturgeon's views as 'Silly and consequently useless' – the editor of the *PM* (Richard Phillips), very correctly, closed the exchange. From this (and other sources) it seems that this disagreement was the root cause of Watkins paying no heed to Sturgeon's lectures at Clarke's Laboratory of Science when the Electrical Society was formed in 1837. Joseph Henry, being in London at this time, however did accept an invitation to attend.

More cautiously perhaps, Watkins took Mary Somerville to task when, in her text *On the Connexion of the Physical Sciences* (1834), she had (in Watkins's opinion) denigrated Saxton's great achievement of the magneto-electric machine. Anonymously, as 'an esteemed correspondent' Watkins wrote:

> Mrs Somerville has acted not very handsomely in describing Mr Saxton's magneto-electric apparatus, inasmuch as she has omitted all mention of that gentleman's name, and spoken of it as though it were merely an improvement on Mr. Faraday's methods, and put forward in detraction of his [Saxton's] justly acquired fame.[85]

Here, as elsewhere, Watkins struck out against the hierarchical scientific establishment, preferring instead to give place and support where it was justly due. One instance of his intense efforts was observed by Joseph Henry in June 1837 when he called upon him after returning from Paris.[86] At the time, Watkins was devising a large battery of thermoelectric antimony–bismuth pairs. In effect, this experiment identified thermoelectricity with other forms of electricity, thereby completing Faraday's project to identify the various seats of electromotive force. In reporting this, Watkins remarked: 'Now that I have pen in hand permit me to state that with thermo-piles I actuate most of the apparatus usually employed for illustrating electro-magnetic phenomena, so that the public teacher may now

85 [F.W.] (1834).

86 Reingold (1979), vol. 3, p. 424.

Optical Instruments

Under this head we have classed not only those Instruments employed for the improvement of our sight, and the examination and measurement of those objects and appearances which from distance, smallness, &c. would otherwise elude our observation; but also those Instruments and Apparatus that are best adapted for the illustration of the properties of light; such for instance as regulate its directness, intensity, state of polarization, colours and interference of rays.

Astronomical and Geodætical Instruments

Astronomical Instruments and Apparatus include constructions for the purposes of observation and computation, and contrivances for exhibiting the motions and phenomena of the heavenly bodies: Geodætical Instruments comprise those in use for the surveying and division of land.

Mechanical Instruments

Instruments thus denominated, comprise all those that are employed for the experimental demonstration of the doctrine of equilibrium of powers of forces,—all those which treat of the generation and communication of motion,—all those that have for their object the illustration of the composition of machines, and also of certain machines themselves.

Hydrodynamical Instruments

These Instruments are commonly employed to illustrate that department of the science of the mechanics of fluids which relate to their forces and motions.

Pneumatical Instruments

The general intention of Instruments thus denominated are for the experimental illustration of the doctrine of elastic fluids, but more particularly the mechanical properties of Air.

Meteorological Instruments

Instruments under this denomination are employed in the investigation of atmospherical changes in general. By their means we are made acquainted with slight variations of pressure, temperature, and humidity of the atmosphere.

Electrical Instruments

These comprise Electrical Machines in general use, with the auxiliary Apparatus for Medical, Philosophical, and Chemical purposes.

Voltaic or Galvanic Instruments

Instruments thus designated are employed to exhibit a peculiar form of electric agency, elicited under particular circumstances, and capable of producing certain effects on bodies, not commonly resulting from the ordinary modes of electrical excitation, such as friction, &c.

Magnetic Instruments

Instruments classed under the foregoing head are employed to exhibit magnetic phænomena, whether produced naturally or artificially; but more especially their relation in respect to each other and their reciprocal action, and the direction they assume when freely suspended.

Electro-magnetical, Electro-dynamical, and Thermo-electrical Instruments

Instruments thus denominated are employed for the exhibition of the relation which subsists between magnetism and electricity.

Chemical Instruments

These Instruments are intended for the experimental investigation of the mutual agencies of the elementary principles of matter, and for the popular illustration of such principles as admit of demonstration by manipulatory practices.

Figure 10.6 Subject divisions of scientific instruments in the 1832 Watkins & Hill catalogue

show by the same apparatus the several rotations, &c. with thermo-electricity, as he does with voltaic electricity.'[87] Thus, another addition was made possible for inclusion in the next trade catalogue, which appeared in 1838.

Each division of experimental science within the Watkins & Hill catalogues was defined according to 'families of instruments'. Such a representation would have been invaluable to a young lecturer starting out in his career, especially also for those who had responsibility for ordering equipment for laboratories. The catalogues of the 1830s – with their 11 defined subject areas – were printed (with updates) at least three times and, by about 1840, the presentation was entirely revamped by splitting the whole gamut of science (but not chemistry) into five subject domains: voltaic and thermoelectric instruments, electrical instruments, hydrostatic, hydraulic, pneumatic, and acoustic instruments, magnetic, electro-magnetic, electro-dynamic and magneto-inductive instruments, and optical instruments. A new version of this five-part arrangement – copiously illustrated with woodcuts – appeared about 1840 and this remained the basis of what could be supplied throughout the 1840s.[88] This same series was reissued in 1856, under the name of Elliott Brothers, which had by then successfully absorbed the business of Watkins & Hill.[89]

87 Watkins (1837), p. 399.

88 Watkins & Hill (1845a), (1845b), (1845c), (1845d), and (1845e).

89 Elliott Brothers (1856a), (1856b), (1856c), (1856d), and (1856e).

Special Orders after 1840

Orders received by Watkins & Hill over the 16-year period from 1840 to 1856 are identified in an isolated ledger, mostly dealing with special business undertaken during Abraham Day's management, although none of these related directly to their current catalogue. Unlike the 1816 'daybook', this ledger contains brief notes on over 300 orders, both institutional and individual. Among the contents were institutional orders from the Admiralty, the Birmingham Mechanics' Institute, the London Institution, the Optical Institute of Munich, the Privy Council, Queen's College (Cork), the Regent Canal Company, the Royal Polytechnic Institution and the Victoria Gallery (Manchester). Of these, the Admiralty requested a 'colladon' apparatus to measure depth by the pressure of the sea;[90] the Privy Council ordered simple sets of chemical apparatus; the Regent Canal Company needed a tube for measuring the rain using a Crosby rain gauge; the Munich Optical Institute (where Fraunhofer worked) ordered an apparatus for the diffraction and dispersion of light; the Royal Polytechnic Institution ordered a 3-foot (Harris) plate electric machine, and, likewise, the Royal Victoria Gallery ordered a 4-foot version of the same machine. Mixed in with institutional orders were others of a more personal nature, mostly from local individuals (amateurs and professionals), many of whom can be identified with certainty due to their later eminence.

Not all orders appeared in the official order book. Over the years, Watkins & Hill received commissions when technical and constructional assistance was required. One such instance came from the Royal Observatory at Greenwich when George Biddell Airy, the Astronomer Royal, requested a 'warning bell apparatus'. On 31 July 1841, he wrote:

> We have lately erected an Electric conductor, and I should be glad if you would construct for us a warning bell apparatus similar to that described by Mr Reader in the Philosophical Transactions for 1792. It will be rather best in our case that all the wires should enter the bottom of the box. So that the form of the instrument will be something like this [here Airy provides a sketch along the following lines] a small bell of brass being suspended by a silk thread to swing between a large ball and a bell ...' [Airy continues by specifying the shape and material of the box, the positioning of wires, that wire diameter (1/10 inch), and that 'quality of wires should admit of soldering to our copper wires'].[91]

---

[90] MHSO: Elliott Archives, Watkins and Hill. The Swiss physicist J.D. Colladon (1802–1893) devised apparatus for measuring sound transmission through water in 1826, then sought to adapt it to measure depth by echo of sound, a precursor of sonar. Editors' note.

[91] 'Correspondence with Tradesmen'. CUL: RGO 6/718, section 24. Letter from Airy [fol. 975r] and acknowledgement from Watkins and Hill [fol. 976].

Abraham Day, then a junior workman at Watkins & Hill, immediately acknowledged Airy's request although, due to the slow pace of work, the apparatus was not made up for over two months. However, the Royal Observatory did receive the following apologetic note:

> 9 Oct. 1841 [Watkins & Hill regretted] ... the order for the Electrical Warning Bell not executed sooner, but circumstances prevented our completing it until within these few days ... If you desire a man from our workshop should form connexions between the bell App^{s} and y^{r} lightning conductor, he shall attend you at any time you may be pleased to appoint.[92]

Alarm bells of various kinds were then becoming popular. Robert Walker, in Oxford, had requested one and so had the Palace of Westminster who required one to be fitted to the Speaker's Chair in the House of Commons, possibly to herald a division upon which Members of Parliament were required to vote.[93]

Orders to and from Karlsruhe

Ideas for a new apparatus were often transmitted by personal contact and correspondence. Friendships were built up between the maker and the user and so ideas were exchanged reciprocally in much the same way that ideas are exchanged informally today. An extant letter to the Watkins family from Wilhelm Eisenlohr, teacher of physics at the Karlsruhe Polytechnic Institute, dated 9 February 1845, not only included an order but it also shows that Eisenlohr kept an account for Watkins for apparatus and materials that he purchased from Germany.[94] His letter is addressed (mistakenly) to 'Frank', probably meaning Watkins's son, Francis Burton Watkins, who was just then dealing with correspondence during his father's serious illness. Clearly, the letter is written in response to earlier queries for Eisenlohr replies:

> Pay my kind respects to all of your family and tell your Father [F.W.] that in respect to Glass [and] Chemicals he will do best to apply to J.G. Aubel in Giessen; the chemist Bullock and Dr Gardener in London will be able to direct him about the importation as they are well known there.

---

92 Ibid. fol. 977.

93 CLO: Papers of the Rev. Robert Walker.

94 Eisenlohr (1799–1872). Letter dated 9 February 1845. MHSO: Elliott Archives, Watkins and Hill.

Thus, we may note that Watkins dealt in Germany through Heinrich Aubel, the technician and janitor at the famous Liebig laboratory at Giessen.[95] His commercial enterprise in the manufacture of potassium and, likewise, the entrepreneurial character of the well-known London chemists Lloyd Bullock and John Gardiner would have brought these individuals into contact. Eisenlohr stayed at Karlsruhe until his dismissal in 1854 and thereafter appears to have founded technical schools in Mannheim and worked at the city observatory.

Included with Eisenlohr's letter to Watkins was a box containing: '2 Diffractive Apps' from Schwerd in Speyer, miscellaneous rings with wire railings, glass tubes, mirrors and 'other trifles', a box with feathers, a box with '6 ground Glass prisms' from Munich, some volumes of music, a parcel with a pipe, a box with plaster casts, and '2 Apps of wood & wire to explain the theory of undulations and a Krank [handle] for them'. The last item, the undulatory apparatus, was a wave machine of Eisenlohr's design for showing the compressions and rarefactions of a progressive wave. Of this, Eisenlohr wrote:

> I beg of Mr Watkins to accept the Undulatory Apps as a mark of my friendship & esteem. He will make it better & handsomer than we can at all events & twice as large. It cannot fail to please physikers if it is put together rightly for which the Directions serve.

The handwritten 'Directions' (with sketched illustration) are extant and very detailed. Assuming that Watkins would find no difficulty assembling the parts, Eisenlohr wished also that Watkins would show it to Charles Wheatstone who, as is well known, was of a musical persuasion and therefore would immediately find the apparatus of interest. In total, Watkins's order for materials from Germany came to the equivalent of £20 14s 5d.

Eisenlohr's letter also included requests. One of these indicates that he had recently visited London where he had noted a Watkins hydroelectric machine on exhibition at the Polytechnic Institution.[96] Of this, he asked:

> If your father can make such an app^{s} for 30 to 40£ I beg you to make me a small sketch. If not write at all events directly whether he has made such a larger one and to which piece and which effects it will give. I have seen the smaller one but they do little. I should like to have a good strong Apparatus. Naturally all the parts must be good and solid and particularly the part where the tubes are secured and must have a very good screw ...

---

95 For the chemical laboratory at Giessen, see Brock (1997).

96 The Polytechnic Institution exhibition hall opened on 6 August 1838 and quickly superseded the Adelaide Gallery of Practical Science. A catalogue for 1845 survives; unfortunately, earlier catalogues once held by the British Library were destroyed during a WWII bombing raid. Gee (1998a), pp. 15–16.

Eisenlohr also wished that Watkins should obtain and send him a chronometer although, on this matter, he was hesitant about its mode of conveyance. It is interesting to note, in passing, that long after this correspondence, Eisenlohr was still in communication with the old firm through orders to Elliott Brothers in 1860.[97]

Shifting Scenes in the London Instrument Trade

Several important changes took place in the second quarter of the nineteenth century, largely arising from the quickening pace of industrialisation. These changes were accompanied by an increased urge to educate or be educated. At an adult level, as already stated, there was a rapid development of Mechanics' Institutes; as a step towards professionalisation, there was increased institutionalised learning, sometimes with mandatory courses in laboratory practice for chemistry; at a juvenile level, science (principally chemistry) continued to be popularised as a rational recreation but now with the availability of boxed sets for self-instruction and, finally there developed the means to introduce chemistry into schools through the agricultural movement.

The constant promotion of chemistry as an engaging and worthwhile practical activity, well-suited to mechanics and, as time would tell, well-suited to schools too, caused newer (non-instrument making) entrepreneurs to enter the customary scientific instrument trade. One such supplier was Robert Best Ede, a Dorking chemist who hitherto supplied toiletries to the Royal Household. He struck up an arrangement with J.J. Griffin in Glasgow to supply portable laboratories as an accompaniment to the seventh edition of *Griffin's Chemical Recreations* (1834). Previous editions of this work had encouraged attendees at Mechanics' Institutes to set up their own home or kitchen laboratories by making ad hoc purchases from local suppliers such as oilmen, tinsmiths, apothecaries and other shops. At a simpler level, the chemistry amusement chest had been around in London since Friedrich Accum's lectures at the Surrey Institution at the turn of the century, although they took off positively for a juvenile market in the wake of Faraday's chemistry lectures for juveniles. In addition to Ede, other suppliers of chemistry sets, amusement chests and youth's laboratories included John Newman (Regent Street), Francis West (Fleet Street), Edward Palmer (Newgate Street), Horne, Thornthwaite and Wood (Newgate Street) and William Edward Statham.[98] Ede's portable laboratories were marketed by John Ward until Ede's death in 1845.[99] Watkins & Hill, too, joined this trend

97 Purchases of a diffraction apparatus, an ear trumpet, a Peltier electrometer and an ebonite rod from Elliott Brothers. Order Book (1860). MHSO: Elliott Archives, Elliott Brothers.

98 Gee (1989).

99 Baddeley (1835); Ward (1845).

about 1840, producing a separate catalogue for chemistry along with Watkins's own text, *Experimental Chemistry*, of which no copy appears extant. By this time, chemistry had developed a professional air and, with the formation of the Chemical Society in 1841, several instrument makers and suppliers, including J.J. Griffin, Richard Knight and Francis Watkins, became members.[100]

The Emergence of J.J. Griffin and William Elliott & Sons

The mid-1840s was a difficult time for Watkins & Hill due to the ill health of both principals. At this time two contrasting and contemporary firms became leaders in the supply of instruments and apparatus, thus expediting the closure of Watkins & Hill. These were the family businesses of John Joseph Griffin in Glasgow and William Elliott in London. In 1847, when Abraham Day took over the daily running of Watkins & Hill, neither of these businesses would have appeared a threat in a competitive sense and the decision of Mary Ann, Francis-III Watkins's widow, to allow Day to manage the business would have been a sensible solution. At least, by allowing the employees to continue, an income could be generated much as before Watkins's death. Francis-III Watkins died in 1847, while his partner, William Hill had died the previous year.[101] Within three years, however, both Griffin and the Elliotts had taken steps that would alter the course of the scientific trade in London and so, inevitably, affect the viability of Watkins & Hill.

Until 1831, Griffin had a fairly free hand in his career as author and journalist. He was well educated, having attended the Andersonian Institution, and had been an outstanding student of chemistry under Andrew Ure. Indeed, such was his passion for chemistry that he wrote *Chemical Recreations* – a remarkable text for its day – very much with those mechanics who attended the Andersonian mechanics' class (latterly the GMI) in mind. Six editions appeared between 1823 and 1826 and hence huge numbers throughout the country learned how to perform experiments at home in a 'kitchen laboratory'. Griffin's contribution as promoter of chemistry education should also be seen in the context of the Anderson–Birkbeck tradition. However, in the 1830s and 1840s, family responsibilities became predominant when, in 1831, his brother, Richard – proprietor of Richard Griffin & Company – died. It then became John's responsibility to manage the family firm in trust for his nephew, Charles, who was then only 13 years old. Not until he separated and moved the apparatus side of the book business to London in 1849 did he trade under his own name, as J.J. Griffin.[102]

[100] Griffin was a founder member of the Chemical Society in 1841; Knight and Watkins joined in 1843 when the Society opened its doors generally to membership.

[101] Gee (1988, thesis), pp. 36–37; Hackmann (2003), p. 26.

[102] Gee (1988, thesis), pp. 229–240.

The supply of apparatus had actually begun in 1837 following a 'chemico-commercial project' (as Griffin called it) to promote the subject by sourcing apparatus in large quantities and therefore selling it cheaply. The very first apparatus catalogue appeared in November of that year; other catalogues of Richard Griffin & Company appeared in 1842–1843 and 1845 after a Continental tour of Prussia, Austria and Bohemia to seek out further suppliers.[103]

Within these catalogues we find that beakers from Bavaria and porcelain ware from Dresden had replaced the customary Wedgwood chemical ware. Glass instruments were imported from Prague, Berlin, Chemnitz and Wittenberg.[104] During this time, Griffin was well-known locally in Glasgow through the Glasgow Philosophical Society, and much appreciated by local and academic chemists including Thomas Thomson at the university, Thomas Graham at Anderson's College and Thomas Clark at the Mechanics' Institute. It is not surprising, therefore, that when Graham left to fill the chemistry post (following the death of Edward Turner) at University College, London, he wrote to Griffin, begging that he should move his business to London. As a Londoner himself (born in Shoreditch), Griffin already had relations and other connections through the book business. Thus, by 1849 – by which time his nephew Charles was already in charge of the book business – Griffin had moved to Marylebone where he opened a 'Chemical Museum' supply house in Baker Street.

Griffin's first London catalogues appeared in the form of 'Scientific Circulars' of which there were six in total. From their content, it becomes clear that Watkins & Hill could no longer compete in this field. Also on opening up the Chemical Museum, Griffin launched another (ninth) edition of *Chemical Recreations* (1849), this time targeted at 'Schoolmasters who desire to teach the elements of chemistry; in popular lectures or experimental lessons'.[105] Within 12 months of arriving in London, Griffin not only gained the support of those within the Chemical Society, but he had expanded his business by taking over the important distribution depôt of John Ward in March 1850. With this expansion, the extra dimensions of philosophical instruments, photographic apparatus and tools of many kinds, were added to later *Scientific Circulars*. By 1854, Griffin's business – now also at Finsbury Square and Bunhill Row – became a focal point for all and sundry, including travellers, explorers and those tooling up before emigration, industrialists whose responsibility it was to respond to the growing legislation, and not least students undertaking mandatory courses in practical chemistry, as well as their teachers.[106]

---

103 Griffin's travel diary. Royal Chemical Society Archives: Griffin's diary is catalogued under Henry Roscoe.

104 Gee and Brock (1991), pp. 52–53.

105 Griffin (1849), p. iii.

106 Gee and Brock (1991), p. 56. 'John Joseph Griffin', www.oxforddnb.com/view/article/11590 accessed 2 July 2011; Gee (1984–1986).

As Griffin's business expanded, changes were afoot at 268 High Holborn where William Elliott had established a successful instrument business, specialising in drawing instruments.[107] William Elliott had followed a traditional route, via an apprenticeship under William Backwell in the Coachmakers' Company from which he was free in 1804, before taking on the challenge of running his own business.[108] It was now time to hand over to his sons. A change in arrangement took place on 24 October 1850, when a partnership was agreed; thereafter, the business became William Elliott & Sons, soon afterward styled Elliott and Sons.[109] Frederick Henry and Charles Alfred Elliott were the sons of their father's second marriage and, at the point of creating the partnership, they were, respectively, 31 and 28 years of age. Charles had worked with his father from boyhood while Frederick, who had received a privileged public school education followed by Christ's College, Cambridge, had developed a career elsewhere. The coming together of these brothers was, for the family, a wise decision for, within three years, their father was dead.

It was at the time of setting up this partnership that the business moved from High Holborn to 56 Strand, in fact, to premises not too distant from Watkins & Hill at Charing Cross, and more or less opposite the philosophical instrument shop of Edward Marmaduke Clarke at 428 Strand.[110]

Until their father's death the business continued much as before, supplying mathematical instruments to educational institutions as well as a range of geodetic instruments.[111] Within the partnership, Frederick appears to have taken the leading role. Certainly, his academic background allowed him membership of the Society of Arts and the Astronomical Society, although not the Chemical Society. As with J.J. Griffin and Watkins & Hill, the Elliott brothers were motivated by the Great Exhibition and, as for many others in the instrument trade, this opportunity to present themselves before the public was pivotal in their development.

---

107 See also: 'William Elliott', www.oxforddnb.com/view/article/48374 accessed 2 July 2011.

108 Clifton (1993b) and Bristow (1993). The company-written article '150 years of instrument making' in *The Elliott Journal*, vol. 1, no. 1, (March 1951) should be treated with caution.

109 A formal 'Deed of Partnership' did once exist according to Wm Elliott's Will. TNA: PROB 11/2177, fol. 603.

110 Kelly's *Post Office Directory* locates 'William Elliott' at 268 High Holborn in 1849 but as 'William Elliott & Sons' at 56 Strand in 1850. For an exact guide to the location of all Elliott premises, see Clifton (1993b) p. 3, Table 2.

111 Elliott and Sons (c.1851) in: Heather (1851), endpapers.

Chapter 11
After Watkins & Hill

A rude awakening to the social and economic importance of science was nowhere felt more positively than at the Exhibition of the Works of Industry of all Nations in 1851. Prince Albert, a visionary and a German, foresaw that nothing should be lost and, indeed, plenty might be gained from Britain making its statement of prowess through the means of an international exhibition. Support for this grand idea came from the offices of the Society of Arts of which he was President, with Henry Cole, a civil servant working under Prince Albert's patronage, providing the background organisation and fund raising, begun in 1849. Ordinary visitors to Paxton's superb Crystal Palace in Hyde Park, particularly those of the artisan and manufacturing classes, might have felt a certain national pride and, if judged by the number of awards and commendations received, there seemed little cause for concern. Others, with wider experience of neighbouring nations and sensitivity to their quickening pace of industrialisation, were very much concerned. Lyon Playfair – one of those fortunate few to have experienced the teaching of continental science under Justus von Liebig at Giessen – saw clearly that there could be no room for complacency. The sentiment was summed up in his introductory lecture to the Government School of Mines for the session 1851–1852 when he lamented:

> ... as surely as darkness follows the setting of the sun, so surely will England recede as a manufacturing nation unless her industrial population become much more conversant with science than they now are.[1]

Indeed, both Germany and France had already instituted technical and scientific instruction while Britain could only boast a system of self-help through the mechanics' institutes. So, whatever prowess she had demonstrated at the Great Exhibition, it was no more than the fruits of her early entry into the Industrial Revolution. The position could not be maintained and remedial action was needed. Notification of the need for action came formally from the Throne, probably through Prince Albert who would have been familiar with the recommendations in the second *Report* of the Commissioners of the Great Exhibition.[2] Queen Victoria's speech at the opening of Parliament

---

1 Playfair (1855). See also: 'Lectures ...' (1853), 2, pp. 417–451.

2 Great Britain. Commissioners for the Exhibition of 1851 (1852), pp. 18–40.

Figure 11.1 The Crystal Palace in Hyde Park, lithograph by Stannard and Dixon, London 1851. © The Trustees of the British Museum, 1949, 1011.94

in November 1852 included a remark that the advancement of the Fine Arts and practical sciences would be readily recognised by her listeners as worthy of the attention of a great and enlightened nation. She went on to say that 'a comprehensive scheme shall be laid before Parliament, having in view the promotion of these objects towards which I invite its aid and co-operation'.[3]

The hint here was of a plan for the future education of the nation. But before considering how the Government actually intervened, we should first reflect upon that aspect of the exhibition that was relevant to Watkins & Hill, the Elliotts, and Griffin in order to detect why the Exhibition had not entirely fulfilled its purpose.

At the Great Exhibition, all things scientific fell under the head of Class X (Philosophical Instruments) for which Playfair had sole responsibility for the grouping of entries.[4] Thus, five divisions had evolved: *A. Instruments for the measurement of space*; *B. Instruments to measure mechanical and physical forces*; *C. Instruments to measure the laws of mechanical and physical science*;

3 House of Lords Official Report, 3rd Series, CXXIII, col. 19.

4 *Official Catalogue* ... (1851), p. 404. See also Hobhouse (2002).

D. Application of mechanical and physical sciences to useful purposes; *E. Chemical and pharmaceutical apparatus*. It was very much intended to invite instrument makers to submit objects of educational value and indeed, as Robert Ellis, compiler of the *Official Catalogue*, remarked: 'A great variety of miscellaneous philosophical apparatus, for popular illustration, and for the purposes of the experimentalist, is also displayed, *and must attract notice.*'[5] [Author's italics]

Sadly, it is clear from the subsequent reports that considerably less attention was paid to the simpler educational items. Jurors laid emphasis on the examining and testing of entries according to their scientific utility rather than their potential as aids in the teaching of science.[6] So, a precision goniometer or an achromatic microscope was far more likely to have received attention than any cheap model or device suited to the schoolroom. This is borne out by the distribution of medals and mentions. Thus, Council Medals (representing novelty or application or material or process, but not confined to excellence of production or workmanship alone) were awarded to Newman & Son for their atmospheric recorder, to Ross for his microscope and equatorial telescope, to George Dollond for his atmospheric recorder, to Ludwig Oertling for large and small delicate balances; Prize Medals (representing high standards of workmanship, according to utility, beauty, cheapness, adaptation to particular markets) were awarded to Elliott & Sons for mathematical drawing instruments, to Horne, Thornthwaite & Wood for photographic apparatus, to Knight & Son for their agricultural cabinets, to Newton & Sons for globes, to Oertling for a balance, to William Simms for astronomical apparatus and a sextant, and Watkins & Hill for a Zamboni dry pile apparatus. J.J. Griffin was rewarded for a calibrated range of alkalimeters, acidometers, hydrometers, chlorimeters, eudiometers, gas tubes and liquid measures.[7] Beyond this, just a few exhibitors received Honourable Mentions: Chadburn Brothers for their 'good and cheap instruments', William Ladd for microscope apparatus, Andrew Pritchard for a microscope and G. Yeates for cheap air pumps.

Generally, the simpler entries expected for schools, such as pumps, chemistry chests, globes and microscopes, were ignored although a few notable examples were submitted: William Newton & Son's orreries, Horne, Thornthwaite & Wood's oxy-hydrogen microscopes, Carpenter & Westley's phantasmagoria lanterns, William Ladd's box of pneumatic experiments, W.E. Statham's chemical cabinets and hydro-pneumatic apparatus, Elliott & Sons's mathematical drawing instruments and Griffin's chemical apparatus suited to parish schools. Watkins & Hill did not submit anything that might have reflected a concern for school science although they did exhibit a syrene (based on Caignard de

5 *Official Catalogue* (1851), p. 405.

6 Glaisher (1852).

7 *Reports by the Juries ...* (1852), p. 295.

la Tour's invention of replicating musical sounds by a series of impulses); this, however, appeared in a sub-class on acoustics dealing with hearing and deafness. The remainder of the Watkins & Hill collection (without any prize) comprised air pumps, thermometers, barometers, an electromagnetic engine, a microscope, a telescope, a delicate astatic galvanometer and a Snow Harris electric machine (with 3 feet diameter glass plate) after a standard design which they had made for at least ten years.[8] The Elliotts, too, submitted meteorological instruments although their transit theodolite was certainly not supplied with school education in mind. It is not the purpose here to analyse all that appeared in Class X but it is important to stress that few exhibitors submitted exhibits with the schoolroom in mind.

Reporting on the exhibits in Class X at the close of the exhibition, James Glaisher concluded:

> The Exhibition, by collecting, within a comparatively small space, almost all the known applications of science throughout the civilised world, is eminently qualified for the attainment of this great end, by promoting the advance in its various branches, and by *infusing a taste for the development of the highest faculties* with which man is endowed.[9] [Author's italics]

Visitors to the Exhibition may well have been inspired although the problem of 'infusing a taste' was not simple for it necessitated a formal and continuous process of instruction, with texts, apparatus, trained teachers and examinations. The days of the simple diffusion of scientific knowledge, through voluntary attendance at the mechanics' institutes were numbered.

State Intervention

The first overt act of the State in the aftermath of discussion resulting from the Great Exhibition was the formation of a Department of Practical Art. Of course, this was not the first utterance on the importance of a practical education relevant to design and manufacture. The School of Mines was an outcome of the Geological Survey of 1832 and was instituted by Henry de la Beche. One of its original functions was to display geological specimens of the kind used for the practical construction of roads and buildings and hence it was called the Museum of Economic Geology (later the Geological Museum of Jermyn Street).

---

8 Hackmann (2003). It is also interesting to note that Watkins and Hill exhibited in two other classes: Class VII *Guns and Weapons*, and Class V *Railway and Marine Mechanisms*, in both cases demonstrating their competence with precision mechanisms.

9 *Reports by the Juries ...* (1852), p. 316.

The Jermyn Street building was opened in 1851 by Prince Albert and took on the title the Government School of Mines and of Science Applied to the Arts. It was an institution for public instruction in mining and its first incumbents included Ramsay (geology), Warington Smyth (mining and mineralogy), Robert Hunt (applied mechanics), Playfair (practical chemistry), Percy (metallurgy) and Edward Forbes (natural history). De la Beche was Director. The title 'Royal School of Mines' did not appear until 1862 when Roderick Murchison was Director. Another important institution, the Royal College of Chemistry, was well established by the mid-century. The Prince Consort assisted in attracting the young August Wilhelm von Hofmann from Bonn to fill the post of Director in 1845. Laboratories were temporarily fitted up in George Street, Hanover Square, where 26 students, many of whom distinguished themselves in later life, enrolled. A building facing Oxford Street was later procured and, by 1853, the Government was responsible for its running. The educational needs of the nation's working classes had been recognised at the Government School of Mines from the moment it was founded and it was from this institution that a pressure group articulated its concerns in the aftermath of the Great Exhibition.[10] In response to their growing concerns, the Government introduced a combined Science and Art Department (later renamed the Department of Science and Art or DSA). This was regularly constituted by minute of the Treasury and placed under the control of the Committee of Council on Education, a body formed in 1839 under the Board of Trade, in the wake of Brougham's Education Bill of 1820, with the express power to dispense parliamentary grant-in-aid for books, buildings and to alleviate the problems of overcrowded parish and dame schools, but not for apparatus of the scientific kind.[11]

These intricacies of government arrangement are of less importance here than it is to note that there were two distinct groupings: the Committee of Council, with Henry Moseley as its Science Inspector for Elementary Schools, and the Department of Science and Art, with Lyon Playfair and Henry Cole who were to dominate the original shape of teacher training in science and lay the foundations for a secondary sector of science schools. Meanwhile, in the wake of the Great Exhibition, Moseley had been commissioned by the Committee of Council in October 1852 to prepare a list of the kind of apparatus which might be used in training schools, and another of apparatus that might fruitfully be employed, with economy and simplicity, in elementary schools. The terms of reference also demanded that he should bear in mind the Committee's generous proposal to defray two-thirds of the cost of apparatus purchased.

When Moseley reported back to the Committee on 5 January 1853, he told the members that:

---

10 *Report of the British Association for the Advancement of Science* (1931), p. 173.

11 Butterworth (1982).

> ... having placed myself in communication with some of the most eminent makers of scientific apparatus in London, with a view to the preparation of such lists, I have found that apparatus for the same object is often made differently by different makers; and that some kinds of apparatus are arranged in sets, so that the same pieces in a set are adapted to serve different uses.[12]

The 'eminent makers' to whom he referred were John J. Griffin, Horne & Thornthwaite and Knight & Co, but Watkins & Hill, and Elliott & Sons, were conspicuously absent. Each of these chosen suppliers was a representative for chemistry and, as Moseley explained, his recommendation was based upon the simple principle that chemistry was 'the science round which all the other experimental sciences are most easily grouped, and to which they are ancillary'. As for the two lists mentioned above, Moseley asked the suppliers what they might recommend for £10, £15 and £20 in elementary schools, and for £100, £130 and £150 in training colleges. Only Griffin extended his more expensive list to include apparatus for the teaching of hydrostatics, electricity and magnetism.

Moseley foresaw that an unchecked two-thirds grant would have been a bad investment had teachers been ill-prepared to use whatever sets were finally decided upon. His remedy was simple: a teacher applying for a £10 set should study George Wilson's *Chemistry* and 'get up the first 87 pages'; for a £15 set, the 'first 261 pages', and for a £20 set, 'the whole book'. A degree of free choice was to be allowed: so, for example, a teacher wishing to obtain a telescope or astronomy slides would be told of some other text.[13] But, being successful in an application also meant that a teacher's knowledge was to be tested on-the-spot by a visiting school inspector. Finally, Moseley informed the Committee:

> ... the services of Mr John Joseph Griffin in this department of popular instruction give him a claim, other circumstances being the same, to a preference. No gentleman has probably laboured more than that gentleman to simplify and to economise the construction of such apparatus.[14]

What Moseley had proposed for the elementary schools was of considerable interest to Playfair and some measure of this common concern is seen through the *First Report* of the Department of Science and Art which was presented to both Houses of Parliament on 1 January 1854. This proposed a two-pronged attack for the development of science education: firstly that it should 'create, exhibit, and distribute the most improved illustrations, models and diagrams

12 Moseley (1853), p. 154.

13 Other recommended texts included: Fownes's *Manual of Chemistry* (1844) and Griffin's *Chemical Recreations* (1849).

14 Moseley (1853), p. 156.

which should be readily accessible ... to all persons ... interested in education ... and to masters in training, and students'; and secondly that it should institute 'a school of the highest class capable of affording the best instruction and the most perfect training, and by assisting in the establishment of local institutions for science instruction ...'[15] The first proposal would have its impact on the instrument-making trade, and the second eventually led to the famous science teacher training establishments at South Kensington.

The overlap in principles of what was desired in the elementary sector by Moseley was not in dispute by Playfair whose efforts were directed towards a new secondary sector. With regard to scientific apparatus of the philosophical kind, Playfair explained:

> Philosophical apparatus illustrative of the several sciences existed in an unconnected way and expensive form, but were rarely so arranged in well thought out consecutive series as to be available to any but the experienced professor who had the knowledge to choose what was suited to his purpose.[16]

By contrast, chemistry was connected and inexpensive, mostly due to Griffin's earlier initiatives but, indeed, the individual machines, engines and apparatus of philosophical science were expensive and hardly suited to individual practical work. Schools were not thought likely to make purchases merely on the basis of their 'availability' or 'excellence' and, therefore, an additional motive of 'cheapness' was needed. Moreover, the matter of arranging some 'well thought out connected series' in natural philosophy was a stumbling block, only to be resolved when physics was perceived to be based upon a unifying principle of energy and its conservation. So, meanwhile, the problem of choice of texts and apparatus was left to 'eminent scientific men' although, at this stage, the problem of how to 'create a taste for it' was thrown open to lecturer–authors and apparatus and instrument suppliers alike. To test what ideas might be forthcoming from the trade, an initiative for an exhibition of 'material helps' was then proposed by the Society of Arts.

An Exhibition of 'Material Helps'

Reverberations over the lack of any system of scientific and artistic instruction echoed around the chamber of the Council of the Society of Arts. Concern for an industrial education was a long-standing affair, although resolved in May 1852

---

[15] House of Commons, First Report of the Department of Science and Art, 1854; (1783) XXVIII.269, p. x.

[16] Ibid., p. xii

when the Earl Granville united the 355 miscellaneous literary and scientific institutions, philosophical societies, athenaeums, and mechanics' institutes, under the auspices of the Society of Arts. Heading the committee of this union was Harry Chester who – wearing another hat – was in a strong position in Government as assistant Secretary to the Committee of the Privy Council on Education. A suitable role for the Institutes' Committee lay in organising an exhibition of educational materials and it was Chester who suggested this. Playfair was on the organising committee although, curiously, not Moseley, although he was a member of the Society of Arts.

While preparations were underway at the Society of Arts for an exhibition in the summer 1854, the theme of 'science education' was also taken up by the managers of the Royal Institution for their Easter Lectures in 1854. Those who had formerly professed on science now professed on science education – with John Tyndall on physics, Charles Daubeny on chemistry, Gordon Latham on language, James Paget on physiology, W.B. Hodgson on economic science, William Whewell on the history of science and Faraday on 'mental education'. Meanwhile, the scientific dealers had responded to the call from the Society's Secretary, Peter Le Neve Foster, for apparatus and materials of 'a cheap and simple kind'. There were, however, no prizes to be won for 'excellence of workmanship' or 'novelty of design' as there has been at the Great Exhibition, nor were there to be any jury reports of the repressive kind. Makers and suppliers were simply encouraged to submit what they could, bearing in mind the educational context.

Three souvenir albums of ephemera (manufacturers and publishers prospectuses) presented at of the 1854 exhibition demonstrate Watkins & Hill's struggle for survival.[17] The absence of any catalogues by the firm in these albums coupled with their not being cited in Yapp's *Official Catalogue* of 1854 for the Royal Society of Arts confirms that they were not present in 1854.[18] The organising secretary, who had specifically requested exhibitors to present 'electric machines, magnetic, electro-magnetic and electro-type apparatus', would surely have attracted the company in earlier times. The impression here, therefore, is one of their withdrawing altogether from this obviously widening educational field. Certainly, it is an indicator that the old firm no longer led the way. By contrast, Elliott Brothers (late Elliott and Sons) were present; not only keen to show off their drawing instruments which they supplied to the Royal Military Academy, Addiscombe College, King's College, Harrow School, Eton College, Marlborough House and Roorkee College in India, founded in 1845–1846, but proud also to announce that they had recently purchased from a Mr Jones of Charing Cross a large circular dividing engine and another

[17] Royal Society of Arts, Manufactures and Commerce (1854b).

[18] Royal Society of Arts, Manufactures and Commerce (1854a), pp. 12–15. See also Layton (1972).

for straight line division from the Admiralty.[19] Indeed, the special short octavo catalogue produced for this exhibition showed clearly that they offered products in the following divisions: firstly, theodolites, levels and transit instruments; secondly, drawing instruments and 'Elliott pencils'; thirdly, telescopes and microscopes; and finally, barometers and philosophical apparatus. In fact, they were already well established, not only with public schools and the military colleges but also as suppliers via the East India Company and as 'opticians to the Ophthalmic Hospital'.

Exhibition Lectures

Sixty lectures were delivered during the Great Exhibition. Those relating to science education often expressed heart-felt advice as well as argued opinions about the nature of science itself. It was in this context that T.H. Huxley made his famous definition of science as 'organised common-sense' and Gustav Hofmann spoke on the value of 'playthings and occupation of early childhood'. In terms of why to include physics and chemistry in the curriculum, Alexander Williamson (University College) considered the 'reasoning powers' offered by these subjects to be more valuable than the 'storehouse of facts' that they contained. Baden Powell argued for the use of astronomical models while Robert Hunt saw meteorology akin to natural history in exercising the powers of observation. In this respect, T. Rymer Jones (King's College), proposed that the microscope should be adopted because it was the most valuable 'tool of observation'. All these arguments are familiar and still valid today.[20]

In the case of Jones's argument for the microscope, W.B. Carpenter, Fullerian Professor in Physiology at the Royal Institution from 1844 (and registrar at the University of London from 1856), responded by urging that a cheap 'educational microscope' should be designed. After all, he said, 'whilst the country Schoolmaster has the best opportunities of turning it to useful account [as an adjunct to Nature Study], it is the city Schoolmaster that, in default of other opportunities, [for whom] its importance as an educational instrument should be the greatest.' Whether a cheap educational microscope was indeed possible was the goal set by a special committee of the Society of Arts to which Robert Field successfully responded with a compound achromatic microscope, rack and fine adjustment, two eyepieces and two object glasses, and all boxed up for three guineas.

---

19 *Elliott Catalogue* [1854], p. 3 in RSA (1854b). The straight-line engine had been built by Ramsden in 1777 and paid for by the Board of Longitude, by whom it had been transferred to Thomas Jones. When the Board of Longitude was wound up, ownership of the engine passed to the Admiralty. Editors' note.

20 Lectures ... (1853).

This 'school microscope' appeared in 1855 at the same time as a cheap alternative for a mere 10s 6d. J.T. Quekett's popular work *A Practical Treatise on the Use of the Microscope* (1848) was already a standard text and, after this exhibition, J. Hogg's *The Microscope, its History, Construction and Application* (1854) and Carpenter's *The Microscope and its Revelations* (1856) became widely known.[21]

After the mid-1850s, several textbooks appeared with their authors making recommendations about apparatus. Thomas Tate was among the first to offer set books: *Electricity for the Use of Beginners* (1854) and *On Magnetism, Voltaic Electricity and Electro-dynamics for the Use of Beginners* (1854). Not to be outdone by Tate in this field, Elliott Brothers, on taking over Watkins & Hill, immediately reprinted Francis-III Watkins's plainly written but long obsolete *Popular Sketch of Electro-magnetism*, issuing it in 1856 with updated footnotes. Similarly, Frederick Newton of Newton & Company reissued his ancestor's, William Newton's, *Use of the Globes* (1854) as a quick and easily achieved introduction to astronomy.[22] Even Griffin brought out a new tenth, two-part, edition of *Chemical Recreations* (1854, 1860) as one last act before handing on the business to his sons. Such reprints could not match those more recent texts which then proliferated. However, members of the instrument trade did respond by launching new apparatus and equipment catalogues specially targeted at schools and schoolmasters. Griffin issued *Apparatus for Instruction in Chemical Science* (1854), Horne, Thornthwaite and Wood published *A Guide to the Purchase and Use of Scientific Instruments* (1856), and other listings for schoolmasters came from Newton & Son (1854), Bland & Long (1854) and Elliott Brothers (1854). Indeed, competition among these suppliers in what then appeared as an expanding market was very apparent. This may well have been a sufficient reason for Elliott Brothers to take over the Watkins & Hill business. At least, on doing so in 1856, they immediately re-issued the obsolescent, but comprehensive, Watkins & Hill illustrated five-part series.[23]

Frederick Temple, who joined the Committee of Council on Education in 1855 from a principalship at the government training school at Kneller Hall, Twickenham, to fill Moseley's position as Her Majesty's Inspector of Schools, involved Thomas Tate, an ex-colleague, mathematics tutor and former teacher at the National Society's Training College in Battersea, in the testing and recommending of apparatus. The outcome of this exercise – a *Special Report on Apparatus for Elementary Schools* – was issued from the Whitehall Education Office on 11 July 1856. However, this step may have caused consternation among the makers and suppliers because the Tate–Temple lists were very prescriptive and it was clear that the customary freedom for the trade to make its own

21 Morrison-Low (2007), pp. 233–248.

22 Gee (1993a).

23 Elliott Brothers (1856a), (1856b), (1856c), (1856d) and (1856e).

recommendations was narrowed. Even so, Elliott Brothers and J.J. Griffin did respond yet again with further catalogues in 1858 – incorporating the expression 'Elementary Schools' in their titles. Yet, at the International Exhibition of 1862, held in London, the suppliers expected to submit examples to the category of 'educational works and appliances' (class XXIX), a category drawn up to cover what was lacking in the Great Exhibition generally ignored this opportunity; their attitude, overtly expressed, was that telescopes, microscopes and other instruments should be placed in the class where they belonged, namely, the class of 'philosophical instruments and processes depending upon their use' (class XIII).[24] Indeed, only Griffin entered class XXIX with a range of apparatus in accordance with Temple's prescriptive lists.[25]

Returning to the Educational Exhibition of 1854, shortly before its close in August 1854, a deputation from the Committee of Institutes, consisting of Lord Ebrington (the chairman following Henry Cole's retirement from this office), and Harry Chester (by this time Vice-President, met with Lord John Russell (Lord President of the Privy Council), to seek the Government's blessing on the idea that there should be a permanent museum to house the contents then on show at St Martin's Hall. The Society of Arts's concern was naturally to do with arts and manufactures although, it argued, this could only be maintained if the education of all classes was improved. Now was the time to put away those religious differences which marked the foundations of the various educational institutions. Tentatively, the Government concurred, pending a vote from the Treasury, and it was agreed that the exhibits should be stored at Gore House (a property in South Kensington purchased with foresight by the Commissioners of the Exhibition of 1851) until a more permanent home was found. Vendors and manufacturers were subsequently canvassed by the Council, by letter dated 22 August 1854, inquiring whether they would be willing to donate items before their impending dispersal. The response was encouraging and ultimately, these items became the basis of an educational museum at South Kensington.

The above account serves to indicate how a permanent museum at South Kensington came into being. Its purpose was twofold: to aid all classes of the public, and especially those engaged in teaching, and to bring together all that was new and worthy of attention in apparatus relating to education, both in the primary and secondary sectors.[26] So, it was here that representatives of the instrument trade and publishers could continue to send examples of what was available, just as they had done at their first coming together at St Martin's Great Hall in the summer

---

24 Timbs (1863), p. 196.

25 *London International Exhibition of 1862* (1862), p. 70 (entry 5598). Medals and honourable mentions were received by Elliott Brothers, J.J. Griffin, and Horne & Thornthwaite, ibid., pp. 197–198.

26 South Kensington Museum (1857), p. iii.

of 1854. The first teachers to venture to South Kensington with the purpose of exploring the exhibits would have come from the established elementary schools. Throughout the 1850s there had been a steady increase in demand for the two-thirds scientific apparatus grant, from a mere £18 in 1853 to over £750 in 1856 and by the year that the museum opened, a sum of £2,345 was tabled as outgoing expenditure, although falling to a meagre £30 19s 10d in 1862, the final year of the grant.[27] But, with the surety of grant-in-aid coming from the Government for the secondary sector, J.J. Griffin and Elliott Brothers responded very positively.

Up to the time of Moseley's departure from the Committee of Council on Education, he had been remarkably successful in instituting a two-thirds grant even if the 'science of common things' for which it was intended was not the kind of disciplined subject science with which the new DSA was concerned. It is sufficient here to remark that Temple's succession in Moseley's position was speedily followed by the relegation of science from its core position in the training colleges to a mere option. But an even greater blow was dealt by 'internal agreements' once the DSA (hitherto under the Board of Trade) fell under the wing of the Committee of Council on Education in 1857. By this time the Government was willing to pay for the kind of proposals stemming from the Department by a system of bonus or premium. All attempts to generate a supply of science masters for the Elementary Division were abandoned and in March 1857, the responsibility for administering the apparatus grant then was transferred from the Education Office in Whitehall to the DSA at South Kensington. The process which followed was then one of winding down the encouragements for elementary schools and so reserving the scheme of 'payment by results' solely for teachers of the Secondary Division whose concerns lay very much with the education of artisans. The diminution in grant offerings ran in parallel with the decline of suitably qualified teachers and, by 1861–1862, the two-thirds grant-in-aid was officially terminated under new regulations following the recommendations of the Newcastle Report. Thus, by the time of the next 'educational exhibition' (actually a division of the International Health Exhibition held in London in 1884), no philosophical instrument makers bothered to enter Class 51 which had specifically been set aside for science teaching apparatus. Nevertheless, the new 50 per cent grant began to support those institutions that participated in the Government scheme. Only 1,330 (January) and 1,239 (May) secondary students were examined under the new DSA system in 1860–1861 but there followed a rapid rise and, by the end of the century, some 150,000 (January) and 114,000 (May) students undertook the DSA examinations with these numbers distributed in the subjects of acoustics, light and heat, magnetism and electricity, inorganic chemistry and practical chemistry.[28]

---

27 Layton (1973), p. 164.

28 The numbers here are broadly interpreted; the actual statistical analysis is far more complex, as shown by Butterworth (1982).

Over the first ten years of the Educational Division of the South Kensington Museum's existence, from 1857 to 1867, eight editions of the *Classed Catalogues* were issued specifying books and apparatus. The first editions laid out what was required of elementary schools but, by 1860, this series specified the requirements of the new and functioning DSA for its 'science schools'. Griffin and Elliott Brothers were the principal chemical suppliers throughout this period although Newton & Co., J. Parkes & Sons, A. Davies & Company, Horne & Thornthwaite, Knight & Sons, and Murray & Heath each submitted apparatus and instrument lists within their own specialisms.

After the 1862 International Exhibition, the South Kensington Museum was removed from the 'Brompton boilers' (an inappropriate iron building shared with the Patent Museum) to space in the southern galleries of the 1862 building and there it continued to receive instruments and apparatus for display until 1878.[29] The Library associated with the Science Collections grew from the steady input by publishers and donors, augmented in 1876 with the inspectorate's collection from the Committee of Council on Education and the old Jermyn Street library of the Museum of Practical Geology in 1886. Thus the Science Library came into being although the apparatus and instrument collection lost its appeal by 1888 and the contents were then returned to the exhibitors.

The Final Years of Watkins & Hill

Having thus examined, albeit briefly, the vast changes taking place in formal education following government intervention, it is now possible to consider the position of Watkins & Hill. Their absence at the Society of Arts exhibition of 'material helps' in 1854 was a clear sign of their reluctance to participate in a future in which they no longer intended to play a significant role. In this respect, their trade was not motivated by the prospects brought about by government grant-in-aid. As a relatively small workforce, with specialised skills in mechanical and electrical work, as evidenced by their entries at the Great Exhibition, most of their work came in the form of individual commissions. That they maintained a presence in the instrument field in this manner is clear from their order book but also noted from their advertising. One example of the latter course comes from R. Smyth and H.E.L. Thuillier's *A Manual of Surveying in India*, a work first published in 1851 and devoted to the revenue surveys in Bengal and the North

[29] Provincial scientific instrument suppliers participating in the DSA grant scheme – such as Mawson & Swan (Newcastle), Mottershead & Co. (Manchester), G. Cussons (Manchester), W. & J. George (Birmingham), Reynolds and Branson (Leeds), and Philip Harris (Birmingham) – did not produce their first apparatus catalogues until the last decades of the nineteenth century.

West Provinces of the Indian subcontinent. In the second edition, published in 1855, there was a substantial section of advertisements by prominent English suppliers generally, although only Watkins & Hill, J.J. Griffin and Elliott Brothers advertised as representatives of the scientific trades.

For the particular readership of Smyth and Thuillier's text – clearly those who specialised in matters of mathematics and measurement – Watkins & Hill indicated they sold 'astronomical, geodetic and mathematical instruments' although they also drew attention to their traditional role as makers of optical and philosophical instruments. Interestingly, Elliott Brothers, who were very much suppliers of geodetic and mathematical instruments, also announced they supplied 'optical and philosophical apparatus'. Of course, their catalogue for the 1854 Educational Exhibition had listed barometers and microscopes although not electrical or philosophical apparatus. However, both firms would be adept at factoring from a network of suppliers and, very likely, from each other. In other words, what they did not make themselves, they would gladly obtain from elsewhere to complete an order. This is confirmed by an order book of 1855 in the Elliott Archives which indicates Watkins & Hill as customers of Elliott Brothers for drawing instruments, a pedometer and a couple of meteorological instruments.[30] We are thus led to assume that a degree of reciprocity already existed between these two instrument houses and so any orders received by either firm from India would surely bring them closer together through co-operation rather than competition.

A further example from the 1850s, signifying their reputation at home rather than abroad, comes from a receipted bill from Watkins & Hill to a Mr Gibbs, probably a tutor in the Royal Household, for a set of German silver drawing instruments contained within a rosewood box engraved with the Prince of Wales's feathers. Since an earlier order book of William Elliott shows that drawing instruments were purchased by Watkins & Hill from Elliotts in the 1830s, then it is also possible that this special set was actually made by Elliott & Sons.[31] Moreover, since the Watkins & Hill bill was made out on 2 November 1854 then it would have been a timely and appropriate present for Albert Edward ('Bertie', as he was affectionately known), the Prince of Wales, later Edward-VII, who was born on 9 November 1841 and so whose thirteenth birthday fell the following week. It is worth noting that this invoice not only shows Watkins & Hill as 'instrument makers to Her Majesty' but also that they

---

30 Bristow (1993), p. 9.

31 For example, a note from the 1830–1831 accounts of William Elliott, states that 'Watkins deposited with Martins [bank, the sum of £360] to meet Harrisons [account]'. MHSO: Elliott Archives, 1.02/2 (scrapbook). Note. The same document shows drafts paid to Hill (14 September 1831 for £176. 5s 0d), to Watkins (27 September 1831 for £100. 0s 0d), and to Harrison several drafts at £50 weekly between 17 August and 15 October 1831. These large sums strongly suggest agreed payments between makers and dealers.

were in the business of 'fitting up' laboratories which, undoubtedly, would be highly remunerative if they obtained a commission. From these examples it would appear that Watkins & Hill were still very much a prominent concern in the mid-1850s.

While considering the inter-trading between the Elliotts and Watkins & Hill, it must be borne in mind that there was a good deal of overlap in the skills of the respective workforces. Given that the Elliotts were on an ascendant path, it was probably expected, and even already arranged, that Elliott Brothers would soon subsume Watkins & Hill. To understand more clearly how such a situation came about, the impact of local changes on the Watkins family become an important issue.

The Plight of Mary Ann Watkins

It is not known how profitable the Watkins & Hill business remained while under the management of Abraham Day. He may have worked hard to maintain goodwill and sustain the high standards of the old firm after the death of the partners. William Hill had died in 1846, and Francis-III Watkins the following year. A little of Day's effort may be deduced from brief but courteous exchanges with Robert Walker at Oxford University, whose custom had begun in 1839 and lasted throughout the period of Day's management. As already stated, Watkins & Hill were very much an independent enterprise offering electrical and mechanical expertise as well as the field of instruments generally. As described in the previous section, they were a fully functioning business in the mid-1850s, and indeed, the Charing Cross shop was still a focal point for scientific persons as may be witnessed by the fact that Richard Taylor – publisher of the *Philosophical Magazine* at Red Lion Court off Fleet Street – still distributed his *Calendar of the Meetings of the Scientific Bodies of London* from Watkins's shop. This sixpenny broadsheet, covering events and meetings of some 33 scientific societies was still available there in 1854–1855.[32] Superficially, therefore, little had changed in the daily affairs at Charing Cross, although all these routine habits were soon to be swept away.

Since the death of the partners and the various realignments of the trade, described in the previous chapter, there had been no attempt to update the five-part catalogue. Of course, it was one thing for Abraham Day to maintain and manage the old firm in order to provide an income for Mary Ann Watkins and work for her faithful employees but it was quite another to envisage a future in which they could regain ground as leaders in the supply of equipment for the

[32] A facsimile of Taylor's 'Calendar' for 1854–1855 (also available at Dollond's in the City) appears in Brock and Meadows (1984), pp. 38–39.

educational field. There is no evidence to suggest that Francis Burton Watkins – Mary Ann's only son and potentially the fourth in line to inherit the business – wished to spend his life in the same trade as his father; nor are there any signs that his mother wished to lead the old firm into a radically different future. Gloomiest of all was the looming prospect of compulsory demolition of the old shop, due to the on-going reconstruction of the district in the vicinity of Charing Cross.

The shops on the north side of the famous statue of Charles-I – including the Golden Cross Tavern, together with the King's Mews to the rear, and those shops on the lower part of St Martin's Lane and West Strand – had already been demolished by the early 1830s, as part of the preparation for Nash's grand design for a large piazza in celebration of the British success at Trafalgar. The loss of these old shops and residences destroyed the 'village feel' at Charing Cross, and more destruction was set to follow with the removal of the remaining buildings on the east and west sides of the statue. The demolition of Northumberland House (the last of the old Strand palaces) – a necessary act in order to create an approach road to the new Thames suspension bridge at Hungerford – was also imminent, but delayed until 1874–1876 when a final wave of demolitions took place. All this devastation had resulted from an Act of 1826 sanctioning moves for the improvement of the area which had become very much run down and neglected.[33] It was, therefore, only a matter of time before the Metropolitan Board of Works ordered that the remaining properties – including the row containing the old Watkins & Hill shop – should be vacated for demolition.[34]

This last factor alone must have weighed heavily upon Mary Ann Watkins. The question of whether to sell up or relocate – possibly a realistic consideration at one time – was no longer a matter of choice; the property had to be vacated. The initiative for a solution to the widow's plight – although not recorded – may have come from the Elliotts rather than Mary Ann herself. Certainly, they wished to obtain a foothold in the emerging market as, indeed, did Griffin, Horne & Thornthwaite, and others too. A takeover of Watkins & Hill would certainly place the aspiring Elliotts on a competitive footing at the very start of a new era when, in an age of government intervention, courses of mandatory and examinable science were urged. Good profits were on the horizon for those, like the Elliotts, who had a sense of the future and who were prepared to cover the full sweep of scientific subjects.

---

[33] On realising that old Charing Cross would soon be lost forever, the artist George Scharf sketched the district for historical record. Barker and Jackson (1990). See also Stamp (1986), pp. 86–99.

[34] In the meantime, Mary Ann Watkins appears to have let the lock-up property associated with 5 Charing Cross to Thomas Scott, estate and land agent (Kelly, *PO Directory*, 1855).

The Elliotts Acquire Watkins & Hill

The Watkins family crisis in the mid-1840s caused by illness of the partners and an insufficient overlap in years between father and son, probably meant that the decision to promote Abraham Day as manager was all that could be done and so the question of whether to relocate into a more commodious accommodation may never have arisen. In any case, the inexorable path leading to closure at 5 Charing Cross brought about by the local scheme for municipal regeneration, points to Mary Ann Watkins opting willingly for the safer course of selling out to the Elliotts. Certainly, the Elliotts took this opportunity to expand, and from this point it may be deduced that the two families completed their negotiations and settlement by late 1855 or early 1856. A plan was in place by mid-1856, whereby customers could expect their orders would be honoured and the workforce was assured of some continuity in the employ of the Elliotts. A printed circular to this effect was issued in July 1856 informing Watkins & Hill customers of the imminent change of ownership.[35] From the Elliotts' standpoint, the purchase of this long-established business – with its customer base, a handful of loyal, experienced employees and a ready-made network of suppliers, and, most importantly, the goodwill of the Watkins family name – immediately placed them alongside the broad-ranging companies of J.J. Griffin and Horne & Thornthwaite. It is possible also, though not certain, that the Elliotts were spurred in their decision-making process by the fact that Horne & Thornthwaite had taken over the Strand shop of Edward Marmaduke Clarke in 1855 whose premises were more or less facing the Elliotts at 56 Strand.[36] In short, the Elliotts had manoeuvred themselves into a position whereby they could profit from the two-thirds grant-in-aid scheme of the Committee of Council on Education. More importantly, at the point of taking over Watkins & Hill, they were well positioned to promote themselves to those institutions which (it was hinted) would ultimately benefit from the 50 per cent grant-in-aid of the DSA when, eventually, it was opened up to institutions (Mechanics' Institutes and so on) in the secondary sector.[37] For Mary Ann Watkins – then nearing 60 years old – a great problem was solved, even if it now meant relinquishing the century-old business house that had been nurtured so conscientiously by her late husband and his ancestors before him.[38]

---

[35] Printed circular (July 1856). CUL: RGO 6/730, *f*.257v.

[36] E.M. Clarke had sold his business on taking up a salaried post as manager of the newly built Royal Panopticon of Science in Leicester Square. Mollan, David and Finucane (2002), p. 63.

[37] This new system under the auspices of the DSA was proposed in 1853 although not actually implemented until March 1860. Thus, Elliott Bros (and the other intending suppliers) had to wait a little longer before they could benefit from this income source. Butterworth (1982), pp. 27–44.

[38] The 1881 Census shows that Mary Ann Watkins was still alive (age 78 years) and living at 25 Robert Street (not far from Charing Cross).

Figure 11.2 Trade card for Elliott Brothers 'Successors to Watkins & Hill'. Private collection

In preparation for their new start in January 1857, the Elliotts prepared (without revision) sets of the Watkins & Hill five-part illustrated catalogue with newly printed covers under their own name.[39] They also (as stated earlier) re-issued Francis Watkins's text, *Popular Sketch of Electro-magnetism or Electro-dynamics*. This work had been very relevant when it first appeared, but was hardly suitable as the sort of text anticipated by those who were about to teach examinable science. Nevertheless, the obsolescent *Popular Sketch* re-appeared with updating footnotes by one 'G.B.H.' covering the more recent contributions of Grove, Faraday, Snow Harris and Bachhoffner. Remnants of the old Watkins & Hill business could still be found in the Elliott catalogue for 1860, when the reprinted Watkins texts appeared alongside the promotion of Bayley's *Slide Rule*, Heather's work on *Mathematical Instruments*, and Haskoll's *Practice of Engineering Fieldwork*.

For a short time during 1856, the Elliotts held possession of the old shop at 5 Charing Cross, thus having begun the take-over during that year while they presumably transferred the stock to 30 Strand, although this property was soon relinquished in favour of 56 Strand.[40] The property at 5 Charing Cross which

[39] Anderson, Burnett and Gee (1990), p. 27.

[40] At the take-over, Frederick Elliott and Charles Alfred Elliott were the joint payees to Westminster for the property at 5 Charing Cross. Poor Rates (St Martin-in-the-Field), WA: F 785.

had been in the possession of the Watkins family for over 100 years was then, unceremoniously, occupied as a warehouse for the Edmiston Pocket Siphonia and Mackintosh Rubber Companies until its demolition in 1875.

The changeover was well organised and smoothly executed, with the continuation of business begun officially on New Year's Day 1857. At the time of the take-over, the Elliotts made the significant appointment of Carl Ludwig Becker, an extremely able mechanic of about 35 years of age, who had undergone his basic training at the factory of A. & G. Repsold in Hamburg and studied under Carl August Steinheil at Munich before arriving in London in 1849.[41] Within two years of the takeover, the coalesced workforces of Watkins & Hill and Elliott Brothers were placed under Becker's management. Just as Francis Watkins had once assisted Lardner, so too, under Becker, special relationships were formed with John Tyndall at the Royal Institution, James Clerk Maxwell at King's College (and later the Cavendish Laboratory), Charles Wheatstone, Edward Frankland and many others.[42] Telegraphy was very much in vogue in those early days and, under Becker, Elliott Brothers became involved in making instruments in connection with the laying of the first trans-Atlantic electric telegraph in 1858. It was probably left to Becker to maintain the slave time-ball – governed by pulses from the Royal Greenwich Observatory along telegraphic wires belonging to the Southern Railway Company – situated in the western tower of their newly acquired premises at 449 Strand in 1873 although this had been in operation since 1852.[43]

Twenty-five years earlier, in the heyday of Francis Watkins's interests in electricity, with the exploration of electromotive power, the magneto-electric machine, electromagnetic induction, and development of chemical and thermoelectric batteries, there had been lively participation and interest shown by engineers, scientists and amateurs in these discoveries. Indeed, such was the fervour among electricians of the period (by men such as George Henry Bachhoffner, John Peter Gassiot, William Leithead, William Sturgeon, Edward M. Clarke and Joseph Saxton) that an Electrical Society was set up (initially at Clarke's Laboratory of Science in the Lowther Arcade) in 1837.[44] But as keen as its members were to explore this new world opened up largely by Faraday, this Society did not last, partly owing to an absence of an immediate commercial application of the various discoveries but also an insufficiency of funds. Indeed,

[41] Bristow (2005), p. 18.

[42] Bristow (1993), p. 10 and Bristow (2005), p. 18. See 'List of apparatus' for Tydall's text *Heat considered as a Mode of Motion* (1863).

[43] Chew and Wilson (1993), p. 37. This famous building with its towers, planned originally by the eminent architect John Nash, now refurbished, stands to this day although no longer containing the Lowther Arcade in which was situated E.M. Clarke's Laboratory of Science and the Adelaide Gallery of Practical Science.

[44] Gee (1993b).

the subject had not yet developed sufficiently to warrant a profession of electrical engineers. By the 1870s, however, the subject had taken great strides, and leaders such as Carl Becker and William Ladd became members of the newly formed Society of Telegraphic Engineers (forerunner of the Institution of Electrical Engineers) when it was set up in 1871.[45]

Despite the inevitable changes that would occur on taking over Watkins & Hill, it is satisfying to note that the Elliott brothers continued their father's tradition in the making and selling of drawing instruments for public schools such as Eton and Harrow, and the military colleges at Sandhurst and Woolwich; it is also satisfying to note that their order book still honoured smaller commissions, such as those by the Reverend Robert Walker at the Clarendon Laboratory, who had been a customer of Watkins & Hill since 1839. Thus, with the assistance of the strengths brought by Becker, the Elliotts obtained many contracts, some with the Admiralty and others overseas through the agency of the East India Company. Charles Elliott retired in 1865 and Frederick Elliott continued until his death in 1873. At this point, the business fell under the joint management of Becker, Willoughby Smith (chief electrician of the Telegraph Construction & Maintenance Company) and Susan Elliott, Frederick's widow. By this time, however, all trace of their major acquisition had vanished although the name of Watkins & Hill lived on through their signature on the many pieces of apparatus still in use. Indeed, a few items of apparatus bearing the Watkins & Hill signature could still be found in the laboratory where the present author studied physics 100 years later![46]

Conclusions

Throughout their 100-year proprietorship at Charing Cross, the three generations of the Watkins family provided a remarkable service, particularly for those whose concerns were with the advance or diffusion of science. Their longevity in business could never have been achieved without exercising a willingness to meet the various challenges and demands that came along. In terms of their commercial survival, each successive generation must therefore have developed a reputation for those quintessential aspects upon which their custom was dependent: namely, the 'making and improving' of awkward newborn articles as conceived by the experimental philosopher, that is, the transforming of an instrument into a robust form, sufficiently foolproof and functional

[45] The Society of Telegraph Engineers formally came into existence on 17 May 1871 at a meeting held in 2 Westminster Chambers, Victoria Street, London. The prime mover in this endeavour was Major Frank Bolton.

[46] This was Goldsmiths College, where Brian Gee took his physics degree. [Editors' note]

for general use; and 'technical support and management', that is, when the instrument maker was called upon to select, install, supervise, maintain, repair and manage.[47] This double-faceted role was clearly made visible through Francis-III Watkins who, on the one hand, brought to commercial fruition a complete range of reliable electromagnetic instruments, while, on the other hand, acted as curator of philosophical instruments at the new university. More generally and as was the custom of all leading instrument makers, the Watkinses must also have made it their business to associate with those whose profession called for improvement and innovation in instrumentation, to seek new processes or methods by which to improve the construction and use of an instrument, and to oblige customers by lending a helping hand or demonstrating or providing explanatory literature where clarity and practical understanding were required. Certainly, the instrument business at Charing Cross was vigorously pursued and those in charge were among the artisan élite who attracted experimentalists from far and wide.

Many of these issues, supporting the instrument-maker's position in a scientific context, rather than a marginal position of technical assistant, can be pursued in the references to this chapter. More generally, it is hoped that the information gathered and situations described about the activities of the Watkins family, now blended into social, historical and scientific contexts, will contribute to an understanding of just a small part of the larger account of the very human activity of scientific endeavour. Collectors of scientific instruments, however, may have gained little from this work, except this: all instruments bearing the signatures of Watkins or Watkins & Hill now have a context in which they can be understood and explained, even if, in truth, not all these instruments were entirely fabricated at the famous address at Charing Cross.

---

[47] Much the same role for the instrument maker was stipulated 100 years later when the microscope makers Watson & Co. (a contemporary of Elliott Bros) listed the 'duties' of their leading employees as follows: (i) to associate with those whose profession, calling for the continual use of scientific instruments, enables them to advise and suggest new designs, or modifications to old; (ii) to keep a constant watch for any new process or method which will serve the purpose of improvement in construction and use of an instrument; and (iii) to be ready with a helping hand for those with an optical difficulty or doubt. Watson (1937), p. 21.

Appendix 1

John Dollond's Refrangibility Experiments (1757–1758)

It took 18 months – from the time of repeating Newton's dubious water–glass prism experiment to solving the problem of chromatic correction – for John Dollond to complete his programme of refrangibility experiments. Of course, he had mulled over the problem long before taking action to test and falsify Newton's assertions but it was not long after achieving success with the compounded crown convex–flint concave objective that he patented the method of making it. He referred to his success at eliminating chromatic aberration simultaneously with correcting for spherical aberration a 'perfect theory'. The outcome was an objective capable of long focal length and, hence, great magnification. As a proto-scientific paper this account appears most lucid; however, the reader should not be taken in by the fluency of style. It does not, for example, indicate from where the materials were obtained, or how the measurements were taken; the paper is also devoid of any mathematical modelling; and, importantly, nor does it reveal how he came by his ideas (practical and theoretical). Thus, as prose, the account is beautifully written; moreover, it lends support to the view proffered by Ramsden, namely, that Dollond was the only person in London at this time with a passion to improve upon the refracting telescope. It certainly was worthy of the Royal Society Copley Medal. However, Dollond concealed much that was formative to his success.

Source: Phil. Trans. R. Soc. Lond. 50 Part 2 (1757), pp. 733–743.

An Account of Some Experiments concerning the different Refrangibility of Light. By Mr. John Dollond. With a Letter from James Short, M.A. F.R.S. Acad. Reg. Suec. Soc.

To the Rev. Dr. Birch, Secret. R.S.
Read June 8, 1758.

DEAR SIR,
I HAVE received the inclosed paper from Mr. Dollond, which he desires may be laid before the Royal Society. It contains the theory of correcting the errors

arising from the different refrangibility of the rays of light in the object-glasses of refracting telescopes; and I have found, upon examination, that telescopes made according to this theory are intirely free from colours, and are as distinct as reflecting telescopes.

I am, DEAR SIR,
Your most obedient humble Servant,
James Short.

Surrey Street,
June 8, 1758.

IT is well known, that a ray of light, refracted by passing thro' mediums of different densities, is at the same time proportionally divided or spread into a number of parts, commonly called homogeneal rays, each of a different colour; and that these, after refraction, proceed diverging: a proof that they are differently refracted, and that light consists of parts that differ in degrees of refrangibility.

Every ray of light passing from a rarer into a denser medium, is refracted towards the perpendicular; but from a denser into a rarer one, from the perpendicular; and the sines of the angles of incidence and refraction are in a given ratio. But light consisting of parts, which are differently refrangible, each part of an original or compound ray has a ratio peculiar to itself; and therefore the more a heterogeneous ray is refracted, the more will the colours diverge, since the ratios of the sines of the homogen[eal] rays are constant; and equal refractions produce equal divergencies.

That this is the case when light is refracted by one given medium only, as suppose any particular sort of glass, is out of all dispute; being indeed self-evident; but that the divergency of the colours will be the same under equal refractions, whatsoever mediums the light may be refracted by, tho' generally supposed, does not appear quite so clearly.

However, as no medium is known, which will refract light without diverging the colours, and as difference of refrangibility seems thence to be a property inherent in light itself, Opticians have, upon that consideration, concluded, that equal refractions must produce equal divergencies in every sort of medium: whence it should also follow, that equal and contrary refractions must not only destroy each other, but that the divergency of the colours from one refraction would likewise be corrected by the other; and there could be no possibility of producing any such thing as refraction, which would not be affected by the different refrangibility of light; or, in other words, that however a ray of light might be refracted backwards and forwards by different mediums, as water, glass, &c. provided it was so done, that the emergent ray should be parallel to the incident one, it would ever after be white; and conversely, if it should come out inclined to the incident, it would diverge, and ever after be coloured. From which it was natural to infer, that all spherical object-glasses of telescopes must be equally affected by the different refrangibility of light, in proportion to their apertures, whatever material they may be formed of.

But it seems worthy of consideration, that not-withstanding this notion has been generally adopted as an incontestable truth, yet it does not seem to have been hitherto so confirmed by evident experiment, as the nature of so important a matter justly demands; and this it was that determined me to attempt putting the thing to issue by the following experiment.

I cemented together two plates of parallel glass at their edges, so as to form a prismatic or wedge-like vessel, when stopped at the ends or bases; and its edge being turned downwards, I placed therein a glass prism with one of its edges upwards, and filled up the vacancy with clear water: thus the refraction of the prism was contrived to be contrary to that of the water, so that a ray of light transmitted thro' both these refracting mediums would be refracted by the difference only between the two refractions. Wherefore, as I found the water to refract more or less than the glass prism, I diminished or increased the angle between the glass plates, till I found the two contrary refractions to be equal; which I discovered by viewing an object thro' this double prism; which, when it appeared neither raised nor depressed, I was satisfied, that the refractions were equal, and that the emergent rays were parallel to the incident.

Now, according to the prevailing opinion, the object should have appeared thro' this double prism quite of its natural colour; for if the difference of refrangibility had been equal in the two equal refractions, they would have rectified each other: but the experiment fully proved the fallacy of this received opinion, by showing the divergency of the light by the prism to be almost double of that by the water; for the object, tho' not at all refracted, was yet as much infected with prismatic colours, as if it had been seen thro' a glass wedge only, whose refracting angle was near 30 degrees.

N.B. This experiment will be readily perceived to be the same as that which Sir Isaac Newton mentions;[1] but how it comes to differ so very remarkably in the result, I shall not take upon me to account for; but will only add, that I used all possible precaution and care in the process, and that I keep the apparatus by me to evince the truth of what I write, whenever I may be properly required so to do.

I plainly saw then, that if the refracting angle of the water-vessel could have admitted of a sufficient increase, the divergency of the coloured rays would have been greatly diminished, or intirely rectified; and there would have been a very great refraction without colour, as now I had a great discolouring; without refraction: but the inconveniency of so large an angle, as that of the vessel must have been, to bring the light to an equal divergency with that of the glass prism,

---

1 Book I, Part *ii*, Prop. 3. Experiment 8 of his Optics.

whose angle was about 60 degrees, made it necessary to try some experiments of the same kind, by smaller angles.

I ground a wedge of common plate glass to an angle of somewhat less than 9 degrees, which refracted the mean rays about 5 degrees. I then made a wedge-like vessel, as in the former experiment, and filling it with water, managed it so, that it refracted equally with the glass wedge; or, in other words, the difference of their refractions was nothing, and objects viewed thro' them appeared neither raised nor depressed. This was done with an intent to observe the same thing over again in these small angles, which I had seen in the prism: and it appeared indeed the same in proportion, or as near as I could judge; for notwithstanding the refractions were here also equal, yet the divergency of the colours by the glass was vastly greater than that by the water; for objects seen by these two refractions were very much discoloured. Now this was a demonstration, that the divergency of the light, by the different refrangibility, was far from being equal in these two refractions. I also saw, from the position of the colours, that the excess of divergency was in the glass; so that I increased the angle of the water-wedge, by different trials, till the divergency of the light by the water was equal to that by the glass; that is, till the object, tho' considerably refracted, by the excess of the refraction of the water, appeared nevertheless quite free from any colours proceeding from the different refrangibility of light; and, as near as I could then measure, the refraction by the water was about 5/4 of that by the glass. Indeed I was not very exact in taking the measures, because my business was not at that time about the proportions, so much as to shew, that the divergency of the colours, by different substances, was by no means in proportion to the refractions; and that there was a possibility of refraction without any divergency of the light at all.

Having, about the beginning of the year 1757, tried these experiments, I soon after set about grinding telescopic object-glasses upon the new principles of refractions, which I had gathered from them; which object-glasses were compounded of two spherical glasses with water between them. These glasses I had the satisfaction to find, as I had expected, free from the errors arising from the different refrangibility of light: for the refractions, by which the rays were brought to a focus, were every-where the differences between two contrary refractions, in the same manner, and in the same proportions, as in the experiment with the wedges.

However, the images formed at the foci of these object-glasses were still very far from being so distinct as might have been expected from the removal of so great a disturbance; and yet it was not very difficult to guess at the reason, when I considered, that the radii of the spherical surfaces of those glasses were required to be so short, in order to make the refractions in the required proportions, that they must produce aberrations, or errors, in the image, as great; or greater, than those from the different refrangibility of light. And therefore, seeing no method of getting over that difficulty, I gave up all hopes of succeeding in that way.

And yet, as these experiments clearly proved, that different substances diverged the light very differently, in proportion to the refraction; I began to suspect, that such a variety might possibly be found in different sorts of glass, especially as experience had already shewn, that some made much better object-glasses, in the usual way, than others: and as no satisfactory cause had as yet been assigned for such difference, there was great reason to presume, that it might be owing to the different divergency of the light by their refractions.

Wherefore, the next business to be undertaken, was to grind wedges of different kinds of glass, and apply them together, so that the refractions might be made in contrary directions, in order to discover, as in the foregoing experiments, whether the refraction and divergency of the colours would vanish together. But a considerable time elapsed before I could set about that work; for tho' I was determined to try it at my leisure, for satisfying my own curiosity, yet I did not expect to meet with a difference sufficient to give room for any great improvement of telescopes; so that it was not till the latter end of the year that I undertook it; when my first trials convinced me, that this business really deserved my utmost attention and application.

I discovered a difference, far beyond my hopes, in the refractive qualities of different kinds of glass, with respect to their divergency of colours. The yellow or straw-coloured foreign sort, commonly called Venice glass, and the English crown glass, are very nearly alike in that respect, tho' in general the crown glass seems to diverge the light rather the least of the two. The common plate glass made in England diverges more; and the white crystal or flint English glass, as it is called, most of all.

It was not now my business to examine into the particular qualities of every kind of glass that I could come at, much less to amuse myself with conjectures about the cause, but to fix upon such two sorts whose difference was the greatest; which I soon found to be the crown, and the white flint or crystal. I therefore ground a wedge of white flint of about 25 degrees, and another of crown of about 29 degrees, which refracted nearly alike; but their divergency of the colours was very different. I then ground several others of crown to different angles, till I got one, which was equal, with respect to the divergency of the light, to that in the white flint: for when they were put together, so as to refract in contrary directions, the refracted light was intirely free from colour. Then measuring the refractions of each wedge, I found that of the white glass to be to that of the crown nearly as 2 to 3; and this proportion would hold very nearly in all small angles. Wherefore any two wedges made in this proportion, and applied together, so as to refract in a contrary direction, would refract the light without any difference of refrangibility.

To make therefore two spherical glasses, that shall refract the light in contrary directions, it is easy to understand, that one must be concave, and the other convex; and as the rays are to converge to a real focus, the excess of refraction must evidently be in the convex; and as the convex is to refract most, it appears from the experiment, that it must be made with crown glass, and the concave with white flint glass.

And further, as the refractions of spherical glasses are in an inverse ratio of their focal distances; it follows, that the focal distances of the two glasses should be inversely as the ratio's [sic] of the fractions of the wedges: for being thus proportioned, every ray of light, that passes thro' this combined glass, at whatever distance it may pass from its axe, will constantly be refracted, by the difference between two contrary refractions, in the proportion required; and therefore the different refrangibility of the light will be intirely removed.

Having thus got rid of the principal cause of the imperfection of refracting telescopes, there seemed to be nothing more to do, but to go to work upon this principle: but I had not made many attempts, before I found, that the removal of one impediment had introduced another equally detrimental (the same as I had before found in two glasses with water between them): for the two glasses, that were to be combined together, were the segments of very deep spheres; and therefore the aberrations from the spherical surfaces became very considerable, and greatly disturbed the distinctness of the image. Tho' this appeared at first a very great difficulty, yet I was not long without hopes of a remedy: for considering the surfaces of spherical glasses admit of great variations, tho' the focal distance be limited, and that by these variations their aberrations may be made more or less, almost at pleasure; I plainly saw the possibility of making the aberrations of any two glasses equal; and as in this case the refractions of the two glasses were contrary to each other; their aberrations, being equal, would intirely vanish.

And thus, at last, I obtained a perfect theory for making object-glasses, to the apertures of which I could scarcely conceive any limits: for if the practice could come up to the theory, they must certainly admit of very extensive ones, and of course bear very great magnifying powers.

But the difficulties attending the practice are very considerable. In the first place, the focal distances, as well as the particular surfaces, must be very nicely proportioned to the densities or refracting powers of the glasses; which are very apt to vary in the same sort of glass made at different times. Secondly, the centres of the two glasses must be placed truly on the common axis of the telescope; otherwise the desired effect will be in a great measure destroyed. Add to these, that there are four surfaces to be wrought perfectly spherical; and any person, but moderately practised in optical operations, will allow, that there must be the greatest accuracy throughout the whole work.

Notwithstanding so many difficulties, as I have enumerated, I have, after numerous trials, and a resolute perseverance, brought the matter at last to such an issue, that I can construct refracting telescopes, with such apertures and magnifying powers, under limited lengths, as, in the opinion of the best and undeniable judges, who have experienced them, far exceed any thing that has been hitherto produced, as representing objects with great distinctness, and in their true colours.

John Dollond

Appendix 2

Fees Charged at Chancery in Procuring Letters Patent

The process of obtaining letters patent was a multi-stage procedure whereby the original petition of the would-be patentee was conveyed between different offices of Chancery until it was ready for acceptance under the Great Seal of Great Britain. During April 1758, Francis Watkins, acting as John Dollond's agent, undertook these onerous duties and paid, from his own pocket, various fees at each stage. This was part of his arrangement with Dollond in order that he should be allowed to join with him in partnership for the future sales of 'patented' achromatic telescopes. The stages set out below were those in force during the mid-eighteenth century although they had evolved over the years since the Clerks Act of 1535. Indeed, the same procedure here was little modified over the next century, that is, until the system of patenting underwent reform by an Act of Parliament in 1852. The fees quoted are typical for the early to mid-eighteenth century, and so are very likely to be close to those paid by Watkins on Dollond's behalf.

Source: Allan Gomme, *Patents of Invention* (Longman, Green, for the British Council, 1946), pp. 16–19.

Stage	Reason	Office fees and other costs*
(i)	To the master in Chancery, praying that a patent be granted	1 s Stamp duty on warrant 6d
(ii)	To the Office of the Secretaries of State for endorsement of the petition	£2 2s 6d
(iii)	To the law officers of the Attorney or Solicitor General for their report	£4 4s 0d
(iv)	To the Office of the Secretaries of State for a warrant and signature of sovereign and Secretary of State requesting that the law officers should prepare a bill for the grant of patent	£7 13s 6d Stamp duty on warrant 7s 6d
(v)	To the law officers of the Patent Bill office for the prepared parchment bill and signature	£2 7s 6d Engrossing £1 1s Stamp duty on warrant 7s 6d

Stage	Reason	Office fees and other costs*
(vi)	To the Office of the Secretaries of State with the parchment copy for the signature of the sovereign. (Referred to as the 'king's bill'.)	£7 13s 6d
(vii)	To the Signet Office for the king's bill to be re-copied exactly (save for changes top and bottom) and sealed with the Signet. (Referred to as the 'signet bill'.)	£5 0s 0d† Stamp duty on warrant 7s 6d
(viii)	To the Office of the Lord Privy Seal for preparation of a writ on parchment	£4 15s 9d† Stamp duty on warrant 7s 6d
(ix)	To Office of the Lord Chancellor for endorsement and date stamp, engrossment‡ on parchment, and seal of the Great Seal	£1 11s 4d† Engrossment £4 6s 4d Sealing and enrolment £7 13s 4d Pursebearer £1 6s 6d For patent box 9s 6d Stamp duty on patent £6

* Stamp duties were introduced in 1694 and those on patents rose steeply, particularly toward the end of the century and into the nineteenth century when a plethora of manufacturing inventions came to the fore: cf. 1694 (£2), 1754 (£6), 1804 (£20) and 1815 (£30).

† An additional charge of one guinea was made per extra parchment skin (a skin would contain about 1,440 words) and additional fees for engrossment (£3 11s 6d) and sealing and enrolment (£1 13s 4d). Additional names in the grant were charged at £2 13s 4d each although this was not necessary in Dollond's case.

‡ The act of writing in the peculiar form of a legal document.

Appendix 3

John Dollond's Patent for the Invention of a New Method of Making Object Glasses of Refracting Telescopes

The form of wording on patents of invention was fairly standardised by the time that John Dollond received his letters patent. In effect, a patent was an 'open letter' from the sovereign – in which the king had written: 'We being willing to give Encouragement to all Arts and Inventions which may be for the Public Good are graciously pleased to condescend to his request,' thus bestowing a privilege on the patentee and allowing him a monopoly for 14 years. The motive by the Crown in awarding patents was a desire to secure benefits for the realm and so to recompense the inventor for what was gained. Such privileges came with conditions, the most important of which concerned the origin of the invention. Dollond had to make a declaration, by affidavit, that he was 'to the best of his knowledge and belief' the 'Sole Inventor'. Any patent could be voided should it be shown that the 'invention is not a new invention' or 'not invented and found out by the said John Dollond'. To validate the letters patent a specification had to be enrolled within three calendar months. Afterwards, the onus was on Dollond to defend any quibble that might arise through dispute or petition or any dealings arising from the rights bestowed upon him. The document presented here is a transcription of the 'king's bill', that is, the copy retained by the Signet Office as a warrant. The tailpiece to this version shows the document to be unfinished. That is, it terminates with a message to the king informing him of the purpose of the document requiring his ascent. The ending to the final (fully engrossed) document (following this copy) would require the Great Seal. Note: the official date of commencement of the patent, is 19 April 1758.

Sources: Three copies of the patent are known: (i) the 'king's bill' (retained by the Signet Office as warrant). TNA SO 7/233; (ii) the final Letters Patent (embellished with Great Seal). Guildhall, City of London (Dollond papers): CLC/444/MS14805/002); and (iii) a quoted copy in the action taken by Peter Dollond against Henry Pyefinch. TNA C 12/2075/4. Note: In the process the bill being copied (from 'kings bill' to 'signet bill' and, finally, to the fully engrossed form), occasional mistakes and variations occurred. For example, one hand might write 'libertys' while another copyist may have preferred 'liberties'; phrases could sometimes be omitted accidentally; and, in the text (below), a pair of parentheses have been included for sense (as in the final form) although they were absent on the original signet office 'kings bill' version. Original spellings are retained, even those in error. This patent is now numbered 721 in Bennet Woodcroft's *Titles of Patents of Invention from March 2, 1617 (14 James I) to July 26, 1853 (16 Victoria)* (London: Queen's Printing Office, 1854).

George the Second by the Grace of God of Great Britain France and Ireland King Defender of the Faith &c. *To all* to whom all these presents Shall come Greeting *Whereas* John Dollond of the parish of Saint Martin in the Fields in the County of Middlesex Optician has by his Petition humbly represented unto us that he hath by Application and Study and at Considerable Expense at last Invented and brought to perfection a new method of making the Object Glasses of Refracting Telescopes by compounding mediums of different Refractive Qualities whereby the Errors arising from the different refrangibility of Light as well as those which are produced by the Spherical Surfaces of the Glasses are perfectly corrected *That* this improvement which has been in vain sought after by Severall others above a Century past will be a great benefit to the publick in General but Especially to the Navy where good Refracting telescopes are of the utmost Importance and as he is the first and Sole Inventor thereof and that no other person has to the best of his knowledge or belief practiced the same he hath therefore most humbly prayed us to Grant to him his Executor Administrator and Assigns our Royal Letters Patent under our Great Seal of Great Britain for the Sole benefit and advantage of the said Invention within that part of our Kingdom of Great Britain called England and Dominion of Wales and Town of Berwick upon Tweed for the Term of fourteen years pursuant to the Statute on such are made and provided *We* being willing to give Encouragement to all Arts and Inventions which may be for the publick Good are graciously pleased to condescend to his request *Know ye* therefore that *We* of our Especial Grace certain knowledge and mere motion *Have given* and Granted and by those presents for us in our Heirs and Successors *Do* give and Grant unto the said John Dollond his Executors Administrators and Assigns our Especial Licence full power Sole priviledge and authority that he the said John Dollond his Executors Administrators and Assigns and any of them by himself and themselves or by his and their Deputy or Deputy Servants or Agents or such others he the said John Dollond his Executors Admons and Assigns shall and lawfully may have and Enjoy the whole profit benefit commodity and advantage from time to time and at all times hereafter during the term of years herein expressed Shall and Lawfully may make use exercise and vend his said Invention within that part of our Kingdom of Great Britain called England and our Dominion of Wales and Town of Berwick upon Tweed as such manner as to him the said John Dollond his Executors Administrators and Assigns or any of them shall in their directions seem moot *And* that he the said John Dollond his Executors Admons and Assigns Shall and Lawfully may have and Enjoy the whole profit benefit commodity and advantage from time to time coming growing accruing and arising by reason of the said Invention and during the term of years herein mentioned *To have* hold Exercise and Enjoy the said licence power priviledges and Advantage herein before granted or mentioned to be Granted unto the said John Dollond his Executors Admons and Assigns for and during and unto the full end and Term of fourteen years from the date of those presents next and immediately Ensuing and to be fully Compleat and Ended according to the Statute in such case made and provided *And* to the End that the said John Dollond his Executors Admons and Assigns and every of them may have and Enjoy the full benefit and the Sole use and Exercise of the said Invention according to our Gracious Intention herein before declared *We* do by those presents for us our Heirs and Successors require and strictly commend all and every person and persons Bodies politick and corporate and all other Subjects whatsoever of what Estate Quality Degree name or condition so ever they be within that said part of our Kingdom of Great Britain called England and Dominion of Wales and Town of Berwick upon Tweed aforesaid that neither they or any of them at any time during the continuance of the said Term of fourteen years hereby Granted either directly or indirectly do make use of put in practice the said Invention of any part of the same so attained unto by the s[d] as aforesaid in any wise Counterfeit Imitate or resemble the same nor shall make or cause to be made any addition thereunto or Subtraction from the same whereby to pretend himself or themselves the Inventor or Inventors Devisor or Divisors thereof without the Licence consent or agreement of the said John Dollond his Executors Admons or Assigns in writing under his and their Hand and Seals first had and obtained in that behalf upon such pains and penalties as can or may be Justly Inflicted on such Offendor for their contempt of this our Royal Command and further to be Answerable to the said John Dollond his Executors Admons and Assigns according to Law for his and their Damage thereby Occasioned *And moreover We* do by these presents for us our Heirs and Successors will and command all and singular the Justices of the Peace Mayors Sheriffs Bayliffs Constables Headboroughs and all other Officers and Ministers whatsoever of us our Heirs and Successors for the time being that they or any of them do not nor shall at any time hereafter during the said Term hereby Granted in any wise Molest Trouble or Hinder to said John Dollond his Executors Amons or Assigns or any of them or his or their Deputys Servants or Agents in or about the due and Lawfull use or Exercise of the aforesaid Invention or anything relating thereto *provided* always and these our Letters Patent are and shall be upon this Condition that if at any time during the said Term hereby Granted it should be made appear to us our House or Successors or any six or more of our or their privy Council that this our Grant is Contrary to Law or prejudicial or Inconvenient to our Subjects in General or that the said Invention is not a new Invention as to the publick use and Exercise thereof in that part of the Kingdom of Great Britain and Dominion of Wales and town of Berwick upon Tweed aforesaid or not Invented and found out by the said John Dollond as aforesaid then upon Signification or Declaration thereof to be made by us our House or Successors under our or their Signet or Privy Seal or by the Lords and those of our or their Privy Council or any Six or more of them under their hands these our Letters Patent shall forthwith cease determine and be utterly Void to all intents and purposes anything herein before contained to the contrary thereof in any wise notwithstanding *provided* also that those Our Letters Patent or anything herein contained shall not Extend or be constructed to Extend to give priviledge unto the said John Dollond his Executors Administrators or Assigns many of them to have or Imitate any Invention or work whatsoever which hath heretofore been found out or Invented by any other of our Subjects whatsoever and publickly used or Exercised in that said part of our Kingdom of Great Britain called England our Dominion of Wales or Town of Berwick upon Tweed aforesaid unto whom the Letters Patent or privilege have been already Granted for the Sole use Exercise and benefit thereof it being our will and pleasure that the said John Dollond and his Exor[s] Admons and Assigns and all and every other person and persons to whom like Letters Patent or priviledges have been already Granted for the sole use and benefit thereof it being Our will and pleasure that the said John Dollond his Executors Admons and Assigns and all and every other person and persons to whom the Letters Patent or priviledges have been already Granted as aforesaid shall distinctly live and practice their Severall Inventions by them Invented and found but according to the full Intent and Meaning of the same Letters Patent and of these present *provided* likewise nevertheless and these our Letters Patent are upon their express condition that if the said John Dollond his Executors or Administrator or any person or persons which shall or may at any Time or Times hereafter during the continuance of this Grant have or Claim any Right Title or Interest in Law or Equity of in or to the Power priviledge and Authority of the sole use and benefit of the said Invention hereby Granted shall make any Transfer or Assignment or any Pretended Transfer or Assignment of the said Liberty and priviledge or any Share or Shares of the Benefit or profit thereof or shall declare any Trust thereof to or for any Number person exceeding the Number of five or shall Open or cause to be Opened any Book or Books for publick subscriptions to be made by any number of persons Exceeding the number of five in Order to the raising any Sum or Sums or Money under pretence of carrying on the said Liberty or priviledge hereby granted or shall by him or themselves or his or their Agent or Servants receive and Sum or Sums of Money whatsoever any number of persons Exceeding in the whole the number of five for such or the like Intents or purposes or shall presume to act as a Corporate Body or shall divide the benefitt of these our Letters Patent or the Liberty & priviledge whereby by us granted into any number of shares exceeding the number of five or shall commit or do or procure to be committed or done any act matter or thing whatsoever during such Time as such person or persons shall have any Right or Title either in Law or Equity in or to the said promises which will be contrary to the true Intent and meaning of a certain Act of Parliament made in the Sixth Year of the Reign of our late Royal Father King George the First Intitled *An Act* for the Better securing Certain powers and priviledges Intended to be Granted by his Majesty by two Charters for the Assurance of Ships and Merchandizes at Sea and for lending Money upon Bottomry and for Restraining several Extravagant and Unwarrantable practices therein Mention'd or in case the said power priviledge or authority shall at any time hereafter become Vested in or in Trust for more than the Number of five person or their representatives at any one time (Reckoning Executors or Administrators in whatever as and for the single person whom they represent as to such Interest as they are or shall be entitled to in Right of such their Testator or Intestate) that then and in any of the said Cases these our Letters Patent and all Liberties and Advantages whatsoever hereby Granted shall latterly cease determine and become Void any thing therein before contained to the contrary thereof in any wise notwithstanding *provided* also that if the said John Dollond shall not particularly Describe and Ascertain the Nature of the said Invention in what manner the same is to be performed by an Instrument in Writing and or his hand or Seal and cause the same to be Inrolled in our High Court of Chancery within three Kalendar months thereof shall be in and by all things good firm Valid sufficient and Effectual in the Law according to the true Intent and Meaning thereof and shall be taken construed and adjudged in the most favourable and beneficial Sense for the best advantage of the said John Dollond his Executors Administrators and Assigns as well in all our Courts of Record as Elsewhere and by all and singular the Officers and Ministers whatsoever of Us our Heirs and Successors in that part of our said Kingdom of Great Britain called England our dominion of Wales and Town of Berwick upon Tweed Aforesaid and Amongst all and Every the subjects of Us our Heirs and Successors whatsoever and whosoever notwithstanding the not full and certain describing the Nature or Quality of the said Invention or of the Materials thereto conducing and belonging *In Witness &c Witness &c* ...[1]

... *May it Please your Most Excellent Majesty*

This contains Your Majestys Grant unto John Dollond of the parish of Saint Martin in the Fields in Your County of Middlesex Optician of the Sole Benefit and Advantage of his Invention of a new method of making the Object Glasses of Refracting Telescopes by compounding mediums of different Refractive Qualities whereby the Errors arising from the Different Refrangibility of Light as well as those which are produced by the spherical surfaces of the Glasses are

[1] This copy of the 'kings bill' was retained by the Signet Office.

perfectly corrected *To hold* to him his Executors Administrators and Assigns within England Wales and Town of Berwick upon Tweed for the Term of fourteen years according to the Statute in that Case made and *provided* that he doth within three calendar months from the date of the Grant hereby Intended Cause a particular Description of the Nature of his said Invention by writing under his hand and seal to be Inrolled in Your High Court of Chancery otherwise Your said Grant to be void and all such Clauses prohibitions and privisous are inserted there in a scroll such in Grants of the like Nature and this Bill is proposed.[2]

By warrant under your Majestys Royal Sign Manual

Ap$^{\ell}$ 12 1758
[signed] *C. Pratt*

... *In witness* whereof we have caused these our Letters to be made patent *Witness ourself at Westminster* the Nineteenth day of April in the Thirty first Year of Our reign by writ of Privy Seal ...[3]

2 The above message to the king was later replaced with the endorsement (engrossed and sealed).

3 Date of commencement.

Appendix 4

Articles of Co-partnership between Mr Francis Watkins and Mr John Dollond 29 May 1758

THE DOLLOND–WATKINS COPARTNERY was set up in order for the partners to trade in common and share profits from the vending (only) of John Dollond's 'New-invented Refracting Telescope', that is telescopes containing the patented combined object glass. Watkins part had arisen as recompense for his 'trouble and expense' in 'soliciting and obtaining' the patent, although to take part, Watkins had to provide a further £200. The articles of agreement set out the formula by which profit-sharing should be carried out and rules for administering the agreement, including rules of closure should the partnership be dissolved for any reason. The duration of the partnership ran alongside the term of the patent and so was for 14 years, dated from 29 May 1758. Marks on the source document show that the patent was sealed on 1 July [1758], witnessed by 'Jn Best' and 'Jn Clark', and stamped 'GR 33' (thirty-third year of the reign of George-II). The agreement clearly states that Dollond was the inventor of a 'new method' (not a new device) by which 'to perfect' (that is, correct) the errors arising from the compounding of different materials of different refrangibility as well as those due to the spherical surfaces.

Source: The transcription here is derived from a copy of the original agreement to be found in the London Metropolitan Archives collection of Dollond papers. Curiously, it lacks signatures of either party. Watkins, too, must have owned a copy although this seems no longer extant. London Metropolitan Archives, Guildhall Library, Dollond Papers: CLC/444/MS14805. Original spellings retained.

Articles of Agreement indented made concluded and agreed upon the Twenty ninth day of May in the Thirty first year of the Reign of our Sovereign Lord George the Second by the Grace of God of Great Britain France and Ireland King Defender of the Faith and in the year of our Lord One thousand seven hundred and fifty eight Between Francis Watkins of the parish of Saint Martin in the Fields in the County of Middlesex optition of the one part and John Dollond of the same parish of the other part as follows that is to say Whereas the said John Dollond hath by Application and Study and at a Considerable Expence

Invented and brought to perfection a New Method of making the Object Glasses of Refracting Telescopes by Compounding mediums of different Refractive Qualities whereby the errors arising from the Different Refrangibility of Light as well as those which are produced by the Spherical Surfaces of the Glasses are perfectly corrected And Whereas the said Francis Watkins hath with Great Trouble and at considerable expence Sollicited and Obtained Letters Patent bearing date at Westminster the Nineteenth Day of April now last past whereby his present Majesty hath Granted unto the said John Dollond his Executors Administrators and Assigns Especial Licence full power sole privilege and authority that the said John Dollond his Executors Administrators and Assigns and every of them by himself and themselves or by his or their Deputy or Deputies Servants or Agents or such others as the said John Dollond his Executors Administrators and Assigns shall at any time agree with no others from time to time and at all times hereafter during the term of years therein and herein Expressed shall and Lawfully may make use exercise and vend his s[d] invention within that part of the Kingdom of Great Britain called England the Dominion of Wales and the town of Berwick upon Tweed such manner as to him the said John Dollond his Executors Administrators and Assigns or any of them shall in their discretions seek work and that the said John Dollond his Executors Administrators and Assigns shall and lawfully may have and enjoy the whole profit Benefit Commodity and Advantage from time to time coming by Owning Arriving and Arising by reason of the said Invention for and during the Term of Years therein and herein mentioned To have hold Exercise and Enjoy the said Licence powers privileges and advantages unto the said John Dollond his Executors Administrators and Assigns for and during and unto the full End and Term of Fourteen Years from the Date of the same Letters Patent next and immediately ensuing and fully compleat and Ended according to the statute in such case made and provided which said Letters Patent the said Francis Watkins hath delivered unto the said John Dollond and Whereas the said John Dollond in Consideration of and as satisfaction for the Trouble and Expense of the said Francis Watkins in soliciting and obtaining the said Letters Patent hath agreed to admit to said Francis Watkins to be an equal partner with him the said John Dollond in the profit benefit comodity and advantage which shall from time to time from the vending the said Invention for and during the remainder yet to come of the said Term of fourteen years *Now These Presents* witness that for the consideration aforesaid and also in consideration of the Great Trust and Confidence which the said Francis Watkins and John Dollond mutually have and repose in each other They the said Francis Watkins and John Dollond each of them separately for himself his representative Executors and Administrators do mutually covenant promise and agree to and with each other of them his Executors and Administrators by these presents in the manner following that is to say *That they* the said Francis Watkins and John Dollond shall and will become partners together in the vending of Refracting Telescopes to be made after the said New Method invented by the said John Dollond and that the said partnership shall commence and be computed from the Day of the Date of these presents and shall continue during the remainder yet to come of the said Letters Patents Granted upon the terms and according to the Agreement herein after Declared and Expressed *And it is hereby Agreed* by and between the said parties that they the said parties respectively shall and will use their utmost endeavours to advance and promote the sale of the said New Invented Telescopes and that the expense of the Tube turning work and the Glassing of such Telescopes shall be first debited out of the Money which shall arise by sale thereof and that one fourth part of the Neat profit arising by such sale or sales shall be taken and retained by the respective Seller whether such seller shall be the said Francis Watkins or the said John Dollond as a recompense for the trouble and hazard of such sale and that the remaining three fourths part being the Neat profit arising from such sales after such Deduction as aforesaid shall be equally divided between them the said Francis Watkins and John Dollond *And it is also hereby agreed* that between the said parties that all such Telescopes shall be sold at the rate of sixty pounds per centum profit and that the said rate shall not at any time during the continuance of the said Copartnership be altered or varied unless with the mutual consent of both the said parties so that the Rates or prices for Glassing of the said Telescopes shall at all times during the continuance of the said Copartnership be established and fixed as follows (that is to say) the Rate or Price of each such Telescope two feet in length or under twelve shillings Three feet in length Sixteen shillings Four feet in length One pound one shilling and Six feet in length One pound twelve shillings which said Rates and Prices shall not be varied or altered unless by mutual consent of both the said parties and that the Tubes and Turning work of the said Telescopes shall be made and done in the common neat manner *And it is hereby Further Agreed* by and between the said parties to these presents that they the said parties respectfully shall and will be true and just to each other in all their accounts Reckonings Dealings and Doings relating to the Business of their said Copartnership and shall and will from time to time write and enter or cause to be written and entered in proper books to be kept by them for that purpose a fair and just account of all such the said telescopes as shall be made and sold by them or either of them and of all such sum and sums of Money as shall be paid and received by them or either of them on account thereof Together with as shall be made and sold by them or either of them and of all such sum and sums of money as shall be paid and Received by them or either of them on account thereof Together with all such Circumstances of Names times and places as shall be heresay or useful to manifest the estate and proceedings of the Business of the said Copartnership and that the said Books of Account shall be kept in some convenient place where to each of the said parties may from time to time have free access to view the same [at] his will and pleasure *And Further* that the said parties Respectively shall and will at the end of Three Calendar Months now next ensuing and so from time to time at the end of Three Succeeding Calendar Months during the continuance of the said Copartnership Account together of and [?]concurring all such of the said New Invented Telescopes shall be made and sold by the said parties either of them and of all the profits arising thereby and that the said parties Respectively shall and may immediately after the Starting and Finishing of every such account take to his use his Respective Share of the Neat profits which shall be then in hand after such Deduction made thereout as aforesaid And it is hereby also Agreed by and between the said parties that in case any different cause of suit or dispute shall arise between them touching any matter or thing in these presents contained or otherwise Relating to the Business of the said Copartnership which cannot be fully decided between themselves the same shall be referred to the hearing and determination of Three Indifferent Persons whereof the Parties in Difference shall name and choose one and the two named and chosen shall forthwith name and choose a third and that they the said parties their representative Executors and Administrators shall perform and fulfill every such order and award as shall be made and set down in writing by and under the hands and seals of the arbitors so named and chosen or any two of them within thirty Days next after [?]Pefferenaure to them made and that every such award shall be made a Rule of Court of King's Bench according to the statute in that Case made and provided And Lastly the said Francis Watkins in consideration of his being admitted by the said John Dollond partner in the profit of the said Invention Hath Remist Released and forever Quit Claimed and by these presents Doth ?Remist Release and forever Quit Claim unto the said John Dollond his Executors Administrators and Assigns all and all manner of action and Suit Debts Sum and Sums of Money and Demands whatsoever which against the said John Dollond his Executors or Administrators set the said Francis Watkins hath or can or may Claim for or in respect of his Solliciting and Obtaining the said recited Letters Patent or otherwise relating thereto to the Day of the Date of these presents In witness whereof the said parties to these presents have hereunto Interchangeably Set their hands and seals the Day and Year first above written.

Appendix 5

Klingenstierna's Admonishment of John Dollond (30 May 1761)

During the Royal Society balloting process for John Dollond's election, it became apparent that Dollond had omitted any acknowledgement to Samuel Klingenstierna who had supplied Dollond with the vital clue which underpinned his refrangibility experiments. When this error of etiquette was realised, Dollond's contribution was reduced in value. Dollond's sense of self-importance did not allow him to remain quiet and through Benjamin Wilson and Torbern Bergman, Klingenstierna came to learn of Dollond's claims. Klingenstierna then wrote to Dollond on 30 May 1761, taking him to task over his despicable behaviour. Two additions were added to the basic letter: (i) *Extract 1* is part of the critique levelled at Dollond, attempting to reason with him over the errors of his way; (ii) *Extract 2* is a further letter (undated) and dispatched with that of 30 May 1761 in which Klingenstierna entreats Dollond to appreciate the need for absolute integrity and openness in all matters of science.

Source: Klingenstierna's correspondence with Dollond is reprinted in full in Nordenmark and Nordström, *Lychnos*, 5 (1939), pp. 313–384.

Extract 1

(i) Concerning the general notion of correcting the aberration of one lens by means of an equal and contrary aberration of another.

> [This] is not owing to you, Sir, as you pretend but to Mr Huygens, who has not contented himself with pointing out or telling us in general Terms that the Thing may be done, but has really effected it in another particular case, having determined the figure of a Concave Eye-Glass, correcting the aberration of a Plano-Convex Object-Glass in a Telescope, as you may learn from the Preface to his *Dioptrica*, where his formulas for determining the Radiuses of the two surfaces of that Concave are to be found. If that was a sufficient reason for denying the name of a Discovery to any thing in my paper, that you could have a claim to, you would be

> in the same case with me in that respect.[1] It would then properly be M. Huygens who should claim the discovery, upon pretense [*sic*] of having pointed out to you and to me the Problem in question. But I am very certain, if that Great and Considering Man lived now, he would applaud to any improvement made by you or by me in the matter, rather than to form us a Chicane about the discovery. And I leave to you, Sir, and to any Man, who will read your Paper, to, judge what help such a pointing out as yours could afford me, towards a Solution of the General Problem of Aberrations, or if I stood in any need of it, having the much better pointing out of Mr Huygens.

(ii) Concerning the misrepresentation that Klingenstierna's paper was nothing but the solution of one single mathematical problem which Dollond had pointed out in the first place. (Here Klingenstierna supposed Dollond to mean the figures of his own object glass although, in fact, the 1760 paper was far more general – a culmination of five years' of study – for which the English equivalent title 'Elements of the Theory of aberrations of Rays of Light refracted in Spherical Lenses, and of the Manner of correcting those Aberrations' was supposed appropriate.)

> ... the Problem, said to be pointed out, being only one particular case of one of the general Problems contained in it. It is unconceivable to me, how an intelligent Person, who has read over a Paper, in order to examine it and to give his opinion about it, as I suppose You have done, should so grossly mistake its meaning;[2] and yet, that must needs be the Case, or something Worse, which I have too good an Opinion of your character to suspect.

(iii) *Concerning Dollond's reason for denying the term 'discovery' to Klingenstierna's mathematical work.* (Rather, Dollond attributes all discoveries to himself, simply by claiming the problems had been solved 'long ago' during the course of his finding the figures of the double object-glasses.)

---

1 [More fully] 'It is properly Mr Huygens that has pointed out to you and to me the Problem in question, and a Mathematician stands not in need of any other pointing out whenever occasion requires. And that may suffice as an Answer to your first reason.'

2 [More fully] '... mistake its meaning. I do not see but that must needs be the case, or something worse: or how an honest Man, being to give an impartial and true account of another Man's Writing to a learned Assembly, should so very unfairly represent it. I see very well, that this Artifice was necessary to you, in order to attribute to yourself my whole discovery at once; but how it may be made consistent with sense, Equity and Honesty, that I confess I see not.'

But suppose you have [solved the problem long ago]; pray Sir, is that a very good Argument, because one particular case of a Problem may have been solved before, the solution being kept a Secret, therefore there is no Discovery at all in a Systematical Writing, designed to explain the whole Theory? It concerns not me, then, or the question before us, whether you have found out by theory the figure of your double Glasses or no. But since you pretend to have done it, and since you use that pretension as an argument of vindicating to you the Discoverys of my Paper, it concerns you, Sir, to make good that pretension. I do not know, Sir, if Mr Maskelyne be a proper Witness or Judge in the case; but that I know from Mr Ferner's Letters, that you have told him yourself, that neither you nor that Gentleman (Mr Maskelyne) have been able to make your Rules agree with mine, and that upon a second examination of our equations, M. Maskelyn had found them to differ from one another, which I take to be a Witness against you, my Rules being certainly right. However, Sir, it seems to be a more natural and unexceptionable Way of Assuring a Discovery to Ones self, to publish it to the World, than to rely on the Testimony of another. You would very much oblidge [*sic*] the Mathematicians by giving your solution and your Method in print, and I for my part desire very earnestly you would no longer keep it a Secret. It is always to the advantage to the Sciences, to see the same Truth found out by different methods and treated in different ways. I hope the more to be complyed with in this my Demand, as you will thereby prevent some very obvious and disadvantageous Reflexions.

Extract 2

Dear Mr Dollond.

The Discovery of the different dissipations of heterogeneous Rays refracted in different Mediums, which undoubtedly belongs to you, and your very happy and ingenious application of that Discovery to the Improvement of Optical Instruments have deservedly done you very much Honour, and I have for my part made me a true pleasure publicldy and privately to acknowledge the Obligations I think the learned World has to you upon that account. I have owned likewise, that without that discovery of yours, I might never have applyed my thoughts to examine the subject of Aberrations. And so, I think, I have done you the Strictest justice in all respects [*More fully in translation*: 'Witness that very Paper, which you now so strangely and undeservedly endeavour to debase.']. If you pretend to share in my discoverys upon any other account, the fairest way you could take, seems to be to produce yours, and to let me be uninquieted. For as to the new method you have taken with me, upon pretenses, that I have proved either to be false or frivolous, to attribute to yourself the inventions contained in my Paper, in Matters where you have given not one Word, that I know, either in Public or to me in particular, that Method, I say, always will be deem'd a very unjust one to any intelligent Judge, and I can foretell you, [it] never will succeed. If you will not believe me, I shall furnish you with some occasions more of trying again, how that Method will do. Mr Clairaut at Paris, is now about to publish something upon the Subject of aberrations and their Corrections; and the Imperial Academy of Petersburg, has proposed as a Problem to be solved for the *next* years price, *how far the Imperfections of Optical Instruments, arising from both aberrations, may be removed by multiplying the Glasses.* And I do not see, but you may have as good a Claim, to whatever shall come abroad upon these occasions, as to any thing in my Paper.

I should not have taken any notice at all, of this your Claim, had not you publicly before the Royal Society endeavoured to detract from a Performance which I had intended as a not unworthy Homage to that Illustrious and Honorable Company, of which 1 have had the Honour to be a Member, above these thirty years. And. for the same Reason I have desired, that this my Answer should be read publicly in the Royal Society before it be delivered to you.

I am, Sir, with much consideration

Your very humble servant

S. Klingenstierna

Appendix 6

Ten Reasons why Telescopes with Combined Object Glasses should be Avoided

Shortly after Dollond and Watkins began to advertise the 'New-invented achromatic telescope' (May 1758), Benjamin Martin ran a campaign in the *Daily Advertiser* in which he stated that no refracting telescope was as good as a reflecting one. At the same time, he made haste to complete an optical text that had been in hand for some time. *New Elements of Optics* appeared in 1759 of which the 'new elements' centred on the subject of aberrations in optical instruments, principally the background theory and the modes of correction. With regard to the method employing combined double convex crown and plano-concave flint glasses (fully illustrated and explained mathematically in the text), Martin dismissed this approach for the ten reasons or difficulties. It should be borne in mind, however, that the object of his dismissal was also intended to accentuate an alternative approach, namely, the catadioptrical achromatic telescope by the broker Caleb Smith, who he proudly presents as an Englishman in contradistinction to Dollond whose name he avoids, except by an aside to a Stranger (thereby suggesting Dollond, the Huguenot).

Source: Benjamin Martin, *New Elements of Optics* (1759), chapter 8.

Remarks on the Nature, Construction, and Effects of the foregoing Achromatic[1] *TELESCOPE.*

288. THE Telescope in Fig. VIII is such, as I think provides against all the Errors of Vision by refracted Light, in the best Manner that any one of this Kind can do; but it is at the same Time a *Memento*, that we ought to expect *no absolute Perfection in the Works of Art*; that is the sole Prerogative of the divine Being, and to be sought for in the *Works of Nature* only.

---

1 From the Greek α *privative*, and χξϖμα, *Colour.* But to say the Truth, no Telescope can shew an Object *without Colours*, if the Object be viewed with a *Lens*; for the ocular Lens only will cause a colour'd Image on the *Retina* in the Fund of the Eye.

289. There has been no Telescope made of the Construction here represented, that I know of: The *Opera Glass* in *Fig.* VII. has been made with a triple Object Glass, but they are put together in a Manner different from that in which they here stand, and which we have shewn to be analogous to the three Prisms.

290. A Telescope, has also been made with a double Object Glass, consisting of a *double Convex Lens,* and a *Plano-Concave,* with its plain Side *outwards*; and five Glasses at the Eye-End all differently placed from those in any of the Telescopes here represented in these Plates.

291. Now with regard to this or any Sort of refracting Telescopes, there are many Imperfections inseparable from them in the Nature of Things, and which therefore all our Art will be in vain applied to elude: For

292. *Firstly,* There is no such thing as a perfect Medium to be found, or such as will refract the Light with a Regularity necessary for perfect Vision. *Glass* of any Kind (being an *artificial Mixture*) is far from being such a Medium; nay *Crystal* itself, the most perfect and natural of all transparent Substances, will admit of no perfect Refraction on many Accounts.

293. *Secondly,* If the Medium were perfect, there is yet no such Thing as a perfect *Politure* of Surface to be effected by Art, which yet is necessary for a true Refraction of Light, and perfect Vision by it.

294. *Thirdly,* The Light incident on a Lens is reflected externally and internally from both the Surfaces, and many Times internally from one Surface to the other, which must necessarily occasion a Confusion in the Refraction in some Degree; for this internal Reflection of Light in a Lens is so considerable, as to occasion a *secondary Focus* in the Lens, in which Objects may be magnified much more than in the *common Focus,* tho' this has pass'd hitherto unobserved.[2]

295. *Fourthly,* The greater the Number of Glasses, the greater the Confusion in Refraction, and *Obscurity* in the Object, by Loss of Light at each Surface by Reflection and Refraction.

296. *Fifthly,* There will ever be an Error from the Figure of the Lenses, as we have shewn, and we can proceed only *one Step* in correcting it, so that there will ever remain an Imperfection of Vision on that Account.

---

[2] I have often magnified Objects this new Way, by a Lens 3 or 4 Inches Diameter, and 12 or 18 Inches common focal Distance, but the secondary Focus not more than 2, or 2½.

297. *Sixthly,* The Error arising from the different Refrangibility of Light admits only of a *palliative Cure*; for I believe there are but few People who will pretend to make a Telescope *in* FACT *as perfect as it is in* THEORY.

298. *Seventhly,* Besides, unluckily for this Telescope, that promises so much Perfection, it *cannot mend one Fault without making another,* for to annihilate the Colours it must be associated with *a Concave Lens*; and such a Conjunction of dissimilar Lenses very much encreases the Error from the Figure of them, as we have amply shewn (from 127 to 137).

299. *Eighthly,* The Aberration of Rays from the Figure is still farther encreased, by our being obliged to have a Concave Lens of such a Form and Position as the Theory forbids in regard to that very Point. A *Plano-Concave* being the best Form, and its Concave Side placed *outwards,* the best Position, (84–88.) But neither of these Particulars can be admitted in this Telescope (266).

300. *Ninthly,* The Image IM [Fig. VIII] should be reversed, or the second Image OP form'd, by two Glasses only, *viz.* EF and GH, as we have shewn, (144.) A Lens interposed between the Image IM and double Object-Glass, appears to me absurd; for if the said Image be free from the *Errors of the Figure* and *different Refrangibility both,* (as we are told) then it is certain it can only do Harm.[3] Yet such a Glass I find in all the Telescopes I have seen of this Construction.

301. *Tenthly,* The Image OP should be view'd with two Eye-Glasses IK, LN, as we have before shewn (140.) which however it is not in those I have seen, tho' constructed with *Seven Lenses.*

302. From these and many other Considerations disadvantageous to refracting Telescopes, which the very best Constructions are liable to, it plainly appears they are far enough from Perfection, or from affording that clear and distinct Vision of Objects which the World has been persuaded to believe, and expect; but this will be farther evinced, when we have first shewn the Nature and Construction of a *reflecting Telescope.*

---

[3] If it be said – it *amplifies the Field of View,* it is granted; but then at the same Time, it will unequally refract the Rays, and, in some Degree, it regenerates the Colours that were before destroy'd.

Bibliography

Manuscripts

Articles of co-partnership between Francis Watkins and John Dollond for making object glasses for refracting telescopes according to the new method of John Dollond, as granted to him by letters patent, 29 [*sic*] May 1758, Guildhall Library: CLC/444/MS14805/001, with CLC/444/MS14805/002 the patent.

Dollond, P. (1789a). 'Account of the late John Dollond's discovery which led to the grand improvement of the refracting telescope', 11pp. Read before the Royal Society 21 May 1789. Royal Society Archives [Letters & Papers, Decade IX, Item 131].

Dollond, P. (1789b). 'An answer to a paper presented to the Royal Society by Mr J. Ramsden relating to the invention of the compound lens in 1729', 10pp. Royal Society Archives [Letters & Papers, Decade IX, Item 146] and quoted by Bennett (1998), p. 5 from a draft letter by Peter Dollond in the Blundell Collection [item 22] at the Museum for the History of Science, Oxford.

Dollond, P. (1789c). 'Some account of the discovery, made by the late Mr. John Dollond, F.R.S. which led to the Grand Improvement of Refracting Telescopes, in order to Correct Some Misrepresentations in Foreign Publications, of that Discovery: with an Attempt to Account for the Mistake in an Experiment made by Sir Isaac Newton on which Experiment, the Improvement of the Refracting Telescope Entirely Depended' (London: Printed for J. Johnson). 15pp. [Originally presented to the Royal Society and read on 21 May 1789. Royal Society Archives [Letters & Papers, Decade IX, Item 131]. *Note:* It was resolved, at the Council of the Society, that the paper should not be published in the *Transactions*; Peter Dollond saw to its appearance privately. His introduction to this paper is dated 1 September 1789. Reprinted in Kelly (1808), pp. 61–77. Peter Dollond's draft of this paper is described in: Bennett (1998).

Pyefinch, H. (1770). Patent No 976 (28 Dec. 1770). TNA: C 210/12.

Ramsden, J. (1789). 'Some observations on the invention of the achromatic telescope', read 18 June 1789. Royal Society Archives [Letters & Papers, Decade IX, Item 138], 6 pp.

Short, J. (1761). 'An account of Mr Klingenstierna's paper entitled *Consideratio circa Logem refractionis Radiorum luminis diversi generis in mediis diverisis, Vid* Newton Opt. lib. 1, part ii. Prop 3. Exp. 8',Read on 16 April 1761. Royal Society Archives [Letters & Papers, Decade IV, Item 66].

Smith, A. (1783). Patent No 1359 (13 March 1783). TNA: C 210/24.

Will of Jeremiah-I Watkins (1764). TNA: PROB 11/899.

Theses and Unpublished Papers

Abrahams, P. (June 2009). 'Printed works related to the history of the telescope'. Typescript online: http://www.europa.com/~telscope/binotele.htm accessed 19 July 2011.

Butterworth, H. (1968). 'The rise of the development of the Department of Science and Art, 1853–1900' (unpublished PhD thesis, University of Sheffield).

Gee, B. (1988). 'The place and contribution of the instrument maker in scientific development, 1820–1850, with special reference to electromagnetism and the diffusion of science' (unpublished PhD thesis, University of Leicester).

Millburn, J.R. (2004). 'William Owen of Fleet Street. An eighteenth-century bookseller and publisher' (unpublished typescript, Museum of the History of Science, Oxford).

O'Brien, T. (1991). 'Newchurch and thereabouts' (unpublished typescript).

Robischon, M.M. (1983). 'Scientific instrument makers in London during the seventeenth and eighteenth centuries'(unpublished PhD thesis, University of Michigan).

Sorrenson, R. (1993). 'Scientific instrument makers at the Royal Society of London, 1720–1780'(unpublished PhD thesis, Princeton University).

Wood, D.O. [n.d.]. 'About the Physical Department, University College, London, and those who worked therein, 1826–1956' (unpublished typescript, University College, London, Library Special Collections).

Printed Sources

Abridgements (1875). *Abridgements of Specifications Relating to Optical, Mathematical and other Philosophical Instruments ... 1636–1866* (London: Office of the Commissioners of Patents for Inventions).

Adams, G. (1747). *Micrographia Illustrata, or, the knowledge of the microscope explain'd, together with an account of a new invented universal single or double microscope* (London: Printed for and sold by the author, and sold by S. Birt, 2nd edn.).

Adams, G. (1777). *A Treatise Describing and Explaining the Construction and Use of New Celestial and Terrestrial Globes* (London: printed for and sold by the author, 4th edn).

Ahlström, O. (1957). 'John Dollond or Chester Moor Hall?', *The Optician*, 37(3), pp. 1–5.

Aldous, V.E. (1999). *My Ancestors were Freemen of the City of London* (London: Society of Genealogists).

Ampère, A. (1823). 'Mémoire sur la théorie mathématique des phénomènes électro-dynamiques uniquement déduite de l'experience ...', *Mémoires de l'Académie Royale des Sciences de l'Institut de France*, 6, pp. 175–387.

Ampère, A. (1826). *Théorie des phénomènes électro-dynamiques, uniquement déduite de l'experience* (Paris: Méquignon-Marvi).

Andersen, H. (1995). *Historic Scientific Instruments in Denmark* (Copenhagen: The Royal Danish Academy of Sciences and Letters).

Anderson, R.G.W., Burnett, J. and Gee, B. (1990). *Handlist of Scientific Instrument-makers' Trade Catalogues 1600–1914* (Edinburgh: National Museums of Scotland in association with the Science Museum London).

Andrade, E.N da C. (1960). *A Brief History of the Royal Society 1660–1960* (London: The Royal Society).

Anon (1738). *The Intelligencer: or, Merchants Assistant. Shewing, in an Alphabetical Manner, all the Names and Places of abode of all the Merchants and Considerable Traders throughout the Cities of London and Westminster, and the Borough of Southwark. With An exact Account of all the Stage-Coaches and Carriers in Great Britain, the Inns where they are set up, and the Time of their Coming in, and Going out which, with Ogilby's Book of the Roads and Fairs, will make A Compleat Pocket-Companion for Merchants and Tradesmen* (London: Printed for W. Meadows, at the Angel in Cornhill; and L. Gilliver and J. Clarke, in Fleet-street; also at their Shop in Westminster-Hall).

Anon (1761). *London and its Environs Described* (London: R. & J. Dodsley).

Anon [John Pote] (1763). *The Foreigner's Guide; or a necessary and instructive companion both for the foreigner and native, in their tour through the cities of London and Westminster* (1729; edition consulted 4th, 1763).

Anon (1771). *The Covenants of an Indenture of Apprenticeship Familiarly Explained and Enforced by Scripture for the use of Apprentices of the City of London* (London: printed for Henry Fenwick, 1771, further editions c.1803 and 1813). GL, pamphlet 9462.

Anon ('By a correspondent') [R.B.] (1829). 'Queries relating to Mr. Hall's original discovery of achromatic telescopes', *Philosophical Magazine and Annals of Philosophy*, 6(33), pp. 233–235.

Anon [R.B.P.] (1917). 'The invention of the achromatic lens', *Notes and Queries*, ser. 12, iii(69), pp. 334–336.

Anon [King, H.C.] (1950). *The House of Dollond, Two Hundred Years of Optical Service, 1750–1950* (Privately printed for Dollond–Aitchison Co.).

Anon [Parsons, H. (1917)]. *The House of Dollond* (London: printed for the company).

Arts and Sciences (1833). 'National Gallery of Practical Science, Adelaide Street', *Literary Gazette*, 878(17), p. 738.

Ashton, T.S. (1961). *An Economic History of England in the Eighteenth Century* (London: Methuen).

Aubert, T. (1951–1952). 'Alexander Aubert, FRS, astronome, 1730–1805', *Notes and Records of the Royal Society of London*, 9(1), pp. 79–95.

Ayscough, J. (1745). *A Description of the Structure of a Microscope: with the method of using it, either with a single or with compound lenses, both for opake as well as transparent objects, made by James Ayscough* (London: Printed for the author).

Ayscough, J. (1750). *A Short Account of the Nature and Use of Spectacles. In which is recommended a glass for spectacles prefereable to any hitherto made use of for that purpose. By __, optician, at the Great Golden Spectacles in Ludgate Street, London* (London: Printed for the author).

Ayscough, J. (1753). *Description of a Single Microscope: and of the apparatus adapted for the solar microscope, the camera obscura, and also for viewing prints in perspective, greatly improved for those uses and translated as Description d'un simple microscope: et des machines adaptées au microscope solaire, la chambre obscure, comme aussi la façon de voir les estampes en perspective, beaucoup perfectionée pour ses usages* (London: Printed for the author).

Ayscough, J. (1754). *A Short Account of the Eye and Nature of Vision Chiefly Designed to Illustrate the Use and Advantage of Spectacles Wherein are Laid Down Rules for Chusing [sic] Glasses Proper for Remedying all the Different Defects of Sight* (London: 3rd edn, privately printed for the author).

Ayscough, J. (1755a). *A Short Account of the Eye and Nature of Vision Chiefly Designed to Illustrate the Use and Advantage of Spectacles Wherein are Laid Down Rules for Chusing [sic] Glasses Proper for Remedying all the Different Defects of SightAs also some reasons for preferring a particular kind of glass, fitter than any other made use of for that purpose. By J. Ayscough, optician* (London: Printed by E. Say, for A. Strahan, in Cornhill; J. Robinson, Ludgate-Street; P. Stevens, facing Stationers-Hall; and Mrs Dodd, without Temple Bar, 4th edn).

Ayscough, J. (1755b). *Description abregée de l'Organe, et du Mechanisme de la Vue. ouvrage destine surtout a dernontrer 1'Usage et l'Utilité des LUNETTES, et a donner des regles pour le choix des Verres propres a supleer aux divers defauts de la Vue, de meme qu'a indiquer les raisons qui doivent faire preferer certans Verres, en particulier, a ceux dont se sert ordinairement* (A La Haye, chez M. Gosse: A Paris, chez M.Bauche: Et à Londres, chez A. Strahan, dans Cornhill; J. Robinson, dans Ludgate-street; P. Stevens, viz-à-viz Stationers-hall; Mr.Dodd, hors de Temple-Bar; et J. Jolliffe, dans St. James's-street).

Ayscough, J. (c.1771). *A Short Account of the Eye and Nature of Vision Chiefly Designed to Illustrate the Use and Advantage of Spectacles Wherein are Laid Down Rules for Chusing [sic] Glasses Proper for Remedying all the Different Defects of Sight* (London: 6th edn, privately printed for John Gilbert) [Not examined].

Ayscough, J. and Mann, J. (c.1750). *The Description of a Pocket Microscope, with the Apparatus Thereto belonging: made by J. Ayscough and J. Mann* (London: At the Sign of Sir Isaac Newton and Two Pair of Golden Spectacles).

Baddeley, W. (1835). 'Ede's portable chemical laboratory', *Mechanics' Magazine*, 23(617), pp. 161–164.

Baker, H. (1742). *The Microscope Made Easy: or, I. The nature, uses and magnifying powers of the best kinds of microscopes described, calculated, and explained: II. An account of what surprizing discoveries have been already made by the microscope* (London: Printed for R. Dodsley, at Tully's Head in Pall-Mall, 1742; other editions: 1743, 1744, 1754 and 1769) [Reprint of the 1769 edition: Science Heritage series (Bristol: Adam Hilger, 1987)].

Barker, F. and Jackson, P. (1990). *A History of London in Maps* (London: Barrie and Jenkins).

Barnard, J. (1737). *A Present for an Apprentice, or a sure guide to win esteem and an estate, with rules for his conduct to his master, and in the world* (London: Printed for T. Cooper at the Globe in Pater-noster Row). Numerous editions of this work ran into the following century.

Barrow, I. (1669). *Optical Works, 1667* (Lectiones xviii) trans. H.C. Fay; ed. A.G. Bennet and D.F. Edgar (London: Worshipful Company of Spectacle makers, 1987).

Barrow, J. (1792). *A Description of a Pocket and Magazine Cases of Drawing Instruments in Which is Explained the Use of Each Instrument* (Printed for J. and W. Watkins optical, mathematical and philosophical instrument makers to their Royal Highnesses the Duke and Duchess of York, and His Royal Highness the Duke of Clarence).

Barty-King, H. (1986). *Eyes Right: The Story of Dollond & Aitchison 1750–1985* (London: Quiller Press).

Beaglehole, J.C. (Ed.) (1962). *The* Endeavour *Journal of Joseph Banks, 1768–1771*, 2 vols (Sydney: Angus and Robertson), vol. 2.

Bechler, Z. (1975).'"A less agreeable matter": The disagreeable case of Newton and achromatic refraction', *British Journal of the History of Science*, 8(2), pp. 101–126. For a commentary on this paper, see Hall (1993).

Beckman, O. (2002a). 'Samuel Klingenstierna and the achromatic lens. Part I. Correspondence with John Dollond', *Journal of the Antique Telescope Society*, No. 22(1), pp. 9–15.

Beckman, O. (2002b). 'Samuel Klingenstierna and the achromatic lens. Part II. Correction of spherical and chromatic aberration', *Journal of the Antique Telescope Society*, No. 23(1), pp. 3–8.

Bellot, H.H. (1929). *University College, London: 1826–1926* (London: London University Press).

Bennett, J.A. (1998). 'Peter Dollond answers Jesse Ramsden', [being an analysis of item 22 in the papers of the Dollond family (MHSO: MSS Blundell)], *Sphæra*, No. 8 (Autumn), pp. 4–5.

Bennett, J. (2006). 'Catadioptrics and commerce in eighteenth-century London', *History of Science*, 44(2), pp. 247–277.

Bernoulli, J. (1771). *Lettres astronomiques: où l'on donne un idée de l'état actuel de l'astronomie practique dans plusieurs villes de l'Europe* (Berlin, chez l'auteur), Lettre cinquième, pp. 65–76.

Bidgood, R. (1981). 'Household and personal possessions in Radnorshire Wills of the 16th, 17th and 18th centuries', *Radnorshire Society Transactions*, 51, pp. 16–28.

Birch, T. (1756–1757). *The History of the Royal Society of London* (London: Royal Society).

Black, J. (1991). *The English Press in the Eighteenth Century* (Aldershot: Gregg Revivals).

Blackstone, H. (1827). *Reports of Cases Argued and Determined in the Courts of Common Pleas and Exchequer Chamber* (London: Joseph Butterworth, 4th edn).

Blunden, E. (1923). *Christs' Hospital* (London: Christophers).

Boas-Hall, M. (1965). *Robert Boyle on Natural Philosophy: An Essay with Selections from his Writings* (Bloomington, IN: Indiana University Press).

Bolitho, H. and Peel, D. (1967). *The Drummonds of Charing Cross* (London: Geo. Allen & Unwin).

Boswell, J. (1953). *The Life of Samuel Johnson* (London: Oxford University Press, reprint of original 1904 edition).

Brewer, J. and Porter, R. (Eds) (1993). *Consumption and the World of Goods* (London: Routledge).

Brewster, D. (1830, reprinted 1999). *The Edinburgh Encyclopaedia; with the Assistance of Gentlemen Eminent in Science and Literature* (Edinburgh: W. Blackwood).

Brewster, D. (1831). *A Treatise on Optics* (London: Longman, Rees, Orme, Brown & Green [Cabinet Cyclopædia Series]).

Brewster, D. (1855). *Memoirs of the Life, Writings and Discoveries of Sir Isaac Newton*, 2 vols (Edinburgh and London: Thomas Constable and Co., Hamilton, Adams and Co.), vol. 1.

Bristow, H.R. (1993). 'Elliott, Instrument Makers of London. Products, customers and developments in the 19th century', *Bulletin of the Scientific Instrument Society*, No. 36, pp. 8–11.

Bristow, H.R. (2005). 'Carl Ludwig Christian Becker: instrument maker and manager for Elliott Brothers', *Bulletin of the Scientific Instrument Society*, No. 84, pp. 18–19.

Bristow, H.R. (2006). 'The take-over of Watkins & Hill in 1857', *Bulletin of the Scientific Instrument Society*, No. 88, pp. 71–76.

Brock, W.H. (1997). *Justus von Liebig, the Chemical Gatekeeper* (Cambridge: Cambridge University Press).

Brock, W.H. and Meadows, A.J. (1984). *The Lamp of Learning: Taylor and Francis and the Development of Science Publishing* (London: Taylor & Francis Ltd.).

Bromley, J. and Child, H. (1960). *The Armorial Bearings of the Guilds of London: A Record of the Heraldry of the Surviving Companies with Historical Notes* (London: Frederick Warne and Co.).

Brown, J. (1979). *Mathematical Instrument-Makers in the Grocers' Company 1688–1800: with notes on some earlier makers* (London: Science Museum).

Browne, W. (Ed.) (1715). *Dr Gregory's Elements of Catoptrics and Dioptrics Translated from the Latin Original with a Large Supplement ...* (London, printed for E. Curll).

Brück, M.T. (2002). *Agnes Mary Clerke and the Rise of Astrophysics* (Cambridge: Cambridge University Press).

Bruijn, J.G. de (1979). 'Van Breda's Instrumentarium', in A.S.H. Breure and J.G. de Bruijn (Eds), *Leven en werken van J.G.S. van Breda (1788–1867)* (Haarlem: Hollandsche Maatschappij der Wettenschappen), pp. 131–145.

Brumwell, S. and Speck, W.A. (2002). *Cassell's Companion to Eighteenth Century Britain* (London: Cassell & Co.).

Bryden, D.J. (1992). 'Evidence from advertising for mathematical instrument making in London, 1556–1714', *Annals of Science*, 49(4), pp. 301–336.

Bryden, D.J. (2003). '"To astonish the ignorant, impose upon the Credulous and amuse the Publick" – A 1707 advertising skirmish between London opticians', *Bulletin of the Scientific Instrument Society*, No. 77, pp. 14–21.

Bryden, D.J. and Simms, D.L. (1992). 'Archimedes and the opticians of London', *Bulletin of the Scientific Instrument Society*, No. 35, pp. 11–13.

Buchanan, P. and Gee, B. (2004). 'Inside the shop of an eighteenth century optician', *Bulletin of the Scientific Instrument Society*, No. 82, pp. 10–15.

Buchanan, P. and Hunt, L.B. (1984). 'Richard Knight (1768–1844): a forgotten chemist and apparatus designer', *Ambix*, 31(2), pp. 57–67.

Buckley, F. (1935). 'Old English Glass: Optical Glasses' *Glass*, 12, pp. 427–8.

Burn, R. (1785). *The Justice of the Peace, and Parish Officer*, 4 vols (London: T. Cadell in the Strand, 15th edn).

Butterworth, H. (1982). 'The DSA examination system', in: MacLeod (Ed.) (1982), pp. 27–44.

Calvert, H.R. (1971). *Scientific Trade Cards in the Science Museum Collection* (London: HMSO).

Campbell, R. (1747). *The London Tradesman, being a compendious view of all the trades, professions, arts, both liberal and mechanic, now practised in the Cities of London and Westminster* (London: T. Gardner; repr. Newton Abbott: David and Charles, 1969).
Caneva, K.L. (1980). 'Ampère, the Etherians, and the Oersted Connexion', *British Journal for the History of Science*, 13(2), pp. 121–138.
Carpmael, W. (1843). *Law Reports of Patent Cases, vol. 1* (London: A. Macintosh), pp. 28–30.
Carpmael, W. (1851). *Law Reports of Patent Cases, vol. 2* (London: A. Macintosh), pp. 117–155.
Cassels, J.W.S. (1979). 'The Spitalfields Mathematical Society', *Bulletin of the London Mathematical Society*, 11(3), pp. 241–258.
Cassels, J.W.S. (1980). 'Addendum: "The Spitalfields Mathematical Society"', *Bulletin of the London Mathematical Society*, 12(5), p. 343.
Cawthorne, H.C. (1929). 'The Spitalfields Mathematical Society (1717–1845)', *Journal of Adult Education*, 3(2), pp. 158–166.
Chambers, N. (2007). *The Scientific Correspondence of Sir Joseph Banks, 1765–1820*, 6 vols (London: Pickering and Chatto).
Champness, R.F. (1968). *A Short History of the Worshipful Company of Spectacle Makers up to the Beginning of the 20th Century* (London: The Company).
Chew, K. and Wilson, A. (1993). *Victorian Science and Engineering Portrayed in the Illustrated London News* (London: Science Museum/Alan Sutton).
Clairaut, A.C. (1762).'Mémoire sur les moyens de perfectionner les lunettes d'approche, par l'usage d'objectifs composés de plusieurs materières différemment réfringentes,' *Histoire de l'Académie Royale des Sciences, année 1756* (Paris: De l'imprimerie royale), pp. 380–437, esp. the section headed 'Consideratio circa legem refractionis Newtonianam radiorum luminis diversi generis in mediis diversis', pp. 405–407 [see below, under Klingenstierna].
Clarke, C. (1773). *A Philosophical Investigation of the Origin, Vicissitude, and Power of Steam Employed in Fire-engines ...* (Newcastle: Printed and sold by T. Slack for the author).
Clarke, E.M. (1836–1837). 'Description of a magneto-electric machine', [Sturgeon's] *Annals of Electricity*, 1(2), pp. 145–155.
Clarke, E.M. (1838). 'A description of a magneto-electric machine', *Silliman's Journal*, 33(2), pp. 213–224.
Clay, R.S. and Court, T.H. (1932). *The History of the Microscope, compiled from original instruments and documents, up to the introduction of the achromatic microscope* (London: Chas. Griffin, repr. Holland Press, 1975).
Clerke, A.M. (1885). *A Popular History of Astronomy during the Nineteenth Century* (Edinburgh: A. & C. Black, 1885, 4th edn 1902).

Clerke, A.M. (1890). 'Hall, Chester Moor', in: Leslie Stephens and Sidney Lee (Eds), *Dictionary of National Biography* 63 vols (London: Smith, Elder & Co., 1885–1900), 24, p. 62.

Clifton, G. (1993a). 'The Spectaclemakers' Company and the origins of the optical instrument-making trade in London', in: Anderson, R.G.W., Bennett, J.A. and Ryan, W.F. (Eds), *Making Instruments Count: Essays on Historical Scientific Instruments Presented to Gerard l'Estrange Turner* (Aldershot: Variorum), pp. 341–364.

Clifton, G. (1993b). 'An introduction to the history of Elliott Brothers up to 1900', *Bulletin of the Scientific Instrument Society*, No. 36, pp. 2–7.

Clifton, G. (1995). [General Ed. G. l'E. Turner], *Directory of British Scientific Instrument Makers 1550–1851* (London: Zwemmer in association with the National Maritime Museum).

Clifton, G. (1997). 'Instrument-making around the Exchange', in: A. Saunders (Ed.) *The Royal Exchange* (London: The London Topographical Society, Publication 152), pp. 230–238.

Clifton, G. (2006). 'Instrument makers in the London Livery Companies', *Bulletin of the Scientific Instrument Society*, No. 88, pp. 36–40.

Clifton, G. and Gee, B. (1997). 'John Newman: a further note', *Bulletin of the Scientific Instrument Society*, No. 53, p. 14.

Cohen, I.B. and Westfall, R.S. (Eds) (1995). *Newton: Texts and Background Commentaries* (London and New York: W. Norton & Co.).

Collins, J. (1710). *The Description and Use of Four Several Quadrants:two great ones and two small ones ...invented and written by ... John Collins, and engrav'd by the curious hand of Mr. Henry Sutton; with additions, not in the former treatise, by John Good* (London: printed for Richard and William Mount, and Tho. Page).

Collyer, J. (1761). *The Parent's and Guardian's Directory, and the Youth's Guide, in the choice of a profession or trade ...* (London: Printed for R. Griffiths in the Strand).

Colyer-Fergusson, T.C. (Ed.) (1906). T*he Registers of the French Church, Threadneedle Street, London*, vol. 3 (London: Publications of the Huguenot Society of London, vol. 16).

Corporation of London Record Office (1996). *Research Guide 1: City Freedom Archives* (London: Corporation of London Record Office).

Court, T.H. and Rohr, M. von (1928–1929a). 'On the development of spectacles in London from the end of the seventeenth century', *Transactions of the Optical Society*, 30(1), pp. 1–21.

Court, T.H. and Rohr, M. von (1928–1929b). 'A history of the development of the telescope from about 1675 to 1830', *Transactions of the Optical Society*, 30(5), pp. 207–260.

Court, T.H. and Rohr, M. von (1930). 'Contributions to the history of the Worshipful Company of Spectacle Makers' [with a chronological list of the more important members, including leading opticians], *Transactions of the Optical Society*, 31(2), pp. 53–90.

Crawford, A. (1977). *A History of the Vintners' Company* (London: Constable).

Crawforth, M.A. (1985). 'Evidence from trade cards for the scientific instrument industry', *Annals of Science*, 42(5), pp. 453–554.

Crawforth, M.A. (1987). 'Instrument makers in the London Guilds', *Annals of Science*, 44(4), pp. 319–377.

Crosland, M. (2007). *Scientific Institutions in France and Britain, c.1700–c.1870* (Aldershot: Ashgate).

Cunningham, G.H. (1927) *London: being a comprehensive survey of the history, tradition and historical associations of buildings and monuments arranged under streets in alphabetical order* (London: Dent).

Daumas, M. (1989). *Scientific Instruments of the Seventeenth and Eighteenth Centuries and Their Makers* (London: Portman Books).

Davies, E. (1905). *A General History of the county of Radnor ... compiled from the Manuscript of the late Rev. Jonathan Williams and other Sources* (Brecknock: Davies & Co.).

Dawes, H. (1996). 'John Frederick Newman (1784–1860): "The ingenious instrument maker, Mr Newman of Lisle Street"', *Bulletin of the Scientific Instrument Society*, No. 50, pp. 11–14.

Death notice (1791). 'Francis Watkins', *Gentleman's Magazine and Historical Chronicle*, 61, p. 1070.

Death notice (1798). 'Walter Watkins', *Gentleman's Magazine and Historical Chronicle*, 68, p. 907.

Death notice (1810). 'Jeremiah Watkins', *Gentleman's Magazine and Historical Chronicle*, 80, p. 186.

De Clercq, P. (1997). *At the Sign of the Oriental Lamp: the Musschenbroek workshop in Leiden, 1660–1750* (Rotterdam: Erasmus Publishing).

De Clercq, P. (2001). 'The papers of the instrument dealer, collector and researcher Thomas Henry Court (1868–1951)', in: Turner, A. (Ed.) (2001), *The Archives of Scholars, Collectors and Dealers: Their Place in the Study of the History of Scientific Instruments. Papers from the XIXth Symposium of the Scientific Instrument Commission, Nuncius. Annali di Storia della Scienza*, XVI, pp. 723–731.

Deadman, H. and E. Scudder (Eds) (1994). *An Introductory Guide to the Corporation of London Records Office* (London: Corporation of London).

Defoe, D. (1726). *The Complete English Tradesman* (London: Printed for Charles Rivington at the Bible and Crown in St Paul's Churchyard).

Desaguliers, J.T. (1717). *Physico-Mechanical Lectures or, an account of what is explain'd and demonstrated in the course of mechanical and experimental philosophy* ... (London: Printed for the author, and sold by him at his house in Channel-Row, Westminster; by Richard Bridger, at the upper End of Hind Court, Fleet-street; and William Vream, within two doors of the Royal Oak in Earl-street near the Seven Dials).

Desaguliers, J.T. (1725, 1734, 1745). *A Course of Mechanical and Experimental Philosophy* (Adorn'd with Thirty-two Copper-plates) (London: n.d. in English and French).

Dollond, J. (1753–1754a). 'A letter from Mr. John Dollond to Mr. James Short, F.R.S. concerning an improvement of refracting telescopes', [dated Vine Court 21 Feb. 1753; read 1 March 1753], *Phil. Trans. R. Soc. Lond.* 48, Part 1, pp. 103–107. [RS, L&P II 421].

Dollond, J. (1753–1754b). 'A new contrivance for measuring small angles' [read 10 May 1753], *Phil. Trans. R. Soc. Lond.* 48, Part 1, pp. 178–181. [RS, L&P II 452].

Dollond, J. (1753–1754c). 'A letter from Mr. John Dollond to James Short, A.M.,F.R.S., concerning a mistake in M. Euler's theorem for correcting the aberrations in the object glasses of refracting telescopes' [dated 11 March 1752; read 23 November 1752], *Phil. Trans. R. Soc. Lond.* 48, (1753–54), Part I, pp. 289–291. [RS, L&P II 437]. Reprinted in: Kelly (1804), p. 23. See also: Speiser, A., Trost, E. and Blanc, C. (Eds) (1962), pp. 38–40.

Dollond, J. (1755). 'An explanation of an instrument for measuring small angles', [dated Denmark Court 4 April 1754; read 25 April 1754], *Phil. Trans. R. Soc. Lond.* 48, Part 2, pp. 551–564.

Dollond, J. (1757). 'An account of some experiments concerning the different refrangibility of light ... with a letter from James Short, MA, FRS, Acad. Reg. Suec. Soc. To the Rev. Dr Birch, Secret. R.S.', [Read June 8, 1758] *Phil. Trans. R. Soc. Lond.* 50, Part 2, pp. 733–743 [RS, L&P III 324].

Dollond, P. (1765). 'An account of an improvement made by Mr. Peter Dollond in his new telescopes; in a letter to Mr James Short, MA, FRS; with a letter of Mr Short's to the Rev Thomas Birch, DD, Secret. R.S.', *Phil. Trans. R. Soc. Lond.*, pp. 54–56.

Dollond, P. (1789). *Some account of the discovery made by ... J. Dollond ... which led to the grand improvement of refracting telescopes ... with an attempt to account for the mistake in an experiment made by Sir Isaac Newton; on which experiment, the improvement ... entirely depended* (London: Printed for J. Johnson in St Paul's Churchyard). Reprinted in Kelly (1808), pp. 61–77.

Elliott and Sons (c.1851). *William Elliott and Sons, List of optical, mathematical and philosophical instruments* (London: William Elliott and Sons, 56 Strand). In J.F. Heather (1851).

Elliott Brothers (1856a). *Descriptive catalogue of electrical instruments and apparatus, manufactured by Elliott Brothers, successors to Watkins and Hill, 5 Charing Cross and 56 Strand, London.*

Elliott Brothers (1856b). *Descriptive catalogue of voltaic and thermo-electric instruments and apparatus, manufactured by Elliott Brothers, successors to Watkins and Hill, 5 Charing Cross and 56 Strand, London.*

Elliott Brothers (1856c). *Hydrostatic, hydraulic, pneumatic, and acoustic instruments, manufactured by Elliott Brothers, successors to Watkins and Hill, 5 Charing Cross and 56 Strand, London.*

Elliott Brothers (1856d). *Descriptive catalogue of optical instruments, manufactured by Elliott Brothers, successors to Watkins and Hill, 5 Charing Cross and 56 Strand, London.*

Elliott Brothers (1856e). *Descriptive catalogue of magnetic, electro-magnetic, electro-dynamic and magneto-electric inductive instruments and apparatus, manufactured by Elliott Brothers, successors to Watkins and Hill, 5 Charing Cross and 56 Strand, London.*

Euler, L. (1747–52). (i) 'Sur la perfection des verres objectifs des lunettes', pp. 1–21; (ii) (Dollond's paper on Euler from Phil.Trans.), pp. 38–40; (iii) 'A Monsieur Dollond,' (Open letter of 1752 in *Phil. Trans.* 48 (1753), pp. 293–296), pp. 41–43. In A. Speiser, E. Trost, and C. Blanc (Eds), (1962).

Euler, L. (1749). 'Sur la perfection des verres objectifs des lunettes', *Mémoires de l'Academie des Sciences de Berlin*, 3, pp. 274–296. See Spieser, Trost and Blanc (Eds), (1962), (ser. 3), 6, pp. 1–21.

Euler, L. (1753–1754a). 'Mr Euler's letter to Mr James Short, F.R.S., [dated Berlin 19 June 1752; read 8 July 1753], *Phil. Trans. R. Soc. Lond.* 48, Part I, p. 292. Ms RS, L&P II 375.

Euler, L. (1753–1754b). Letter from L. Euler: 'A Monsieur Monsieur Dollond' [dated Berlin 16 June 1752; read 8 July 1753], *Phil. Trans. R. Soc. Lond.* 48, Part I, pp. 293–296. See also: Spieser, Trost and Blanc (1962), vol. 6, pp. 41–43.

Euler, L. (1769). 'Considérations sur les nouvelles lunettes d'Angleterre de Mr.Dollond et sur le principe qui en est le fondement' [read 16 Sept. 1762], *Mém. de l'Academie des Sciences*, Berlin, 18 (1769), pp. 226–248. See also: Spieser, Trost and Blanc (1962), vol. 8, pp. 101–121.

Evans, W., Ross, C. and Werner, A. (1995). *Whitefriars Glass. James Powell & Son of London* (London: Museum of London).

[F.W.] (1834). 'An esteemed correspondent', *Mechanics' Magazine*, 21, p. 96.

Farey, J. [prepared by] (1829) Select Committee of 1829 in: *British Parliamentary Papers*, 1829 (332) III. [Containing cases at law respecting patent rights.]

Fauque, D. (1985). 'Alexis-Marie Rochon (1741–1817), savant astronome et opticien', *Revue d'Histoire des Sciences*, 38(1), pp. 3–36.

Ferguson, J. (1768). *The Young Gentleman and Lady's Astronomy: familiarly explained in ten dialogues between Neander and Eudosia* (London: printed for A. Millar and T. Cadell).

Ferguson, J. (1823). *Lectures on Select Subjects –; with notes, and an additional volume containing the most recent discoveries in the arts and sciences by David Brewster* (Edinburgh: Stirling & Slade, and Bell & Bradfute, 3rd edn), 2 vols.

FitzHugh, T.V.H. (1998). [rev. S. Lumas for Society of Genealogists] *The Dictionary of Genealogy* (London: A. & C. Black).

Fitzroy, A. (1928). *The History of the Privy Council* (London: John Murray).

Foster, J. (2000). *Alumni Oxoniensis: the members of the University of Oxford 1500–[1886]: their parentage, birthplace, and year of birth, with a record of their degrees: being the Matriculation Record of the University, Part 2: 1715–1886*, vol. 4 (Bristol: Thoemmes), p. 1508.

Fuss, N.I. (1783). *Éloge de monsieur Léonard Euler: lu á l'Académie Impériale des Sciences, dans son assemblée de 23 octobre 1783; avec une liste complette des ouvrages de m. Euler* (St Petersburg: [n.p.]).

Gee, B. (1969). 'Georg Simon Ohm', *Physics Education*, 4, pp. 106–113.

Gee, B. (1970). 'André Marie Ampère (1775–1836)', *Physics Education*, 5, pp. 359–369.

Gee, B. (1983). 'Through a hole in a window shutter', *Physics Education*, 18, pp. 93–97, 140–143.

Gee, B. (1984–1986). 'John Ross Griffin', in D. Jeremy (Ed.), *Dictionary of Business Biography*, 6 vols (London: Butterworth), 2, pp. 664–665.

Gee, B. (1989). 'Amusement chests and portable laboratories: Practical alternatives to the regular laboratory', in James (1989), pp. 37–59.

Gee, B. (1990a). 'Joseph Henry's trade with instrument makers in London and Paris', *Bulletin of the Scientific Instrument Society*, No. 25, pp. 19–24.

Gee, B. (1990b). 'Faraday's plight and the origins of the magneto-electric spark', *Nuncius. Annali di Storia della Scienza*, V, pp. 43–69.

Gee, B. (1990c). 'Instrument makers in physics history', in John Roche (Ed.), *Physicists Look Back* (Bristol: Adam Hilger), pp. 202–222.

Gee, B. (1991). 'Electromagnetic engines. Pre-technology and development immediately following Faraday's discovery of electromagnetic rotations', *History of Technology*, 13, pp. 41–72.

Gee, B. (1992). 'The Newtons of Chancery Lane and Fleet Street revisited. Part I: A question of establishment', *Bulletin of the Scientific Instrument Society*, No. 35, pp. 3–14.

Gee, B. (1993a). 'The Newtons of Chancery Lane and Fleet Street revisited. Part II: The Fleet Street business and other genealogy', *Bulletin of the Scientific Instrument Society*, No. 36, pp. 12–14.

Gee, B. (1993b). 'The early development of the magneto-electric machine', *Annals of Science*, 50(2), pp. 101–133.

Gee, B. (1996). 'John Newman: A second look', *Bulletin of the Scientific Instrument Society*, No. 51, pp. 22–25.

Gee, B. (1998a). 'The spectacle of science and engineering in the metropolis. Part I: E.M. Clarke and the early West End exhibitions', *Bulletin of the Scientific Instrument Society*, No. 58, pp. 11–18.

Gee, B. (1998b). 'The spectacle of science and engineering in the metropolis. Part II: E.M. Clarke and the Royal Panopticon of Science and Art', *Bulletin of the Scientific Instrument Society*, No. 59, pp. 6–13.

Gee, B. and Brock, W.H. (1991). 'The case of John Joseph Griffin, from artisan-chemist and author-instructor to business-leader', *Ambix*, 38(1), pp. 29–61.

Gee, B. and Clackson, S.G. (1992). 'The origin of practical work in the English school science curriculum', *School Science Review*, 73(265), pp. 79–83.

Gee, B. and Hackmann, W. (1998). 'Icons and heroes of an electrical age', *Bulletin of the Scientific Instrument Society*, No. 59, pp. 1–2.

George, M.D. (1965, reprint 1984). *London Life in the Eighteenth Century* (Harmondsworth: Penguin).

Gerhold, D. (1994). *Courts of Equity. A guide to chancery and other legal records* (Isle of Wight: Pinhorns).

Gibson, J. (Ed.) (2001). *Bishops' Transcripts and Marriage Licences, Bonds and Allegations* (Bury, Lancs.: Federation of Family History Societies, 5th edn).

Gibson, J. and Churchill, E. (2002). *Probate Jurisdictions: Where to look for a Will* (Bury, Lancs.: Federation of Family History Societies, 5th edn).

Glaisher, J. (1852). 'Philosophical instruments and progress, as represented at the Great Exhibition', in *Lectures on the Results of the Great Exhibition of 1851*, delivered before the Society of Arts, Manufactures and Commerce (London: David Bogue), pp. 321–402.

Goldsmith, M. (1947). *Soho Square* (London: Sampson Low, Marston & Co.).

Gomme, A.A. (1946). *Patents of Invention, Origin and Growth of the Patent System in Britain* (London: Longman, Green & Co.).

Good, J. (1706). *Multum in parvo: or tables exactly calculated for the latitude of London ... [Universall Dialling]* (London: printed for, and sold by Richard Mount, and Company).

Goodison, N. (1977). *English Barometers, 1680–1860: a history of domestic barometers and their makers and retailers* (Woodbridge: Antique Collectors' Club).

Grant, R. (1852). *A History of Physical Astronomy, from the Earliest Ages to the Middle of the Nineteenth Century* (Repr. New York: Johnson, 1966).

Gray, S. (1731–1732). 'A letter to Cromwell Mortimer, MD, Secretary of the Royal Society, containing several Experiments concerning Electricity', *Phil. Trans. R. Soc. Lond.*, 37(417–426), pp. 18–44.

's Gravesande, W.J. (1726). *Mathematical Elements of Natural Philosophy confirmed by Experiment, or an Introduction to Sir Isaac Newton's Philosophy*, trans. J.T. Desaguliers (London: printed for J. Sennex, J. Innys, ... T. Longman, 2nd edn, 1726).

Great Britain. Commissioners for the Exhibition of 1851 (1852). *Report of the Commissioners*, 2 vols (London: Spicer Brothers; W. Clowes & Sons).

Gregory, D. (1684). *Catoptricae et dioptricae sphaericae elementa* (Oxonii: E Theatro Sheldoniano).

Gregory, D. (1994). *Radnorshire: a historical guide* (Llanrwst: Gwasg Carreg Gwalch).

Griffin, J.J. (1849). *Chemical Recreations: a popular compendium of experimental chemistry, for the use of beginners* (London: John J. Griffin, 9th edn).

Grimwade, A. (1982). *London Goldsmiths 1697–1897: their marks and lives, from the original registers at Goldsmiths' Hall and other sources* (London: Faber).

Grove, W.R. (1853). [Note on improving refracting telescopes]. *Monthly Notices of the Royal Astronomical Society*, 13, pp. 80–81.

Gwynn, R.D. (1970–1976). 'The distribution of the Huguenot refugees in England, II: London and its environs', *Proceedings of the Huguenot Society of London*, 22(6), pp. 509–568.

Gwynn, R.D. (2001). *Huguenot Heritage: the history and contribution of the Huguenots in Britain* (Brighton: Sussex Academic Press). Revised reprint of 1985 (Routledge) edition.

Haag, E. and Haag, E. (1846–1859). *La France protestante: ou, Vies des protestants français qui se sont fait un nom dans l'histoire depuis les premiers temps de la réformation jusqu'à la reconnaissance du principe de la liberté des cultes par l'Assemblée nationale; ouvrage précéde d'une notice historique sur le protestantisme en France ...* 10 vols. (Paris; Genève: J. Cherbuliez).

Hackmann, W.D. (1978). *Electricity from Glass, the history of the frictional electrical machine, 1600–1850* (Alpen aan den Rijn: Sijthoff & Noordhoft).

Hackmann, W. (2003). 'Sir William Snow Harris's Plate Electrical Machine by Watkins & Hill – the End of the Line', *Bulletin of the Scientific Instrument Society*, No. 77, pp. 26–31.

Hakfoort, C. (1995). *Optics in the Age of Euler: conceptions of the nature of light, 1700–1795* (Cambridge: Cambridge University Press).

Hall, A.R. (1993). *All was Light: an introduction to Newton's Opticks* (Oxford: Clarendon Press).

Hall, A.R. and Simpson, A.D.C. (1996). 'An account of the Royal Society's Newton Telescope', *Notes and Records of the Royal Society of London*, 50(1), pp. 1–11.

Hans, N. (1998). *New Trends in Education in the Eighteenth Century* (London: Routledge). Originally published in London in 1951 by Routledge and Kegan Paul.

Harben, H.A. (1918). *A Dictionary of London: being notes topographical and historical relating to the streets and principal buildings in the city of London* (London: Jenkins).

Harte, N. and North, J. (1991). *The World of UCL, 1828–1990* (London: University College London).

Hartley, H. (1960). *The Royal Society: its origins and founders* (London: The Royal Society).

Hayward, P.A. (1987). *Hayward's Patent Cases, 1600–1883: a compilation of the English patent cases for those years*, 11 vols (Abingdon: Professional Books).

Heal, A. (1935). *The London Goldsmiths 1200–1800: a record of the names and addresses of the craftsmen, their shop-signs and trade-cards* (Cambridge: Cambridge University Press; facsimile edition, Newton Abbott: David and Charles).

Heal, A. (1947). *The Signboards of Old London Shops* (London: Batsford).

Heather, J.F. (1851). *A Treatise on Mathematical Instruments: including most of the instruments employed in drawing, for assisting the vision, in surveying and levelling, in practical astronomy, and for measuring the angles of crystals: in which their construction, and the methods of testing, adjusting and using them, are concisely explained* (London: John Weale).

Heilbron, J.L. (1982). *Elements of Early Modern Physics* (Los Angeles: University of California Press).

Helden, A. van (1975). 'The historical problem of the invention of the telescope', *History of Science*, 13(4), pp. 251–263.

Hellman, C.D. (1932). 'John Bird (1709–1776), mathematical instrument maker', *Isis*, 17(1), pp. 127–153.

Hendry, J. (1980). 'Newton's theory of colour', *Centaurus*, 23(3), pp. 230–251.

Herschel, J.F.W. (1821). 'On the aberrations of compound lenses and object-glasses', *Phil. Trans. R. Soc. Lond.*, 111, pp. 222–267.

Hewish, J. (1980). *The indefatigable Mr Woodcroft. The legacy of invention* (London: British Library Reference Division and Science Reference Library).

Hewish, J. (2000). *Rooms near Chancery Lane. The patent office under the commissioners, 1852–1883* (London: The British Library).

Heymann, H.J. (1952). 'Samuel Klingenstierna (1698–1765)', in: S. Lindroth (Ed.) *Swedish Men of Science (1650–1950)* (Stockholm: Swedish Institute/Almqvist & Wiksell), pp. 59–65.

Hills, R. L. (2002). *James Watt: his time in Scotland 1736–1774* (Ashbourne: Landmark).

Hobhouse, H. (2002). *The Crystal Palace and the Great Exhibition. Art, science and productive industry. A history of the Royal Commission for the Exhibition of 1851* (London: Athlone Press).

Hodgson, R. (1864). 'The achromatic object-glass', *Monthly Notices of the Royal Astronomical Society*, 24, pp. 195–196.

Holbrook, M. [with additions and revisions by R.G.W. Anderson and D.J. Bryden] (1992). *Science Preserved. A directory of scientific instruments in collections in the United Kingdom and Eire* (London: HMSO).

Holden-Macmichael, J. (1906). *The Story of Charing Cross and its Immediate Neighbourhood* (London: Chatto & Windus).

Horwitz, H. (1995). *Chancery Equity Records and Proceedings, 1600–1800* (London: PRO Handbook No. 27, HMSO).

Howse, D. (1975). *Greenwich Observatory*. Vol. 3: *The Buildings and Instruments* (London: Taylor & Francis)

Howse, D. (1989a). *Nevil Maskelyne: the Seaman's Astronomer* (Cambridge: Cambridge University Press).

Howse, D. (1989b). 'Prices of Lalande's astronomical instruments', *Bulletin of the Scientific Instrument Society*, No. 21, pp. 9–10.

Hulme, E.W. (1896). 'The history of the patent system under the prerogative and at common law', *Law Quarterly Review*, 12(46), pp. 141–154.

Hulme, E.W. (1898). *English Patent Law, its History, Literature, and Library* (London: Library Bureau). Reprinted in *The Library*, 10 (February–April 1898), pp. 42–45.

Hulme, E.W. (1900). 'The history of the patent system under the prerogative and at common law: a sequel', *Law Quarterly Review*, 16(55), pp. 44–56.

Hulme, E.W. (1902). 'On the history of the patent law in the seventeenth and eighteenth centuries', *Law Quarterly Review*, 18, pp. 280–288.

Hulme, E.W. (1917). 'Privy Council law and practice of letters patent for invention from the Restoration to 1794', *Law Quarterly Review*, 33(1), pp. 63–75 and pp. 181–195.

Humphreys, A.L. (1914). 'History of glass-making and the glass trade', *Notes and Queries*, ser. 11, ix, pp. 312–313.

Hunter, M. and Schaffer, S. (Eds) (1989). *Robert Hooke: new studies* (Woodbridge, Suffolk: Boydell Press).

Hutchinson, K. (1981). 'Euler, Dollond, Klingenstierna and the refutation of Newtons stance on chromatic aberration', (typescript) in: J.N. Crossley (Ed.), *Proceedings of the First Australian Conference on the History of Mathematics* (Clayton, Victoria, Australia: Monash University Department of Mathematics), pp. 136–144.

Hutchinson, K. (1991). 'Idiosyncrasy, achromatic lenses, and early Romanticism', *Centaurus*, 34(2), pp. 125–171.

Hyde, R. (1982). [Notes in] *The A to Z of Georgian London* (London: London Topographical Society, Publication No. 126).

Inkster, I. and Morrell, J. (Eds) (1983). *Metropolis and Provinces: Science in British Culture 1790–1850* (London: Hutchinson).

International Exhibition of 1862 (1862). *The Illustrated Catalogue of the Industrial Department: British Division*, vol.11 (London: printed for Her Majesty's Commissioners).

Ilardi, V. (2007). *Renaissance Vision from Spectacles to Telescopes* (Philadelphia, PA: Memoirs of the American Philosophical Society, vol. 259).

Inwood, S. (1998). *A History of London* (London: Macmillan).

Jackson, M.W. (2000). *Spectrum of Belief: Fraunhofer and the craft of precision optics* (Cambridge, MA: MIT Press).

Jaecks, D.H. (1986). 'Developments in 18th century optics and early instrumentation', in: Stock and Orna (1986), pp. 51–65.

Jaecks, D.H. (2010). 'An investigation of the eighteenth century achromatic telescope', *Annals of Science*, 67(2), pp. 149–186.

James, F.A.J.L. (Ed.) (1989). *The Development of the Laboratory: essays on the place of experiment in industrial civilization* (Basingstoke: Macmillan).

Jenkins, E.J. (1985). '*A Magnificent Pile*'. *A centenary history of the Leeds Central High School* (Leeds: City of Leeds School).

Jenkins, F.A. and White, H.E. (1957). *Fundamentals of Optics* (New York: McGraw Hill, 3rd edn).

Jones, E. (Ed.) (2001). *The Welsh in London 1500–2000* (Cardiff: University of Wales Press on behalf of the Honourable Society of Cymmrodorian).

Jungnickel, C. and McCormmach, R. (1996). *Cavendish* (Philadelphia, PA: the American Philosophical Society, vol. 220).

Kelly, J. (1804, 1808). *The Life of John Dollond FRS, Inventor of the Achromatic Telescope* (London, 1st edn, privately printed pamphlet (8pp), repr. from *Phil. Mag.* 18 (Feb. 1804) pp. 47–52; 2nd printing R. Taylor). Also in: *Phil. Trans. R. Soc. Lond.* (Abridged), 10, (1809), p. 341. Third edition (1808) [for W.M. Thistleton, 123pp] includes reprints of letters and papers by John and Peter Dollond, Short, Euler and Maskelyne.

Kelly, T. (1957). *George Birkbeck: Pioneer of adult education* (Liverpool: Liverpool University Press).

King, H.C. (1938). 'John and Peter Dollond: the significance of their optical work', *Dioptric Review*, 40(3), pp. 177–201.

King, H.C. (1955, 1980). *The History of the Telescope* (London: Charles Griffin & Co.; repr. New York: Dover).

King, S. (1840). 'On a large achromatic object-glass of a Telescope worked by Mr. Dollond, the flint glass of which was prepared by the late Dr. Ritchie', *Monthly Notices of the Royal Astronomical Society*, 5, pp. 65–66.

King, H.C and Millburn, J.R. (1978). *Geared to the Stars: the evolution of planetariums, orreries and astronomical clocks* (Bristol: Adam Hilger).

Kiralfy, A.K.R. (1958). *Potter's Historical Introduction to English Law and its Institutions* (London: Sweet & Maxwell, 4th edn).

Klingenstierna, S. (1754). 'Anmärkning vid Brytnings-Lagen af särskilta slags Ljus-strälar, då de gå ur ett genomskinande medel in i åtskilliga andra', *Kungliga vetenskapsakademiens Handlingar [Transactions of the Royal Swedish Academy]*, 16, pp. 297–300.

Klingenstierna, S. (1755). 'Consideratio circa legem refractionis Newtonianam radiorum luminis diversi generis in mediis diversis', [Note of 18 Jan. 1755 to Dollond]. Reprinted facsimile of MS and Swedish trans. (pp. 36–39) of Nordenmark and Nordström (1938); and *Hist. Mém. Acad. R. Paris* (1756), pp. 380–437 [read 1761], in: Clairaut (1762), pp. 405–407.

Klingenstierna, S. (1759–60) 'De aberratione luminis, in superficiebus et lentibus sphæricis refractorum', (dated Stockholm, April 20, 1760, read April 2, 1761), *Phil. Trans. R. Soc. Lond.* 51, Part II, pp. 944–977. Ms RS L&P IV 62.

Lalande, J.J. [La Lande, Joseph Jérôme le Français de] (1764). *Astronomie* 2 vols (Paris: Chez Desaint & Saillant).

Lalande, J.J. [La Lande, Joseph Jérôme le Français de] (1792). *Astronomie* 3 vols (Paris: Chez la Veuve Desaint ... De l'Imprimerie de P. Didot l'aîne, 3rd edn).

Lalande, J.J. [La Lande, Joseph Jérôme le Français de] (1801). 'History of astronomy for the year 1800', *Philosophical Magazine*, 9(34), pp. 105–118.

Langford, P. (1989). *A Polite and Commercial People: England 1727–1783* (Oxford: Clarendon Press).

Laurance, L. (1920). *General and Practical Optics* (London: School of Optics, 3rd edn). [A text for the examination candidates of the Worshipful Company of Spectacle Makers].

Law, F.W. (1979). *The Worshipful Company of Spectacle Makers: a history* ([London: Worshipful Company of Spectacle Makers]).

Layton, D. (1972). 'The Educational Exhibition of 1854', *Journal of the Royal Society of Arts*, 120, pp. 156–160, 262–264, 321–323.

Layton, D. (1973). *Science for the People: the origins of the school science curriculum in England* (London: Allen and Unwin).

Lectures... (1853). *Lectures on the Results of the Great Exhibition of 1851: delivered before the Society of Arts, Manufactures, and Commerce, at the suggestion of H.R.H. Prince Albert, President of the Society*, 2 vols (London: David Brogue).

Leigh, S. (1819). *New Picture of London: or ... a brief and luminous guide to the stranger: on all subjects connected with general information, business, or amusement* (London: W. Clowes, 3rd edn).

Leight, S. and Taylor, S. (Eds) (1997). *The Livery Companies of the City of London* (London: Corporation of London).

Lettsom, W.G. (1881). 'The inventor of the achromatic telescope' (letter dated 17 August 1881) in *Astronomical Register*, 19, pp. 232–233.

Lilly, J. (1771). *A Collection of Modern Entries: or, select proceedings in the Courts of King's Bench, Common Pleas, and Exchequer ...* (London: printed by W. Strahan and M. Woodfall for J. Worrall and B. Tovey, 4th edn).
Lillywhite, B. (1972). *London Signs: a reference book of London signs from the earliest times to about the mid-nineteenth century* (London: Allen and Unwin).
Lindberg, S.G. (Ed.) (1956). *Bengt Ferrner: Resa in Europa, en astronom, industriespion och teaterhabitue genom Denmark, Tyskland, Holland, England och Italien – 1758–1762* (Uppsala: Almquist & Wiksell). ['Travels in Europe', diary of Bengt Ferrner], Lychnos-bibliotek, No. 14.
Lloyd, A. (1998). *The Chequered History of Lady Hawkins' School Kington, Herefordshire, founded 1632; with reference to the Vaughans of Hergest* (Kington: Printed for the author).
London. Royal Commission for the Great Exhibition of the works of industry of all nations (1851). *Official and Descriptive Catalogue*, 4 vols (London: Spicer Bros and W. Clowes & Sons).
London. Royal Commission for the Great Exhibition of the works of industry of all nations (1851). *Reports by the Jurors* (London: Spicer Bros and W. Clowes & Sons).
London (1851). *Medals and Honourable Mentions awarded by the International Juries* (London: G.E. Eyre & W. Spottiswood).
London (1852). *Report of the Juries on the Subjects in the Thirty Classes into Which the Exhibition was Divided* (London: W. Clowes & Sons).
London. International Exhibition (1862). *Jurors' Reports* (*Class 29. Educational Works and Appliances*) (London: Bell and Daldy).
London County Council (1935). *Survey of London, vol. 16. Charing Cross (the Parish of St. Martin-in-the-Fields, Part I)* (London: for the LCC).
London County Council (1937). *Survey of London vol. 18. The Strand (the Parish of St. Martin-in-the-Fields, Part II)* (London: for the LCC).
Lysons, D. (n.d.). Collectanea; or, a collection of advertisements and paragraphs from the newspapers, relating to various subjects ... for the collector [Cuttings from London newspapers between 1660 and 1825], vol. III.
Lysons, D. (1795). *The Environs of London: an historical account of the towns, villages and hamlets, within twelve miles of that capital*, 4 vols (London: Cadell).
MacLeod, C. (1988). *Inventing the Industrial Revolution. The English patent system, 1660–1800* (Cambridge: Cambridge University Press).
MacLeod, R.M. (Ed.) (1982). *Days of Judgement: science, examinations and the organization of knowledge in late Victorian England* (Driffield: Nafferton).
Mann, J. and Ayscough, J. (c.1747). *A Description of a Pocket Microscope with the Apparatus Thereto Belonging: adapted for viewing both opake and transparent objects. And likewise of the part called The Solar Apparatus; as greatly improved and made by* [J.M. and J.A.] (London: At the Sign of Sir Isaac Newton and Two Pair of Golden Spectacles near the West End of St Paul's, London), 9pp.

Mann, J. and Ayscough, J. (c.1747). *A Description of the Solar, or Camera-obscura Microscope, as Made and Sold by* – (London: At the Sign of Sir Isaac Newton and Two Pair of Golden Spectacles near the West End of St Paul's), 14pp.

Marriner, S. (1980). 'English bankruptcy records and statistics before 1850', *Economic History Review*, 33(3), pp. 351–366.

Martin, B. (1743). *A Course of Lectures in Natural and Experimental Philosophy, Geography and Astronomy ...* (Reading: J. Newbery & C. Micklewright, printed and sold by J. Newbery and C. Micklewright: also by Mess. Ware, Birt, Astley, Austen, Robinson, Dodsley and Needham, in London; Mr. Fletcher in Oxford; Mr. Thurlbourn in Cambridge; Mess. Ward and Chandler at York and Scarborough; Mr. Collins in Salisbury; Mr. Frederick at Bath; Mr. Craighton in Ipswich; and Mr. Wimpey in Newbury).

Martin, B. (1747). *Philosophia Britannica: or a new and comprehensive system of the Newtonian philosophy, astronomy and geography ...* (Reading: Micklewright & Co.for the author; and for M. Cooper in Pater-noster-row, London; R. Raikes at Gloucester; B. Collins at Salisbury; and J. Leake, and W. Frederick at Bath).

Martin, B. (Ed.) (1755–1764). *The General Magazine of Arts and Sciences, Philosophical, Philological, Mathematical, and Mechanical: ...* (London: for W. Owen).

Martin, B. (1759–1763). *The Young Gentleman and Lady's Philosophy ...*, 2 vols (London: W. Owen).

Martin, B. (1759, 1769). *New Elements of Optics; or, the theory of aberrations, dissipation, and colours of light: of the general and specific refractive powers and densities of mediums; the properties of single and compound lenses; and the nature, construction and use of refracting and reflecting telescopes and microscopes* (London: For the author). [In three parts: (i) telescope types, (ii) spherical aberrations and (iii) chromatic aberrations].

Maskelyne, N. (1761). 'Theorem of the aberration of the rays of light refracted through a lens, on account of the imperfection of the spherical figure', *Phil. Trans. R. Soc. Lond.*, 52, pp. 17–21.

Matthew, H.C.G. and Harrison, B. (Eds) (2004). *Oxford Dictionary of National Biography* (Oxford: Oxford University Press; updated and expanded: http:// www.oxforddnb.com/.

Maxted, I. (1977). *The London Book Trades 1775–1800. A preliminary check list of members* (Folkestone: Wm. Dawson).

Maxwell-Lyte, H. (1926). *Historical Notes on the Use of the Great Seal of England* (London: HMSO).

McConnell, A. (1994). 'From craft workshop to big business – the London scientific instrument trade's response to increasing demand 1750–1820', *London Journal*, 19(1), pp. 36–53.

McConnell, A. (2007). *Jesse Ramsden (1735–1800). London's leading scientific instrument maker* (Aldershot: Ashgate).

McKenzie, D.F. (Ed.) (1978). *Stationers' Company Apprentices 1701–1800* (Oxford: Oxford Bibliographical Society).

Medawar, P. (1963). 'Is the scientific paper a fraud?' [BBC Broadcast] in: *The Listener*, pp. 377–378.

Milbourn, T. (1888). *The Vintners' Company: their muniments, plate, and eminent members, with some account of the ward of Vintry* (London: printed for the Vintners' Company).

Millburn, J.R. (1976). *Benjamin Martin. Author, instrument-maker, and 'country showman'* (Leyden: Noordhoff).

Millburn, J.R. (1983). 'The London evening courses of Benjamin Martin and James Ferguson, eighteenth-century lecturers on experimental philosophy', *Annals of Science*, 40(5), pp. 437–455.

Millburn, J.R. (1985). 'Instrument makers of Fleet Street from Archival Sources', *Bulletin of the Scientific Instrument Society*, No. 9, pp. 8–10.

Millburn, J.R. (1986a). *Benjamin Martin: supplement* (London: Vade-Mecum Press).

Millburn, J.R. (1986b). *Retailer of the Sciences: Benjamin Martin's Scientific Instrument Catalogues, 1756–82* (London: Vade-Mecum Press).

Millburn, J.R. (1988). 'Historical patents and the instrument collector', *Bulletin of the Scientific Instrument Society*, No. 18, pp. 3–7.

Millburn, J.R. (1991). 'Instrument makers and the royal arms', *Bulletin of the Scientific Instrument Society*, No. 21, pp. 3–7 and No. 28, pp. 2–7.

Millburn, J.R. (1995a). 'Gunnery instruments and their makers in the eighteenth century', *Journal of the Ordnance Society*, 7, pp. 27–44.

Millburn, J.R. (1995b). 'The Meandering Microcosm', *Antiquarian Horology*, 22(2), pp. 160–161.

Millburn, J.R. (2000) *Adams of Fleet Street, Instrument Makers to King George III* (Aldershot: Ashgate).

Mills, A.A. (1981). 'Newton's prisms and his experiments on the spectrum', *Notes and Records of the Royal Society of London*, 36(1), pp. 13–35.

Mills, A.A. and Jones, M.L. (1989). 'Three lenses by Constantine Huygens in the possession of the Royal Society of London', *Annals of Science*, 46(2), pp. 173–182.

Mills, A.A. and Turvey, P.J. (1979). 'Newton's telescope: an examination of the reflecting telescope attributed to Sir Isaac Newton in the possession of the Royal Society', *Notes and Records of the Royal Society of London*, 33(2), pp. 135–155.

Mitchell, B.R. [in collaboration with Deane, P.] (1962), *Abstract of British Historical Statistics* (Cambridge: Cambridge University Press).

Moens, W.J.C. and Fergusson, T.C.C. (Eds) (1896–1916). 'The registers of the French Church, Threadneedle Street, 1600–1840', *Huguenot Society Publications*, 9, 13, 16, 23 (Lymington, Hants: the Society).

Moll, G. (1830). 'Electromagnetic experiments', *American Journal of Science*, 19(2), pp. 329–337.

Mollan, R.C., Davis, W. and Finucane, B. (Eds) (2002). *Irish Innovators in Science and Technology* (Dublin: The Royal Irish Academy).

Montucla, J.-É. (1802). *Histoire des Mathematiques, dans laquelle on rend compte de leurs progrès ... jusqu'à nos jours ...* (4 vols, 1799–1802 or years VII–X of the new republic) (Paris: Henri Agasse), vol 3, pp. 448–449 and note 3, p. 448. [This edition was completed and edited by J.J. Le Français de Lalande.]

Moore, K. with additions by M. Sampson (1995). *A Guide to the Archives and Manuscripts of the Royal Society* (London: The Royal Society).

Morrison-Low, A.D. (2007). *Making Scientific Instruments in the Industrial Revolution* (Aldershot: Ashgate).

Mortimer, T. (1763). *The Universal Director: or, the nobleman and gentleman's true guide to the masters and professors of the liberal arts and sciences ...* (London: printed for J. Coote).

Morton, A.Q. and Wess, J.A. (1993). *Public and Private Science. The King George III Collection* (Oxford: Oxford University Press in association with the Science Museum, London).

Morus, I.R. (2005). *When Physics Became King* (Chicago: University of Chicago Press).

Moseley, H. (1853). 'Special report on grants to aid in the purchase of apparatus for instruction in science', *Minutes of the Committee of Council on Education* (London: George E. Eyre and William Spottiswoode), pp. 154–214.

Murdoch, T. (1985). *The Quiet Conquest. The Huguenots 1685–1985.* Museum of London exhibition in association with the Huguenot Society of London, 15 May–31 October 1985 (London: Museum of London in association with A.H. Jolly (Editorial) Ltd).

Neale, J. (1747). *Directions for Gentlemen, who have Electrical Machines, How to Proceed in Making their Experiments ... With a compleat Series of Experiments, and the Management of them, as shewn in his Course.* (London, Printed for the Author, in the Year MDCCXLVII [1747]).

'New Publications' (1829). 'New Publications', *Mechanics' Magazine*, 11(289–290), pp. 29–31; pp. 45–48.

[Newbery, J.] (1761). *The Newtonian system of philosophy adapted to the capacities of young gentlemen and ladies, ... being the substance of six lectures read to the Lilliputian society, by Tom Telescope...* (London: Printed for J. Newbery).

Newton, I. (1671–1672). 'A new theory of light and colors', *Phil. Trans. R. Soc. Lond.*, 6(69–80), pp. 3075–3087.

Newton, I. (1672). 'Mr Isaac Newton's answer to some considerations upon his Doctrine of Light and Colors ...', *Phil. Trans. R. Soc. Lond.*, 7(81–91), pp. 5084–5103.

Newton, I. (1704) *Opticks: or, a treatise on reflections, refractions, inflections and colours of light* (New York: Dover, 1952, based on the 4th edn. (1730), edited by E.T. Whittaker.)

Nordenmark, N.V.R. and Nordström, J. (1938). 'Om Uppfinningen av den Akromatiska och aplanatiska Linsen', *Lychnos*, 4, pp. 1–52 [in Swedish].

Nordenmark, N.V.R. and Nordström, J. (1939). 'Om Uppfinningen av den Akromatiska Linsen', *Lychnos*, 5, pp. 313–384 [in Swedish, containing a summary in English and a reprint of Veritas (1791)].

'Notice ...' (1828). 'Notices respecting new books', *Philosophical Magazine...*, 2nd series, 4, pp. 220–222.

O'Connell, S. [with contributions by R. Porter, C. Fox and R. Hyde] (2003). *London 1753* (London: The British Museum).

Official Catalogue ... (1851). See London: Royal Commission for the Great Exhibition ... of 1851.

Owen, C. (1914a). 'History of glass-making and the glass trade', *Notes and Queries*, ser. 12, iii, p. 230.

Owen, C. (1914b). 'The invention of the achromatic lens', *Notes and Queries*, ser. 12, iii, pp. 334–336.

P[ézenas, Esprit], L[e].P[ère] (1767). *Cours complet d'optique; traduit de l'anglois de Robert Smith [1738], contentant la théorie, la pratique & les usages de cette science. Avec des additions considérables sur toutes les nouvelles découvertes qu'on a faites en cette matière depuis la publication de l'ouvrage anglois*, 2 vols (Avignon: Chez La veuve Girard & François Sequin, Jean Aubert; Se vend a Paris: chez Charles-Antoine Jombert, Charles Saillant).

Pacey, A. (1974). *The Maze of Ingenuity: ideas and idealism in the development of technology* (London: Allen Lane).

Patent Commissioners (1856). *Letters Patent and Specifications of Letters Patent, 1617–1852* (London: Printed by G.E. Eyre and W. Spottiswoode, printers to the Queen's Most Excellent Majesty).

Pedersen K.M. (2001). 'Thomas Bugge's journal of a voyage through Germany, Holland and England, 1777' in: J. Lützen (Ed.) *Around Casper Wessel and the Geometric Representation of Complex Numbers* (Copenhagen: for the Royal Danish Academy of Sciences and Letters), pp. 29–46.

Pedersen, K.M. and de Clercq, P. (2009). *Thomas Bugge's Journal of a Voyage through Germany, Holland and England in 1777* (Aarhus: Aarhus University Press).

Phillips, H. (1964). *Mid-Georgian London: a topographical and social survey of central and western London about 1750* (London: Collins).

Playfair, L. (1855). 'The study of abstract science essential to the progress of industry' in *British Eloquence: the literary, political and sacred oratory of the nineteenth century. Literary addresses.* Second series (London and Glasgow: R. Griffin & Co.).

Plumley, N. (1976). 'The Royal Mathematical School within Christ's Hospital', *Vistas in Astronomy*, 20(1), pp. 51–59.

Porter, R. (1994). *London. A social history* (London: Hamish Hamilton).

Priestley, J. (1767). *A History of the Present State of Electricity: with original experiments* (London: printed for J. Dodsley in Pall-Mall; J. Johnson and B. Davenport in Pater-noster Row, and T. Cadell (successor to Mr Millar) in the Strand).

Prosser, R.B. (1917). 'The invention of the achromatic lens', *The Observatory*, 40(516), pp. 297–301; also in: *Optician and Scientific Instrument Maker*, 53(1369), pp. 203–205; reprinted from *Notes and Queries*, 12 (June 1917), pp. 334–336.

'Public meeting' (1823). 'Public meeting for the establishment of the London Mechanics' Institute', *Mechanic's Magazine*, 1(12), pp. 177–192.

R.B. (1829). [R.B.] 'Queries relating to Mr. Hall's original discovery of achromatic telescopes', *Philosophical Magazine and Annals of Philosophy*, 6(32), pp. 233–235.

Ranyard, A.C. (1881). 'The invention of the achromatic telescope', *Astronomical Register*, 19(224), pp. 194–201 and pp. 279–280.

Ranyard, A.C. (1886a). 'Note with respect to the invention of the achromatic telescope', *Monthly Notices of the Royal Astronomical Society*, 46, pp. 460–461.

Ranyard, A.C. (1886b). 'Report of a meeting of the RAS held on 9 Apr.1886', *The Observatory*, 9(110), pp. 177–187.

Reed, G. (1984). 'Dollond vs. Hall, through an achromatic lens', *Griffith Observer* (Los Angeles) 48(5), pp. 2–11.

Rees, A. (1819). *The Cyclopaedia; or, Universal dictionary of arts, sciences, and literature* 45 vols (London: Longman, Hurst, Rees, Orme & Brown ...).

Reports by the Juries ... (1852). See London. (1852).

Reingold, N. (Ed.) (1979 and 1981).*The Papers of Joseph Henry*, vols 3 and 4 (Washington DC: Smithsonian Institution Press).

Robison, J. (1795). [Art.] 'Optics' in *Encyclopaedia Britannica*, 3rd edn, vol. 13 (Edinburgh: Archibald Constable & Company), pp. 231–364.

Roche J. (Ed.) (1990), *Physicists Look Back: studies in the history of physics* (Bristol: Adam Hilger).

Rochon, A.M. de (1798a). 'Qui traite du Platine, de son utilité dans les arts, du perfectionnement du Flintglass, & des avantages des Télescopes sur les lunettes achromatiques', *J. Phys. Chem. d'Histoire Naturelle et des Arts*, [Messidor an VI] (tome Quatrieme), 47, pp. 3–15.

Rochon, A.M. de (1798b). 'Observations on platina, and its utility in the arts, together with some remarks on the advantages which reflecting [telescopes] have over achromatic telescopes', [translated from Rochon (1798a)], *Philosophical Magazine*, 2 (n.s.), pp. 19–27 and pp. 170–177.

Rochon, A.M. de (1801a). 'Memoire sur les verres achromatiques, adapté à la mesure des angles, et sur les avantages que l'on peut retirer de la double réfraction pour la mesure précise des petits angles', *J. Phys. Chem. d'Histoire Naturelle et des Arts*, 52, pp. 169–199.

Rochon, A.M. de (1801b). 'Memoir on achromatic glasses, adapted for the measurement of angles; and on the advantages which may be derived from double refraction for the accurate measurement of small angles by C. Rochon' [read by Delambre during the third quarter of Year IX of the Institut de France], *Philosophical Magazine*, 10(38), pp. 179–180.

Rochon, A.M. de (1824). 'Rochon, Alexis-Marie de', [Art.] in: *Biographie Universelle, Ancien et Moderne*, 38, pp. 337–347.

Rohr, M. von (1926–1927). 'Contributions to the history of opticians in the first half of the nineteenth century (with special reference to spectacle history)', *Transactions of the Optical Society*, 28(3), pp. 117–148.

Rolt, L.T.C. and Allen, J.S. (1977). *The Steam Engine of Thomas Newcomen* (Hartington, Derbyshire: Moorland Publishing Company).

Rothenberg, M. (Ed.) (1992). *The Joseph Henry Papers*, vol. 6 (Washington DC: Smithsonian Institution Press).

Royal Society of Arts, Manufactures and Commerce (1854a). *Official Catalogue of the Exhibition held in St Martin's Hall, July 4, 1854*, edited by G.W. Yapp (London: Longman, Brown, Green, and Longman).

Royal Society of Arts, Manufactures and Commerce (1854b). *A Collection of Prospectuses of the Educational Exhibition of 1854...* [being an album of trade catalogues and other miscellanea].

Rudd, M.E. (1999). 'A reply to the comments of Dr. Henry King regarding the paper by Rolf Willach on achromatic lenses', *Journal of the Antique Telescope Society*, 16(1), pp. 8–10.

Rudd, M.E., Jaecks, D.H., Willach, R., Sorrenson, R. and Abrahams, P (2000). 'New light on an old question: Who invented the achromatic telescope?', *Journal of the Antique Telescope Society*, 19(1), pp. 3–12.

Rudé, G. (2003). *Hanoverian London 1714–1808* (London: Secker & Warburg, 1971; Sutton, 2003).

Schaffer, S. (1989). 'Natural philosophy and public spectacle in the eighteenth century', *History of Science*, 21(1), pp. 1–43.

Schaffer, S. (1995). 'Glass Works', in: Cohen and Westfall (1995), pp. 202–217.

Schwartz, L.D. (1992). *London in the Age of Industrialisation: entrepreneurs, labour force and living conditions, 1760–1850* (Cambridge: Cambridge University Press).

Scouloudi, I. (Ed.) (1985). *Returns of Strangers in the Metropolis, 1593, 1627, 1635, 1639. A study of an active minority* (Quarto series of the Huguenot Society of London, vol. LVII) (London: Huguenot Society of London).

Scouloudi, I. (Ed.) (1987). *Huguenots in Britain and their French Background, 1550–1800* (London: Macmillan).

Seaborne Davies, D. (1934). 'The early history of the patent specification', *Law Quarterly Review*, 50, pp. 86–109, pp. 260–274.

Shapiro, A. (Ed.) (1984). *The Optical Papers of Isaac Newton*, vol. 1 (The optical lectures, 1670–1672) (Cambridge: Cambridge University Press).

Shapiro, A.E. (1979). 'Newton's "achromatic" dispersion law: theoretical background and experimental evidence', *Archive for the History of Exact Science*, 21(2), pp. 91–128. [For a commentary on this paper, see Hall (1993)].

Short, J. (1753–1754). 'A letter from James Short, F.R.S. to Peter Duval, Esq. F.R.S.', [dated and read 9 April 1752], *Phil. Trans. R. Soc. Lond.*, 48, Part 1, pp. 287–288.

Simpson, A.D.C. (1985). 'Richard Reeve – the "English Campani" and the origins of the London telescope-making tradition', *Vistas in Astronomy*, 28(1–2), pp. 357–365.

Simpson, A.D.C. (1989). 'Robert Hooke and practical optics: technical support at a scientific frontier', in: Hunter and Schaffer (1989), pp. 21–62.

Simpson, A.D.C. (1992). 'James Gregory and the reflecting telescope', *Journal of the History of Astronomy*, 23(2), pp. 77–92.

Simpson, A.D.C. (2009). 'The beginnings of commercial manufacture of the reflecting telescope in London', *Journal for the History of Astronomy*, 40(4), pp. 421–466.

Smith, A. (1783). *Visus Illustratus; or, The sight rendered clear and distinct: being an enquiry and examination into the cause of the inefficacy or defect of the present mode of constructing spectacles ... and to illustrate the use ... of spectacles formed upon a new construction, ...* (London: Printed for the author, and sold by Egerton; and Walker, 1783).

Smith, C. (c. 1735). *The Description, Use and Excellency of a New Instrument, or Sea Quadrant, invented by Caleb Smith ...* (London: 'to be had of the Inventor ... Royal Exchange; or at the Crown Coffee House, King-street, near Guildhall').

Smith, C. (1740). 'A new method of improving and perfecting catadioptrical telescopes, by forming the speculums of glass instead of metal', *Phil. Trans. R. Soc. Lond.*, 41(4), pp. 326–340. [dated April 5, 1739]. Also in: *Phil. Trans.* (Abridged) no. 456 (Jan. 1740), p. 113.

Smith, R. (1738). *A Compleat System of Opticks in Four Books, viz. A Popular, a Mathematical, a Mechanical, and a Philosophical Treatise ...* in four books (2 vols),Printed for the author and sold there [Cambridge] by Stephen Austen at the Angel and Bible in St Paul's Church-yard, and Robert Dodsley at Tulley's Head in Pall Mall). (Reprint: Thoemmes Continuum, 2003).

Sorrenson, R. (1999). 'George Graham, visible technician', *British Journal for the History of Science*, 32(2), pp. 203–222.

Sorrenson, R. (2001). 'Dollond & Son's pursuit of achromaticity, 1758–1789', *History of Science*, 39(1), pp. 31–55.

South Kensington Museum (1857). *Catalogue of the Educational Division of the South Kensington Museum* (London: HMSO).

Speiser, A, Trost, E. and Blanc, C. (Eds) (1962). *Leonhardi Euleri Opera Omnia, tertia* (ser. 3), *'Opera Physica Miscellanea Epistolae', sextum* [vol. 6], (Societatis Scientiarum Naturalium Helveticae). Note: the *Opera Omnia* (76 vols) comprises almost all works and correspondence listed in the Euler archives 'Eneström Index' including those employed in this work: (i) E118 'Sur la perfection des verres objectifs des lunettes', (ser. 3, vol. 6), pp. 1–21;(ii) E210 Dollond–Euler correspondence, pp. 38–43; (iii) E216 'Examen d'une controverse sur la loi de refraction des rayons de differentes couleurs par rapport à la diversité des milieux transparents par lesquels ils sont transmis' (ser. 3, vol. 5), pp. 172–184; E380 'Considérations sur les nouvelles lunettes d'Angleterre de Mr. Dollond et sur le principe qui en est le fondement' (ser. 3, vol. 8), pp. 102–121.

Spitalfields Mathematical Society (1804). *A Catalogue of Books, belonging to the Mathematical Society, Crispin Street, Spitalfields* (London: for James Whiting at the Cicero Press).

Stamp, G. (1986). *Changing Metropolis: earliest photographs of London, 1839–1879* (Harmondsworth: Penguin).

Stewart, L. (1999). 'Other centres of calculation, or, where the Royal Society didn't count: commerce, coffee-houses and natural philosophy in early modern London', *British Journal for the History of Science*, 32(2), pp. 133–153.

Stewart, L. and Weindling, P. (1995). 'Philosophical threads: natural philosophy and public experiment among the weavers of Spitalfields', *British Journal for the History of Science*, 28(1), pp. 37–62.

Stock, J.T. and Orna, M.V. (Eds) (1986). *The History and Preservation of Chemical Instrumentation* (Dordrecht: Kluwer/D. Reidel).

Talbot, S. (1996). 'Jesse Ramsden FRS: his "optical testament"', *Bulletin of the Scientific Instrument Society*, No. 50, pp. 27–29.

Talbot, S. (2002a). 'Caleb Smith's Sea Quadrants of c.1735 as constructed by Thomas Heath and George Adams Snr of London', *Bulletin of the Scientific Instrument Society*, No.73, pp. 2–7.

Talbot, S. (2002b). 'The astroscope by James Mann of London: the first commercial achromatic refracting telescope *c.*1735', *Bulletin of the Scientific Instrument Society*, No.75, pp. 6–8.

Tallis, J. (1847). *London Street Views 1838–1840,* together with the revised and enlarged views of 1847 by Peter Jackson (London: Nattali and Maurice, 1969).

Taylor, E.G.R. (1966). *The Mathematical Practitioners of Hanoverian England 1714–1840* (Cambridge: Cambridge University Press).

Telescope, Tom (1761). See [Newbery, J.].

Thomson, G.S. (1940). *The Russells in Bloomsbury, 1669–1771* (London: Jonathan Cape).

Thorne, J. (1876). *Handbook to the Environs of London: alphabetically arranged, containing an account of every town and village, and of all places of interest, within a circle of twenty miles round London* (London: Murray; Godfrey Cave Associates, reprint 1983).

Timbs, J. (1863). *The Industry, Science, and Art of the Age: or, the International Exhibition of 1862* (London: Lockwood & Co.).

Tomalin, C. (1997). *Jane Austen: a life* (London: Viking).

Tuckwell, W. (1900). *Reminiscences of Oxford* (London: Cassell).

Turnbull, H.W. (Ed.) (1959). *The Correspondence of Isaac Newton vol 1: 1661–1675* (Cambridge: Cambridge University Press for the Royal Society).

Turner, G. l'E. (1969). 'James Short, F.R.S., and his construction of reflecting telescopes', *Notes and Records of the Royal Society of London*, 24(1), pp. 91–108.

Turner, G. l'E. (1974–1975). 'Henry Baker, FRS, Founder of the Bakerian Lecture', *Notes and Records of the Royal Society of London*, 29(1), pp. 53–79.

Turner, G. l'E. (1976). 'The London trade in scientific instrument-making in the 18th century', *Vistas in Astronomy*, 20(1), pp. 173–182.

Turner, G. l'E. (1989). *The Great Age of the Microscope* (Bristol: Adam Hilger/IOP).

Turner, G. l'E. (2000). 'The Government and the English optical glass industry, 1650–1850' *Annals of Science*, 57(4), pp. 399–414.

Venn, J. and Venn, J.A. (1922–1954). *Alumni Cantabrigiensis: a biographical list of all known students, graduates and holders of office at the University of Cambridge*, 10 vols (Cambridge: Cambridge University Press).

Veritas [Anon.] (1790), Communication to the editor [Mr Urban] under a pseudonym (30 September 1790) in *Gentleman's Magazine and Historical Chronicle*, 60(2), pp. 890–891.

Wackerbarth, A.D. (1868). 'Questions relative to the original inventor of the achromatic telescope', *Monthly Notices of the Royal Astronomical Society*, 28(7), pp. 202–205.

Wallis, J. (1811). *London: being a complete guide to the British capital* (London: Sherwood, Neely and Jones).

Ward, J. (1845). *Ward's Condensed Catalogue of Chemical, Philosophical, Electrical and Electrotype Apparatus (manufactured by first rate workmen on the premises)* 6th edn (London: John Ward, operative chemist, etc., 79 Bishopsgate St. Within).

Watkins, F. (1747). *A Particular Account of the Electrical Experiments hitherto made publick, with [a] Variety of new ones; and full Instructions for performing them: To which is annex'd The Description of a compleat Electrical Machine, and its Apparatus, with the Way of using it.* Illustrated with Cuts. Printed for, and sold by the Author, at Sir Isaac Newton's Head, Charing-Cross. MDCCXLVII (London). Optician to their Royal Highnesses the Prince and Princess of Wales. ([Sold by] W. Owen, Bookseller and Publisher, at Homer's Head, Temple Bar).

Watkins, F. (1754). *L'Exercise du Microscope, contenant un Abregé de tout ce qui a été ecrit par les meilleurs Autheurs touchant les Objets le plus curieux: avec Les Precautions qu'on doit prendre pour faire les Observations avec Succes. Auquel est ajouté La Description d'un Microscope, qu'on peut appeller* universel, *d'autant qu'on y trouve les Proprietez de toutes les differentes Sortes qui ayent encore parues.* Construit sur un nouveau Plan, par François Watkins (à Londres. Imprimé pour François Watkins, Opticien, à Charing Cross). [97pp, 3 pl.].

Watkins, F. (1828, 1856). *A Popular Sketch of Electro-magnetism or Electrodynamics* ... (London: John Taylor, Bookseller and Publisher to the University of London and Watkins and Hill, Charing Cross); (2nd edn [reprinted and updated by] Elliott Brothers, successors to Watkins and Hill).

Watkins, F. (1833)'.'On the magnetic powers of soft iron', *Phil. Trans. R. Soc. Lond.*, 123, pp. 333–342.

Watkins, F. (1837). 'On the thermo-electric spark, as obtained from a single pair of metallic elements', *Philosophical Magazine*, series 3, 11, pp. 398–399.

Watkins, F. (1838). 'On the decomposition of water by thermo-electricity', *Philosophical Magazine*, series 3, 13, p. 541.

Watkins, J. & W. (1796). *A Short Account of the Azimutal[sic] or Invariable Compass, etc. [Edited with an introduction by J. and W. W.]* (London).

Watkins and Hill (1828). *Description of apparatus constructed and sold by Watkins and Hill, for illustrating the most striking phenomena of electromagnetism.* In: Watkins (1828).

Watkins and Hill (1832). *A descriptive catalogue of optical, mathematical, philosophical, chemical instruments and apparatus, constructed and sold by Watkins and Hill* ... no. 5 Charing Cross, London.

Watkins and Hill (1836). *A descriptive catalogue of optical, mathematical and chemical instruments and apparatus, constructed and sold by Watkins and Hill, London.*

Watkins and Hill (1838). *A new and enlarged catalogue of optical, mathematical, philosophical and chemical instruments and apparatus, constructed and sold by Watkins and Hill, No. 5 Charing Cross, London.*

Watkins and Hill (1845a). *Descriptive catalogue of voltaic and thermo electric instruments and apparatus, constructed and sold by Watkins and Hill, No. 5 Charing Cross, London.*

Watkins and Hill (1845b). *Descriptive catalogue of electrical instruments and apparatus, constructed and sold by Watkins and Hill, No. 5 Charing Cross, London.*

Watkins and Hill (1845c). *Descriptive catalogue of hydrostatic, hydraulic, pneumatic and acoustic instruments, constructed and sold by Watkins and Hill, No. 5 Charing Cross, London.*

Watkins and Hill (1845d). *Descriptive catalogue of magnetic, electro-magnetic, electro-dynamic, and magneto-inductive instruments and apparatus, constructed and sold by Watkins and Hill, No. 5 Charing Cross, London.*

Watkins and Hill (1845e). *Descriptive catalogue of optical instruments and apparatus, constructed and sold by Watkins and Hill, No. 5 Charing Cross, London.*

Watson (1937). *Watson Centenary 1837–1937* ([n.p.]: W. Watson & Son Ltd.)

Webb, C. (1998). *London Livery Company Apprenticeship Registers, vol. 14: Spectaclemakers' Company 1666–1800 [and] Loriners' Company 1722–31, 1759–1800* (London: Society of Genealogists).

Webb, D. (1990). 'VII. Guide books to London before 1800: a survey', *London Topographical Record*, 26, pp. 138–152.

Webster, T. (1844). *Reports and Notes of Cases on Letters Patent for Invention*, 2 vols (London: Blenkarn).

Westfall, R.S. (1980). *Never at Rest: a biography of Isaac Newton* (Cambridge: Cambridge University Press).

Westminster Poll Book (1749) *A Copy of the Poll for a Citizen for the City and Liberty of Westminster ... 22nd November to 8th December 1749.* (London: Printed for J. Osborn at the Golden-Ball, in Pater-Noster Row ... MDCCXLIX).

Willach, R. (1996). 'New light on the invention of the achromatic telescope objective', *Notes and Records of the Royal Society of London*, 50(2), pp. 195–210.

Willach, R. (1997). 'The early history of the achromatic telescope objective', *Journal of the Antique Telescope Society*, 12(1), pp. 4–13. [This paper is a re-worked (more comprehensive) version of Willach (1996)].

Woolrich, A.P. (1997). 'John Farey Junior (1791–1851), engineer and polymath', *History of Technology*, 19, pp. 111–142.

Worster, B. (1722). *Compendious and Methodical Account of the Principles of Natural Philosophy: as they are explain'd and illustrated in the course of experiments, perform'd at the Academy in Little Tower Street. By Benjamin Worster, A.M.* (London: printed for the author).

Zuidervaart, H. (2011). 'De opmerkelijke geschiedenis van de Haagse instrumentmakersfirma 'Eastland & Regenboog'(c.1768—1839), of hoe het beruchte Engelse patent voor de achromatische telescoop in Nederland werd ontdoken', *Studium. Tijdschrift voor Wetenschaps- en Universiteitsgeschiedenis/ Revue d'Histoire des Sciences et des Universités*, 4(3), pp. 171–180.

Index

For Product Safety Concerns and Information please contact our EU representative GPSR@taylorandfrancis.com Taylor & Francis Verlag GmbH, Kaufingerstraße 24, 80331 München, Germany

Batch number: 10725971

Printed by Printforce, the Netherlands